Hossein Abbaspour | Matilde Marcolli | Thomas Tradler (Eds.)

Deformation Spaces

Aspects of Mathematics

Edited by Klas Diederich

The texts published in this series are intended for graduate students and all mathematicians who wish to broaden their research horizons or who simply want to get a better idea of what is going on in a given field. They are introductions to areas close to modern research at a high level and prepare the reader for a better understanding of research papers. Many of the books can also be used to supplement graduate course programs.

Volumes of the series are listed on page 175

www.viewegteubner.de

Hossein Abbaspour | Matilde Marcolli
Thomas Tradler (Eds.)

Deformation Spaces

Perspectives on algebro-geometric moduli

A Publication of the Max-Planck-Institute
for Mathematics, Bonn

**VIEWEG+
TEUBNER**

Bibliographic information published by the Deutsche Nationalbibliothek
The Deutsche Nationalbibliothek lists this publication in the Deutsche Nationalbibliografie;
detailed bibliographic data are available in the Internet at http://dnb.d-nb.de.

Dr. Hossein Abbaspour
Département de mathématiques
Université de Nantes
2, rue de la Houssinière - BP 92208
F-44322 Nantes Cedex 3
France

abbaspour@univ-nantes.fr

Prof. Dr. Matilde Marcolli
Mathematics Department
California Institute of Technology
1200 E.California Blvd.
Pasadena, CA 91125
USA

matilde@caltech.edu

Dr. Thomas Tradler
New York City College of Technology
Department of Mathematics N-711
300 Jay Street
Brooklyn, NY 11201
USA

ttradler@citytech.cuny.edu

Prof. Dr. Klas Diederich (Series Editor)
Fachbereich Mathematik
Bergische Universität Wuppertal
Gaußstraße 20
D-42119 Wuppertal

klas@math.uni-wuppertal.de

Mathematics Subject Classification
58H15 Deformations of structures, **53D55** Deformation quantization, star products, **18D50** Operads, **14F42** Motivic cohomology; motivic homotopy theory, **13D10** Deformations and infinitesimal methods, **2G05** Deformations of complex structures, **32G15** Moduli of Riemann surfaces, Teich-müller theory, **14D23** Stacks and moduli problems

1st Edition 2010

Editorial Office: Ulrike Schmickler-Hirzebruch | Nastassja Vanselow

Softcover re-print of the Hardcover 1st edition 2010
Vieweg+Teubner Verlag is a brand of Springer Fachmedien.
Springer Fachmedien is part of Springer Science+Business Media.
www.viewegteubner.de

Cover design: KünkelLopka Medienentwicklung, Heidelberg
Printed on acid-free paper

ISBN 978-3-8348-2669-5

Preface

In recent years the Hochschild and cyclic complex and their algebraic structures have been intensively studied from different perspectives. Some of these algebraic gadgets have been around since the early work of Gerstenhaber on the deformations of associative algebras, while others, such as cyclic homology, were introduced by Connes in the early development of noncommutative geometry. More recent developments from this perspective include the theory of Hopf cyclic (co)-homology of Hopf algebra.

Various algebraic structures of Hochschild and Cyclic (co)-homology, such as Batalin-Vilkovisky and Gerstenhaber algebras, received a topological reincarnation by the works of Chas and Sullivan and other authors on free loop space. These compelling ideas, such as an action of the moduli space of surfaces possibly with various compactifications, have been considered in several different settings. The algebraic analogue of these constructions on Hochschild and cyclic complexes (of Frobenius algebras) are usually known under the name of Deligne conjecture. This theory develops parallel to symplectic field theory and Gromov-Witten invariants. As an algebraic theory, this corresponds to a deformation problem over PROPs or properads as opposed to operads, which naturally include genus, or in physics terminology, the correct "h-bar" terms.

The editors had organized two workshops in July 2007 and August 2008 at the Max-Planck-Institut für Mathematik in Bonn with a generous support from the Hausdorff Center. Participants of these workshops were mainly algebraic topologist, noncommutative geometers, and specialists in deformations theory. The aim of these workshops was to bring together the mathematicians who work on deformations of algebraic and geometric structures and Hochschild and cyclic complexes. As it is clear from the volume, the subject of these activities was influenced by physics and the new perspectives that it offers.

This volume collects a few self-contained and peer-reviewed papers by the participants, which present topics in algebraic and Motivic topology, quantum field theory, algebraic geometry, noncommutative geometry and the deformation theory of Poisson algebras. The papers were contributed by some of the speakers of the workshops, and we hope they provide a reasonable view of some of the activities held at the workshops.

We would like to thank the staffs of the Hausdorff Center and Max-Planck-Institut in Bonn for their valuable assistance in organizing the workshops.

August 2009,
Hossein Abbaspour, Matilde Marcolli, and Thomas Tradler

Contents

On the Hochschild and Harrison (co)homology of C_∞-algebras and applications to string topology

Grégory Ginot

ABSTRACT. We study Hochschild (co)homology of commutative and associative up to homotopy algebras with coefficient in a homotopy analogue of symmetric bimodules. We prove that Hochschild (co)homology is equipped with λ-operations and Hodge decomposition generalizing the results in [**GS1**] and [**Lo1**] for strict algebras. The main application is concerned with string topology: we obtain a Hodge decomposition compatible with a non-trivial BV-structure on the homology $H_*(LX)$ of the free loop space of a triangulated Poincaré-duality space. Harrison (co)homology of commutative and associative up to homotopy algebras can be defined similarly and is related to the weight 1 piece of the Hodge decomposition. We study Jacobi-Zariski exact sequence for this theory in characteristic zero. In particular, we define (co)homology of relative A_∞-algebras, i.e., A_∞-algebras with a C_∞-algebra playing the role of the ground ring. We also give a relation between the Hodge decomposition and homotopy Poisson-algebras cohomology.

The Hochschild cohomology and homology groups of a commutative and associative k-algebra A, k being a unital ring, have a rich structure. In fact, when M is a symmetric bimodule, Gerstenhaber-Schack [**GS1**] and Loday [**Lo1**] have shown that there are λ-operations $(\lambda^k)_{k\geq 1}$ inducing so-called γ-rings structures on Hochschild cohomology groups $HH^*(A, M)$ and homology groups $HH_*(A, M)$. In characteristic zero, these operations yield a weight-decomposition called the Hodge decomposition whose pieces are closely related to (higher) André-Quillen (co)homology and Harrison (co)homology. These operations have been widely studied for their use in algebra, geometry and their intrinsic combinatorial meaning.

The Hochschild (co)homology of the singular cochain complex of a topological space is a useful tool in algebraic topology and in particular in string topology. In fact, Chas-Sullivan [**CS**] have shown that the (shifted) homology $H_{*+d}(LM)$, where $LM = \mathrm{Map}(S^1, M)$ is the free loop space of a manifold M of dimension d, is a Batalin-Vilkovisky-algebra. In particular, there is an associative graded commutative operation called the loop product. When M is simply connected, there is an isomorphism $H_{*+d}(LM) \cong HH^*(C^*(M), C^*(M))$ which, according to Cohen-Jones [**CJ**], identifies the cup-product with the loop product. Alternative

---

2000 *Mathematics Subject Classification.* Primary 16E45, 18G55; Secondary 13D03, 17B63, 55P35.

Key words and phrases. C_∞-algebras, Hochschild (co)homology, string topology, Harrison (co)homology, A_∞-algebras, homotopy Poisson algebras.

proofs of this isomorphism have also been given by Merkulov [**Mer**] and Félix-Thomas-Vigué [**FTV2**]. This isomorphism is based on the isomorphism $H_*(LX) \cong HH^*(C^*(X), C_*(X))$, where X is a simply connected space, and the fact that the Poincaré duality should bring a "homotopy isomorphism" of bimodules $C_*(X) \to C^*(X)$. Since $HH^*(C^*(X), C^*(X))$ is a Gerstenhaber algebra, it is natural to define string topology operations for Poincaré duality topological space X using the Hochschild cohomology of their cochain algebra $C^*(X)$. To achieve this, one needs to work with homotopy algebras, homotopy bimodules and homotopy maps between these structures, even in the most simple cases. This was initiated by Sullivan and his students, see Tradler and Zeinalian papers [**Tr2, TZ, TZ2**]. For example, they show that for nice enough spaces, $HH^*(C^*(X), C^*(X))$ is a BV-algebra.

In fact the cochain complex $C^*(X)$ is "homotopy" commutative since the Steenrod $\cup_1$-product gives a homotopy for the commutator $f \cup g - g \cup f$. This fact motivates us to study Hochschild (co)homology of commutative up to homotopy associative algebras (C_∞-algebras for short) in order to add λ-operations to the string topology picture for nice enough Poincaré duality spaces. These λ-operations have to be somehow compatible with the other string topology operations. We achieve this program in Section 5. In particular we prove that if X is a triangulated Poincaré duality space, the Hochschild cohomology of its cochain algebra is a BV-algebra equipped with λ-operations commuting with the BV-differential and filtered with respect to the product, see Theorem 5.7.

Besides string topology there are other reasons to study C_∞-algebras and cohomology theories associated to their deformations, i.e. Hochschild and Harrison. Actually, associative structures up to homotopy (A_∞-structures for short), introduced by Stasheff [**St**] in the sixties, have become more and more useful and popular in mathematical physics as well as algebraic topology. A typical situation is given by the study of a chain complex with an associative product inducing a graded commutative algebra structure on homology. The quasi-isomorphism class of the algebraic structure usually retains more information than the homology. In many cases, it is possible to enforce the commutativity of the product at the chain level at the price of relaxing associativity. For instance, in characteristic zero, according to Tamarkin [**Ta**], the Hochschild cochain complex $C^*(A, A)$ of any associative algebra A has a C_∞-structure. The same is true for the cochain algebra of a space [**Sm**], also see Lemma 5.7 below. It is well-known that these structures retain more information on the homotopy type of the space than the associative one, for instance see [**Ka**]. Moreover the commutative, associative up to homotopy algebras are quite common among the A_∞-ones and deeply related to the theory of moduli spaces of curves [**KST**]. In fact an important class of examples is given by the formal Frobenius manifolds in the sense of Manin [**Ma**].

In this paper, we study Hochschild (co)homology of C_∞-algebras with value in general bimodules. The need for this is already transparent in string topology, notably to get functorial properties. Note that we work in characteristic free context (however usually different from 2) in order to have as broad as possible homotopy applications. In particular we do not restrain ourself to the rational homotopy framework. In characteristic zero, a similar approach (but different application) to string topology has been studied by Hamilton-Lazarev [**HL**], see Section 6 for details.

To construct λ-operations, we need to define homotopy generalizations of symmetric bimodules. Notably enough, the appropriate symmetry conditions are not the same for homology and cohomology. These "homotopy symmetric" structures are called C_∞-bimodules and C_∞^{op}-bimodules structures respectively. We define and study λ-operations and Hodge decomposition for Hochschild (co)homology of C_∞-algebras. In particular, the λ-operations induce an augmentation ideal spectral sequence yielding important compatibility results between the Gerstenhaber algebra and γ-ring structures in Hochschild cohomology. In characteristic zero, the Hodge decomposition of Poisson algebras is related to Poisson algebras homology [**Fr1**]. We generalize this result in the homotopy framework.

We study Harrison (co)homology of C_∞-algebras and prove that, if the ground ring k contains the field $\mathbb{Q}$ of rational numbers, the weight 1 piece of the Hodge decomposition coincides with Harrison (co)homology. For strictly commutative algebras, the result is standard [**GS1, Lo1**].

It is well-known that, in characteristic zero, for unital flat algebras, Harrison (co)homology coincides with André-Quillen (co)homology (after a shift of degree). In that case, a sequence $K \to S \to R$ gives rise to of the change-of-ground-ring exact sequence, often called the Jacobi-Zariski exact sequence. We obtain a homotopy analogue of this exact sequence. Our approach is to define relative A_∞ and C_∞-algebras, i.e., A_∞ and C_∞-algebras for which the "ground ring" is also a C_∞-algebra. We define Hochschild and Harrison (co)homology groups for these relative homotopy algebras. These definitions are of independent interest. Indeed, recently, several categories of strictly associative and commutative ring spectra have arisen providing exciting new constructions in homotopy theory, for instance see [**EKMM, MMSS**]. Our constructions of Hochschild and Harrison (co)homology of relative homotopy algebras are algebraic, chain complex level, analogues of topological Hochschild/André-Quillen (co)homology of an R-ring spectrum, where R is a commutative ring spectrum.

Here is the plan of the paper. In section 1 we recall and explain the basic property of A_∞-algebras and their Hochschild cohomology. We give some details, not so easy to find in the literature, for the reader's convenience. In Section 2 we recall the definition of C_∞-algebras, introduce our notion of a C_∞-bimodule, generalizing the classical notion of symmetric bimodule, and then of Harrison (co)homology. We also study some basic properties of these constructions. In Section 3 we establish the existence of λ-operations, Hodge decompositions in characteristic zero and study some of their properties. In Section 4, we study the homotopy version of Jacobi-Zariski exact sequence for Harrison (co)homology and establish a framework for the study of A_∞-algebras with a C_∞-algebra as "ground ring". In the last section we apply the previous machinery to string topology and prove that there exists λ-operations compatible with a BV-structure on $HH^*(C^*(X), C^*(X))$ for X a triangulated Poincaré duality space. The last section is devoted to some additional remarks (without proof) and questions.

Acknowledgement : The author would like to thank Ralph Cohen, Andrei Lazarev, Jim Stasheff and Micheline Vigué for helpful discussions and suggestions.

Notations :

- in what follows k will be a commutative unital ring and $R = \bigoplus R^j$ a $\mathbb{Z}$-graded k-module. All tensors products will be over k unless otherwise stated and/or sub scripted.

- We use a cohomological grading for our k-modules with the classical convention that a homological grading is the opposite of a cohomological one. In other words $H_i := H^{-i}$ as graded modules. A (homogeneous) map of degree k between graded modules V^*, W^* is a map $V^* \to W^{*+k}$.

- When $x_1, \ldots, x_n$ are elements of a graded module and σ a permutation, the *Koszul sign* is the sign $\pm$ appearing in the equality $x_1 \ldots x_n = \pm x_{\sigma(1)} \ldots x_{\sigma(n)}$ which holds in the symmetric algebra $S(x_1, \ldots, x_n)$.

- We use Sweedler's notation $\delta(x) = \sum x^{(1)} \otimes x^{(2)}$ for a coproduct δ.

- A *strict* up to homotopy structure will be one given by a classical differential graded one.

- The algebraic structures "up to homotopy" appearing in this paper are always uniquely defined by sequences of maps $(D_i)_{i \geq 0}$. Such maps will be referred to as *defining maps*, for instance see Remark 1.5.

1. Hochschild (co)homology of an A_∞-algebra with values in a bimodule

In this section we recall the definitions and fix notation for A_∞-algebras and bimodules as well as their Hochschild (co)homology. For convenience of the reader, we also recall some "folklore" results which might not be found so easily in the literature and are needed later on.

1.1. A_∞-algebras and bimodules.
The *tensor coalgebra* of R is $T(R) = \bigoplus_{n \geq 0} R^{\otimes n}$ with the deconcatenation coproduct

$$\delta(x_1, \ldots, x_n) = \sum_{i=0}^{n} (x_1, \ldots, x_i) \otimes (x_{i+1}, \ldots, x_n).$$

The suspension sR of R is the graded k-module $(sR)^i = R^{i+1}$ so that a degree $+1$ map $R \to R$ is equivalent to a degree 0 map $R \to sR$.

Let V be a graded k-module. The *tensor bimodule* of V over the tensor coalgebra $T(R)$ is the k-module $T^R(V) = k \oplus T(R) \otimes V \otimes T(R)$ with structure map

$$\delta^V(x_1, \ldots, x_n, v, y_1, \ldots, y_m) = \sum_{i=1}^{n} (x_1, \ldots, x_i) \otimes (x_{i+1}, \ldots, x_n, v, y_1, \ldots, y_m)$$

$$\oplus \sum_{k=1}^{m} (x_1 \ldots, x_n, v, y_1, \ldots, y_i) \otimes (y_{i+1}, \ldots, y_m).$$

If D is a coderivation of $T(R)$, then a coderivation of $T(V)$ to $T^R(V)$ over D is a map $\Delta : T^R(V) \to T^R(V)$ such that

$$(1.1) \qquad (D \otimes \mathrm{id} + \mathrm{id} \otimes \Delta) \oplus (\Delta \otimes \mathrm{id} + \mathrm{id} \otimes D) \circ \delta^V = \delta^V \circ \Delta.$$

We denote $A^\perp(R) = \oplus_{n \geq 1} sR^{\otimes n}$ the coaugmentation

$$0 \to k \to T(sR) \to A^\perp(R) \to 0$$

and abusively write δ for its induced coproduct.

DEFINITION 1.1.   • *An A_∞-algebra structure on R is a coderivation D of degree 1 on $A^\perp(R)$ such that $(D)^2 = 0$.*
- *An A_∞-bimodule over R structure on M is a coderivation D_M^R of degree 1 on $A_R^\perp(M) := T_{sR}(sM)$ over D such that $(D_M^R)^2 = 0$.*
- *A map between two A_∞-algebras R, S is a map of graded differential coalgebras $A^\perp(R) \to A^\perp(S)$.*
- *A map between two A_∞-bimodules M, N over R is a map of graded differential bicomodules $A_R^\perp(M) \to A_R^\perp(N)$.*

Henceforth, R-bimodule will stand for A_∞-bimodule over an A_∞-algebra R.

NOTATION 1.2. We will denote "$a \otimes m \otimes b \in A_R^\perp(M)$" a generic element in $A_R^\perp(M)$. That is, $a, b \in A^\perp(R)$, $m \in M$ and $a \otimes m \otimes b$ stands for the corresponding element in $A^\perp(R) \otimes sM \otimes A^\perp(R) \subset A^\perp(R)$.

REMARK 1.3. These definitions are the same as the definitions given by algebras over the minimal model of the operad of associative algebras and their bimodules and goes back to the pioneering work [St].

REMARK 1.4. Coderivations on $A^\perp(R)$ are the same as coderivations on $T(sR)$ that vanishes on $k \subset T(sR)$.

It is well-known that a coderivation D on $T(sR)$ is uniquely determined by a simpler system of maps $(D_i : sR^{\otimes i} \to sR)_{i \geq 0}$. The maps $\tilde{D}_i$ are given by the composition of D with the projection $T(sR) \to sR$. The coderivation D is the sum of the lifts of the maps $\tilde{D}_i$ to $A^\perp(R) \to A^\perp(R)$. More precisely, for $x_1, \ldots, x_n \in sR$,

$$(1.2) \quad D(x_1, \ldots, x_n) = \sum_{i \geq 0} \sum_{j=0}^{n-i} \pm x_1 \otimes \ldots \otimes \tilde{D}_i(x_{j+1}, \ldots, x_{j+i}) \otimes \ldots \otimes x_n$$

where $\pm$ is the sign $(-1)^{|D_i|(|x_1| + \cdots + |x_j|)}$. Furthermore, there are isomorphisms of graded modules $\operatorname{Hom}(sR^{\otimes i}, sR) \ni \tilde{D}_i \longmapsto D_i \in s^{1-i} \operatorname{Hom}(R^{\otimes i}, R)$ defined by

$$(1.3) \qquad D_i(r_1, \ldots, r_i) \quad = \quad (-1)^{i|\tilde{D}_i| + \sum_{k=1}^{i-1}(k-1)|r_k|} \tilde{D}_i(sr_1, \ldots, sr_i).$$

Note that the signs are given by the Koszul rule for signs. It follows that a coderivation D on $T(sR)$ is uniquely determined by a system of maps $(D_i : R^{\otimes i} \to R)_{i \geq 0}$. Such a coderivation D is of degree k if and only if each D_i is of degree $k + 1 - i$. According to Remark 1.4 above, a coderivation D on $A^\perp(R)$ is one on $T(sR)$ such that $D_0 = 0$.

REMARK 1.5. We call the maps $(D_i : R^{\otimes i} \to R)_{i \geq 0}$ the *defining maps* of the associated coderivation $D : T(sR) \to T(sR)$. We also use similar terminology for all other kind of coderivations appearing in the rest of the paper.

Similarly, a coderivation D_R^A on $A_R^\perp(M)$ (over D with $D_0 = 0$) is given by a system of maps $(D_{i,j}^M : R^{\otimes i} \otimes M \otimes R^{\otimes j} \to M)_{i,j \geq 0}$. All of these properties are formal consequences of the co-freeness of the tensor coalgebra (in the operadic setting). Also a very detailed and down-to-earth account is given in [Tr1].

REMARK 1.6. Given a coderivation D of degree 1 on $A^\perp(R)$ defined by a system of maps $\left(D_i : R^{\otimes i} \to R\right)_{i \geq 0}$, it is well-known [St] that the condition $(D)^2 = 0$ is equivalent to an infinite number of equations quadratic in the D_i's. Namely, for $n \geq 1$, $r_1, \ldots r_n \in R$,

$$(1.4) \quad \sum_{i+j=n+1} \sum_{k=0}^{i-1} \pm D_i(r_1, \ldots, r_k, D_j(r_{k+1}, \ldots, r_{k+j}), r_{k+j+1}, \ldots, r_n) = 0.$$

In particular, if $D_1 = 0$, Equation (1.4) implies that D_2 is an associative multiplication on R.

There are similar identities for the defining maps $(D_{i,j}^M : R^{\otimes i} \otimes M \otimes R^{\otimes j} \to M)_{i,j \geq 0}$ of an A_∞-bimodule [**Tr1**]. It is trivial to show that, when $D_1 = 0$ and $D_{00}^M = 0$, D_{10}^M and D_{01}^M respectively endows M with a structure of left and right module over the algebra (R, D_2).

EXAMPLE 1.7. Any A_∞-algebra (R, D) is a bimodule over itself with structure maps given by $D_{i,j}^R = D_{i+1+j}$.

REMARK 1.8. Similarly to coderivations, a map of graded coderivation $F :$ $A^\perp(R) \to A^\perp(S)$ is uniquely determined by a simpler system of maps $(F_i : R^{\otimes i} \to R)_{i \geq 1}$, where F_i is induced by composition of F with the projection on S. The details are similar to those of Remark 1.4 and left to the reader. The maps F_i are referred to as the defining maps of F.

1.2. Hochschild (co)homology. Let (R, D) be an A_∞-algebra and M an R-bimodule. We call a coderivation from $k \oplus A^\perp(R) = T(sR)$ into $A_R^\perp(M)$ a *coderivation of R into M*. By definition it is a map $f : T(sR) \longrightarrow A_R^\perp(M)$ such that

$$\delta^M \circ f = (\mathrm{id} \otimes f + f \otimes \mathrm{id}) \circ \delta.$$

As in Remark 1.5, such a coderivation is uniquely determined by a collection of maps $(f_i : R^{\otimes i} \to M)_{i \geq 0}$ where the f_i are induced by the projections onto sM of the map f restricted to $sR^{\otimes i}$ (for instance see [**Tr1**]).

DEFINITION 1.9. *The Hochschild cochain complex of an A_∞-algebra (R, D) with values in an R-bimodule M is the space $s^{-1}\mathrm{CoDer}(R, M)$ of coderivations of R into M equipped with differential b given by*

$$b(f) = D_M \circ f - (-1)^{|f|} f \circ D.$$

It is classical that b is well-defined and $b^2 = 0$ see [**GJ2**]. We denote $HH^*(A, M)$ its cohomology which is called the Hochschild cohomology of R with coefficients in M.

EXAMPLE 1.10. Let (R, m, d) be a differential graded algebra and (M, l, r, d_M) be a (differential graded) A-bimodule. Then R has a structure of an A_∞-algebra and M a structure of A_∞-bimodule over R given by maps:

$$(1.5) \qquad D_1 = d, \qquad D_2 = m \quad \text{and} \quad D_i = 0 \text{ for } i \geq 3;$$

$$(1.6) \qquad D_{0,0}^M = d_M, \quad D_{1,0}^M = l, \quad D_{0,1}^M = r \text{ and } D_{i,j}^M = 0 \text{ for } i + j \geq 1.$$

The converse is true: if R, M are, respectively, an A_∞ algebra and a R-bimodule with $D_{i \geq 3} = 0$ and $D_{i,j}^M = 0$ $(i + j \geq 1)$, then the identities (1.5), (1.6) define a differential graded algebra structure on R and a bimodule structure M. We call this kind of structure a *strict* homotopy algebra or a *strict* homotopy bimodule.

We have seen that the k-module $\mathrm{CoDer}(R, M)$ is isomorphic to the k-module $\mathrm{Hom}\left(\bigoplus_{n\geq 0} R^{\otimes n}, M\right)$ by projection on sM, i.e. the map $f \mapsto (f_i : R^{\otimes i} \to M)_{i\geq 0}$. Thus the differential b induces a differential on $\mathrm{Hom}\left(\bigoplus_{n\geq 0} R^{\otimes n}, M\right)$, which for a homogeneous map $f : R^{\otimes n} \to M$, is given by the sum $b(f) = \alpha(f) + \beta(f)$ where $\alpha(f) : R^{\otimes n} \to M$ and $\beta(f) : R^{\otimes n+1} \to M$ are defined by

$$
\begin{aligned}
\alpha(f)(a_1, \ldots, a_n) &= (-1)^{|f|+1} d_M(f(a_1, \ldots, a_n)) \\
&+ \sum_{i=0}^{n-1} (-1)^{i+|a_1|+\cdots+|a_i|} f(a_1, \ldots, d(a_{i+1}), \ldots, a_n) \\
\beta(f)(a_0, \ldots, a_n) &= (-1)^{|f||a_0|+|f|} l(a_0, f(a_1, \ldots, a_n)) \\
&+ (-1)^{n+|a_0|+\cdots+|a_{n-1}|+|f|} r(f(a_0, \ldots, a_{n-1}), a_n) \\
&- \sum_{i=0}^{n-1} (-1)^{i+|a_0|+\cdots+|a_i|+|f|} f(a_0, \ldots, m(a_i, a_{i+1}), \ldots, a_n).
\end{aligned}
$$

Hence $\alpha+\beta$ is the differential in the standard bicomplex giving the usual Hochschild cohomology of a differential graded algebra [Lo2]. Consequently Definition 1.9 coincides with the standard one for strict A_∞-algebras, that is the one given by the standard complex.

It is standard that the identity $(D)^2 = 0$ restricted to A yields that $(D_1)^2 = 0$ and $|D_1| = |D| = 1$. Therefore, (R, D_1) is a chain complex whose cohomology will be denoted $H^*(R)$. Moreover the linear map $D_2 : R^{\otimes 2} \to R$ passes to the cohomology $H^*(R)$ to define an associative algebra structure. Similarly D_{00}^M is a differential on M and $H^*(M)$ has a bimodule structure over $H^*(R)$ induced by D_{10}^M and D_{01}^M. The link between the cohomology of $H^*(A)$ and the one of A is given by the following spectral sequence.

PROPOSITION 1.11. *Let (R, D) be an A_∞-algebra and (M, D^M) an R-bimodule with R, M, $H^*(R)$, $H^*(M)$ flat as k-modules. There is a converging spectral sequence*

$$
E_2^{p,q} = HH^{p+q}(H^*(R), H^*(M))^q \implies HH^*(R, M).
$$

The subscript q in $HH^*(H^*(R), H^*(M))^q$ stands for the piece of internal degree q in the group $HH^*(H^*(R), H^*(M))$ (the internal degree is the degree coming from the grading of $H^*(R)$).

Proof: There is a decreasing filtration of cochain complex $F^{*\geq 0} C^*(R, M)$ of $\mathrm{CoDer}(R, M)$ where $F^p C^*(R, M)$ is the subspace of coderivation f such that

$$
f(R^{\otimes n}) \subset \bigoplus_{p+i+j\leq n} R^{\otimes i} \otimes M \otimes R^{\otimes j}.
$$

The filtration starts at F_0 because any coderivation f is determined by maps $R^{\otimes i\geq 1} \to M$. It is thus a bounded above and complete filtration. Hence, it yields a cohomological converging spectral sequence computing $HH^*(R, M)$. The maps D_i and $D_{j,k}^M$ lower the degree of the filtration unless $i = 1$, $j = k = 0$. Consequently the differential on the associated graded is the one coming from the inner differentials D_1 and $D_{0,0}^M$. It follows by Künneth formula, that

$$
E_1^{**} \cong \mathrm{CoDer}\left(A^\perp(H^*(R)), A^\perp_{H^*(R)}(H^*(M))\right).
$$

The differential on the E_1^{**} term is induced by D_2, D_{10}^M, D_{01}^M. These operations give $H^*(R)$ a structure of associative algebra and $H^*(M)$ a bimodule structure. Hence the differential d^1 on E_1^{**} is the same as the differential defining the Hochschild cohomology of the graded algebra $H^*(R)$ with values in $H^*(M)$. Now, Example 1.10 implies that $E_2^{*,*} = HH^*(H^*(R), H^*(M))$.   ∎

The Hochschild cohomology $HH^*(R, R)$ of any A_∞-algebras R has the structure of a Gerstenhaber algebra as was shown in [**GJ2**]. The product of two elements $f, g \in C^*(R, R)$ (with defining maps (f_n), (g_m)) is the coderivation $\mu(f, g)$ defined by

$$(1.7) \quad \mu(f, g)(a_1, \ldots, a_n) = \sum_{j \geq 2, r_1, r_2 \geq 0} \pm (a_1 \otimes \ldots \otimes D_j(\ldots f_{r_1}, \ldots, g_{r_2}, \ldots) \otimes \ldots a_n).$$

In the formula the sign $\pm$ is the Koszul sign. There is also a degree 1 bracket defined by $[f, g] = f \widetilde{\circ} g - (-1)^{(|f|+1)(|g|+1)} g \widetilde{\circ} f$ where

$$f \widetilde{\circ} g(a_1, \ldots, a_n) = \sum_{i,j} \pm (a_1 \otimes \ldots \otimes f_i(\ldots g_j, \ldots) \otimes \ldots a_n).$$

PROPOSITION 1.12. *Let R be an A_∞-algebra and take $M = R$ as a bimodule. Then $(HH^*(R, R), \mu, [,])$ is a Gerstenhaber algebra and the spectral sequence $E_{m \geq 2}^{**}$ is a spectral sequence of Gerstenhaber algebras.*

Proof: The fact that the product μ and the bracket $[,]$ make $HH^*(R, R)$ a Gerstenhaber algebra is well-known [**GJ2, Tr2**]. Also see Remark 1.13 below for a sketch of proof.

The product map $\mu : F^p C^* \otimes F^q C^* \to F^{p+q} C^*$ and bracket $[,] : F^p C^* \otimes F^q C^* \to F^{p+q-1} C^*$ are filtered maps of cochain complexes. Thus both operations survive in the spectral sequence. At the level E_0 of the spectral sequence, the product μ boils down to

$$\mu(f, g)(a_0, \ldots, a_n) = \sum a_0 \ldots \otimes D_2(f(\ldots), g(\ldots)) \otimes \ldots \otimes a_n$$

which, after taking the homology for the differential d_0, identifies with the usual cup product in the Hochschild cochain complex $\mathrm{Hom}(H^*(R)^{\otimes *}, H^*(R))$ through the isomorphism between coderivations and homomorphisms. Similarly the bracket coincides with the one introduced by Gerstenhaber in the Hochschild complex of $H^*(R)$. The Leibniz relation hence holds at level 2 and on the subsequent levels. ∎

REMARK 1.13. Actually, the product structure is the reflection of a A_∞-structure on $C^*(R, R)$. It is easy to check that the maps $\gamma_i : C^*(R, R)^{\otimes i} \to C^*(R, R)$ defined by

$$\gamma_i(f^1, \ldots, f^i)(a_1, \ldots, a_n) = \sum_{j \geq i, r_1, \ldots, r_i \geq 1} \pm (a_1 \otimes \ldots \otimes D_j(\ldots f_{r_1}^1, \ldots, f_{r_2}^2, \ldots$$

$$\ldots, f_{r_i}^i, \ldots) \otimes \ldots \otimes a_n)$$

together with $\gamma_1 = b$, the Hochschild differential, give a A_∞-structure to $C^*(R, R)$. Thus the map $\mu = \gamma_2$ gives an associative algebra structure to $HH^*(R, R)$. Moreover it is straightforward to check that the Jacobi relation for $[,]$ is satisfied on $C^*(R, R)$. The Leibniz identity and the commutativity of the product are obtained as in Gerstenhaber fundamental paper [**Ge**].

The Hochschild homology of an A_∞-algebra R was first defined in [**GJ1**]. Let M be an R-bimodule and $b : M \otimes T(sR) \to M \otimes T(sR)$ be the map

$$b(m, a_0, \ldots, a_n) = \sum_{p+q \leq n} \pm D_{p,q}^M(a_{n-p+1}, \ldots, a_n, m, a_1 \ldots, a_q) \otimes a_{q+1} \cdots a_{n-p}$$

$$+ \sum_{i+j \leq n} \pm m \otimes a_1 \otimes \cdots D_{j+1}(a_i, \ldots, a_{i+j}) \otimes a_{i+j+1} \otimes \cdots a_n.$$

DEFINITION 1.14. *The Hochschild homology $HH_*(R, M)$ of an A_∞-algebra (R, D) with values in the bimodule (M, D^M) is the homology of $(M \otimes T(sR), b)$.*

The fact that $b^2 = 0$ follows from a straightforward computation or from Lemma 1.16 below.

REMARK 1.15. Recall that we use a cohomological grading for R, M. Thus a cycle $x \in M^i \subset M \otimes T(sR)$ gives an element $[x] \in HH_{-i}(R, M)$ in homological degree $-i$.

Given a bimodule M over R, there is a map $\gamma_M : M \otimes T(sR) \to T^{sR}(sM)$ defined by

$$\gamma_M = \tau \circ (s\,\mathrm{id} \otimes \delta)$$

where τ is the map sending the last factor of $M \otimes T(sR) \otimes T(sR)$ to the first of $T(sR) \otimes M \otimes T(sR)$.

LEMMA 1.16. *Given any coderivation ∂ of $T^{sR}(sM)$ over D, there is a unique map $\overline{\partial} : M \otimes T(sR) \to M \otimes T(sR)$ that makes the following diagram commutative:*

$$\begin{array}{ccc} M \otimes T(sR) & \xrightarrow{\gamma_M} & T^{sR}(sM) \\ \overline{\partial} \downarrow & & \downarrow \partial \\ M \otimes T(sR) & \xrightarrow{\gamma_M} & T^{sR}(sM). \end{array}$$

Proof: The map $\overline{\partial}$ is the sum $\sum \overline{\partial}^{[i]}$, where $\overline{\partial}^{[i]}$ takes value in $M \otimes sA^{\otimes i}$. By induction on i, it is straightforward that

$$\gamma_M(\overline{\partial}^{[i]}) = \sum_{i=0}^{n} \left(\sum_{p+q=n-m} a_{j+1} \otimes \ldots \otimes \partial_{p,q}(a_{n-p+1}, \ldots, a_n, m, a_1, \ldots \right.$$

$$\ldots, a_q) \otimes a_{q+1} \otimes \ldots \otimes a_i + \sum_{j=i+1}^{m-1} \pm a_i \otimes \ldots \otimes D_{n-m+2}(a_j,$$

$$a_{j+1}, \ldots, a_{j+n-m+1}) \otimes \ldots \otimes a_n \otimes m \otimes a_1 \ldots \otimes a_i$$

$$+ \sum_{j=0}^{i-n+m-1} \pm a_i \otimes \ldots \otimes a_n \otimes m \otimes a_1 \ldots \otimes D_{n-m+2}(a_j, \ldots$$

$$\left. \ldots, a_{j+n-m+1}) \otimes \ldots \otimes a_i \right).$$

It follows that the map $\overline{\partial}$ exists and satisfies

$$\overline{\partial}(m, a_0 \otimes \ldots \otimes a_n) = \sum_{p+q \leq n} \pm \partial_{p,q}(a_{n-p+1}, \ldots, a_n, m, a_1 \ldots, a_q) \otimes a_{q+1} \otimes \cdots$$

$$\cdots \otimes a_{n-p} + \sum_{i+j \leq n} \pm m \otimes a_1 \otimes \cdots \otimes D_{j+1}(a_i, \ldots$$

$$\cdots, a_{i+j}) \otimes a_{i+j+1} \otimes \cdots \otimes a_n.$$

EXAMPLE 1.17. For $\partial = D_M^R$, $(D_M^R)^2$ is the trivial coderivation, hence $\overline{(D_M^R)}^2 = \overline{(D_M^R)}^2 = 0$ and $\overline{D_M^R}$ is a codifferential. Moreover $D_M^R \circ \gamma_M = \gamma_M \circ b$, thus $\overline{D_M^R} = b$ and $b^2 = 0$.

EXAMPLE 1.18. Let (R, d, m) be a differential graded algebra and (M, d_M, l, r) a strict R-bimodule. The only non-trivial defining maps are $D_1 = d$, $D_2 = m$, $D_{00}^M = d_M$, $D_{01}^M = r$, $D_{10}^M = l$. Hence one has

$$b(m \otimes a_1 \otimes \ldots \otimes a_n) = d_M(m) \otimes a_1 \otimes \cdots \otimes a_n + \sum \pm m \otimes \cdots da_i \ldots \otimes a_n$$
$$+ r(m, a_1) \otimes a_2 \cdots \otimes a_n + \pm l(a_n, m) \otimes a_1 \cdots \otimes a_n$$
$$+ \sum \pm m \otimes a_1 \cdots m(a_i, a_{i+1}) \cdots \otimes a_n$$

which is the usual Hochschild boundary for a differential graded algebra. Thus Definition 1.14 is equivalent to the standard one for strict algebras and bimodules.

THEOREM 1.19. *Let R be an A_∞-algebra and M an R-bimodule, flat as k-modules. There is a converging spectral sequence*

$$E_{pq}^2 = HH_{p+q}(H^*(R), H^*(M))_q \Longrightarrow HH_{p+q}(R, M).$$

The subscript q in $HH_n(A, B)_q$ stands for the piece of $HH_n(A, B)$ of internal homological degree q (thus of internal cohomological degree $-q$).

Proof: Consider the filtration $F_{p \geq 0} C_*(R, M) = \bigoplus_{i \leq p} M \otimes sR^{\otimes i}$ dual to the filtration of Proposition 1.11. It is an exhaustive bounded below filtration of chain complex thus it gives a converging homology spectral sequence. Now the result follows as in the proof of Proposition 1.11.   ∎

2. C_∞-algebras, C_∞-bimodules, Harrison (co)homology

In this section we introduce the key definition of C_∞-bimodules and also recall the Harrison (co)homology of C_∞-algebras for which there is not so much published account.

2.1. Homotopy symmetric bimodules.
Commutative algebras are associative algebras with additional symmetry. Similarly a C_∞-algebra could be seen as a special kind of A_∞-algebra. Indeed, this is the point of view we adopt here. The *shuffle product* makes the tensor coalgebra $(T(V), \delta)$ a bialgebra. It is defined by the formula

$$\mathrm{sh}(x_1 \otimes \ldots, \otimes x_p, x_{p+1} \otimes \ldots \otimes x_{p+q}) = \sum \pm x_{\sigma^{-1}(1)} \otimes \ldots \otimes x_{\sigma^{-1}(p+q)}$$

where the summation is over all the (p, q)-shuffles, that is to say the permutation of $\{1, \ldots, p+q\}$ such that $\sigma(1) < \cdots < \sigma(p)$ and $\sigma(p+1) < \cdots < \sigma(p+q)$. The sign $\pm$ is the sign given by the Koszul sign convention. A $(p_1, \ldots, p_r)$-shuffle is a permutation of $\{1, \ldots, p_1 + \cdots + p_r\}$ such that $\sigma(p_1 + \cdots + p_i + 1) < \cdots < \sigma(p_1 + \cdots + p_i + p_{i+1})$ for all $0 \leq i \leq r - 1$.

A B_∞-*structure* on a k-module R is given by a product M^B and a derivation D^B on the (shifted tensor) coalgebra $A^\perp(R)$ such that $(A^\perp(R), \delta, M^B, D^B)$ is a differential graded bialgebra [**Ba**]. A B_∞-algebra is in particular an A_∞-algebra whose codifferential is D^B.

DEFINITION 2.1.   • *A C_∞-algebra is an A_∞-algebra (R, D) such that the coalgebra $A^\perp(R)$, equipped with the shuffle product and the differential D, is a B_∞-algebra.*
   • *A C_∞-map between two C_∞-algebras R, S is an A_∞-algebra map $R \to S$ which is also a map of algebras with respect to the shuffle product.*

In particular there is a faithful functor from the category of C_∞-algebras to the category of A_∞-algebras. Moreover a A_∞ algebra defined by maps $D_i : R^{\otimes i} \to R$ is a C_∞-algebra if and only if, for all $n \geq 2$ and $k + l = n$, one has

$$(2.8) \qquad D_n(\operatorname{sh}(x_1 \otimes \ldots \otimes x_k, y_1 \otimes \ldots \otimes y_l)) = 0.$$

EXAMPLE 2.2. According to Example 1.10 and identity (2.8), any differential graded commutative algebra (R, m, d) has a natural C_∞-structure given by $D_1 = d$, $D_2 = m$ and $D_i = 0$ for $i \geq 3$.

REMARK 2.3. Definition 2.1 is taken from [**GJ2**]. In *characteristic zero*, a more classical and equivalent one is to say that a C_∞-algebra structure on R is given by a degree 1 differential on the cofree Lie coalgebra $C^\perp(R) := \operatorname{coLie}(sR)$. The equivalence between the two definitions follows from the fact that $\operatorname{coLie}(sR) = A^\perp(R)/\operatorname{sh}$ is the quotient of $A^\perp(R)$ by the image of the shuffle multiplication $\operatorname{sh} : A^{\perp \geq 1}(R) \otimes A^{\perp \geq 1}(R) \to A^\perp(R)$. For arbitrary characteristic, Definition 2.1 is slightly weaker (see Example 2.4 below) than the one given by the operad theory, namely by a (degree 1) codifferential on $C^\perp(R)$.

EXAMPLE 2.4. Since the universal enveloping coalgebra of a cofree Lie coalgebra $\operatorname{coLie}(V)$ is the tensor coalgebra $(T(V), \delta, \operatorname{sh})$ equipped with the shuffle product, a degree 1 differential on the cofree Lie coalgebra $C^\perp(R) := \operatorname{coLie}(sR)$ canonically yields a C_∞-algebra structure on R. We call such a C_∞-structure a *strong C_∞-algebra* structure. Note that strict C_∞-algebras are strong C_∞-algebras. Over a ring k containing $\mathbb{Q}$, all C_∞-algebras are strong (Remark 2.3).

A bimodule over a C_∞-algebra is a bimodule over this C_∞-algebra viewed as an A_∞-one. However this notion does not capture all the symmetry conditions of a C_∞-algebra. In the following sections we will need up to homotopy generalization of symmetric bimodules.

DEFINITION 2.5. *A C_∞-bimodule structure on M is a bimodule over (R, D) such that the structure maps D_{ij}^M satisfy, for all $n \geq 1$, $a_1, \ldots, a_n \in R$, $x \otimes m \otimes y \in A_R^\perp(M)$, the following relation*

$$(2.9) \quad \sum_{i+j=n} \pm D_{(i+|x|)(j+|y|)}^M\big(\operatorname{sh}(x, a_1 \otimes \ldots \otimes a_i), m, \operatorname{sh}(y, a_{i+1} \otimes \ldots \otimes a_n)\big) = 0.$$

The sign $\pm$ is the Koszul sign of the two shuffle products multiplied by the sign $(-1)^{(|a_1| + \cdots + |a_i| + i)(|m| + 1)}$. With Sweedler's notation associated to the coproduct structure of $T(sR)$, identity (2.9) reads as

$$(2.10) \qquad \sum \pm D^M\big(\operatorname{sh}(x, a^{(1)}), m, \operatorname{sh}(y, a^{(2)})\big) = 0.$$

EXAMPLE 2.6. Let (R, m, d_R) be a graded commutative differential algebra and (M, l, r, d_M) a graded differential R-bimodule. Then M has a bimodule structure as explained in the previous section. Moreover this bimodule structure is a C_∞-bimodule structure if and only $l(m, a) = (-1)^{|a||m|} r(a, m)$ for all $m \in M, a \in R$, that is, M is symmetric in the usual sense.

EXAMPLE 2.7. If (R, D) is a C_∞ algebra such that $D_1 = 0$, then $D_2 : R \otimes R \to R$ is associative (Remark 1.6) and graded commutative by Equation (2.8). Thus (R, D_2) is a graded commutative algebra. Furthermore, if (M, D^M) is a C_∞-bimodule such that $D_{00}^M = 0$, Equation (2.10) implies that D_{01}^M and D_{10}^M give a structure of symmetric bimodule over (R, D_2) to M.

EXAMPLE 2.8. Any C_∞-algebra is a C_∞-bimodule over itself. This follows from identity (2.8) and the observation that, for any $a \otimes r \otimes b \in A_R^\perp(R)$ and $x \in A^{\perp *\geq 1}(R)$, one has

$$\sum \pm \left(\mathrm{sh}(a, x^{(1)}) \otimes r \otimes \mathrm{sh}(b, x^{(2)}) \right) = \mathrm{sh}(a \otimes r \otimes b, x).$$

A C_∞-bimodule is a left (and right by commutativity) module over the shuffle bialgebra. The module structure is the map $\rho : A_R^\perp(M) \otimes A^\perp(R) \to A_R^\perp(M)$ given by the composition

$$A_R^\perp(M) \otimes A^\perp(R) \xrightarrow{\mathrm{id} \otimes \delta} A_R^\perp(M) \otimes A^\perp(R) \otimes A^\perp(R) \xrightarrow{(\mathrm{sh} \otimes \mathrm{id} \otimes \mathrm{sh}) \circ \tau} A_R^\perp(M)$$

where the map τ

$$A^\perp(R) \otimes M \otimes A^\perp(R) \otimes A^\perp(R) \otimes A^\perp(R) \xrightarrow{\tau} A^\perp(R) \otimes A^\perp(R) \otimes M \otimes A^\perp(R) \otimes A^\perp(R)$$

is the permutation of the two $A^\perp(R)$ factors sitting in the middle.

PROPOSITION 2.9. A R-bimodule M is a C_∞-bimodule if and only if $A_R^\perp(M)$ is a differential module over the shuffle bialgebra $A^\perp(R)$. That is to say if the following diagram commutes

$$\begin{array}{ccc}
A_R^\perp(M) \otimes A^\perp(R) & \xrightarrow{\rho} & A_R^\perp(M) \\
D^M \otimes \mathrm{id} + \mathrm{id} \otimes D \downarrow & & \downarrow D^M \\
A_R^\perp(M) \otimes A^\perp(R) & \xrightarrow{\rho} & A_R^\perp(M).
\end{array}$$

Proof: The compatibility with the coalgebra structure is part of the definition of a R-bimodule. It remains to prove the compatibility with the product. Let's denote by $x \bullet y$ the shuffle product $\mathrm{sh}(x, y)$. First we have to check that ρ defines an action of $(A^\perp(R), \mathrm{sh})$. Using that sh is a coalgebra map it is equivalent, for all $u, x, y \in A^\perp(R)$, $m \in M$, to :

$$(u \bullet x^{(1)}) \bullet y^{(1)} \otimes m \otimes (v \bullet x^{(2)}) \bullet y^{(2)} = u \bullet (x^{(1)} \bullet y^{(1)}) \otimes m \otimes v \bullet (x^{(2)} \bullet y^{(2)})$$

which holds by associativity of sh.

Since D is both a coderivation and derivation, one has, for all $a, b, x \in A^\perp(R)$, $m \in M$,

$$D^M \left(\rho(a, m, b, x) \right) = \sum a^{(1)} \bullet x^{(1)} \otimes D_{**}^M \left(a^{(2)} \bullet x^{(2)}, m, x^{(3)} \bullet b^{(1)} \right) \otimes b^{(2)} \bullet x^{(4)}$$
$$+ \rho \left(D^M \otimes \mathrm{id} + \mathrm{id} \otimes D \right) (a, m, b, x).$$

The sum is over all decompositions $\delta^3(x) = \sum x^{(1)} \otimes x^{(2)} \otimes x^{(3)} \otimes x^{(4)}$ such that $x^{(2)}$ or $x^{(3)}$ is not in k. But then the difference $D^M \circ \rho - \rho \circ \left(D^M \otimes \mathrm{id} + \mathrm{id} \otimes D \right)$ is 0 if and only if D^M satisfies the defining conditions of a C_∞-bimodule.   ■

The strict notion of a symmetric bimodule is self-dual. However this is not true for its homotopy analog. Thus we will also need the dual version of a C_∞-bimodule, that we call a C_∞^{op}-bimodule.

DEFINITION 2.10. *Let (R, D) be a C_∞-algebra. A C_∞^{op}-bimodule structure on M is an R-bimodule structure, such that the structure maps D_{ij}^M satisfy for all $n > 1$, $a_1, \ldots, a_n \in R$, $x \otimes m \otimes y \in A_{\bar{R}}^\perp(M)$:*

$$\sum_{i+j=n} \pm D_{(j+|x|)(i+|y|)}^M \big(\mathrm{sh}(x, a_{i+1} \otimes \ldots \otimes a_n), m, \mathrm{sh}(y, a_1 \otimes \ldots \otimes a_i)\big) = 0.$$

As in Definition 2.5, the sign is given by the Koszul rule for signs.

EXAMPLE 2.11. *If M is a strict symmetric bimodule (over a strict algebra) then it is a C_∞^{op}-bimodule.*

EXAMPLE 2.12. A C_∞-algebra has no reason to be a C_∞^{op}-bimodule in general. However its dual is always an C_∞^{op}-bimodule. More precisely, let (R, D) be a C_∞-algebra and write $R^\star = \mathrm{Hom}(R, k)$ for the dual module of R. Then the maps $D_{kl}^{R^\star} : R^{\otimes k} \otimes R^\star \otimes R^{\otimes l} \to R^\star$ given by

$$D_{kl}^{R^\star}(r_1, \ldots, r_k, f, r_{k+1}, \ldots, r_{k+l})(m) = \pm f\big(D_{k+l+1}(r_{k+1}, r_{k+l}, m, r_1, \ldots, r_k)\big)$$

yields an A_∞-bimodule structure on $R^\star$, see [**Tr1**] Lemma 3.9. The equation

$$\sum_{i+j=n} \pm D_{(j+|x|)(i+|y|)}^{R^\star} \big(\mathrm{sh}(x, a_{i+1} \otimes \ldots \otimes a_n), f, \mathrm{sh}(y, a_1 \otimes \ldots \otimes a_i)\big) = 0.$$

is equivalent to

$$\sum_{i+j=n} \pm D_{(j+|x|)(i+|y|)}^{R} \big(\mathrm{sh}(y, a_1 \otimes \ldots \otimes a_i), m, \mathrm{sh}(x, a_{i+1} \otimes \ldots \otimes a_n)\big) = 0.$$

which is satisfied because R is a C_∞-bimodule (Definition 2.5).

The argument of Example 2.12 can be generalized to prove

PROPOSITION 2.13. *The dual $M^\star = \mathrm{Hom}(M, k)$ of any C_∞-bimodule is a C_∞^{op}-bimodule. The dual of any C_∞^{op}-bimodule is a C_∞-bimodule.*

The (operadic) notion of strong C_∞-algebras (Example 2.4) gives rise to the notion of *strong C_∞-bimodules* which form a nice subclass of the C_∞-bimodules because, under suitable freeness assumption, they are also C_∞^{op}-bimodules, see Proposition 2.16 below. Let (R, D) be a strong C_∞-algebra.

DEFINITION 2.14. *A strong C_∞-bimodule structure on M over a strong C_∞-algebra (R, D) is a structure of strong C_∞-algebra on $R \oplus M$, given by a codifferential D^M on $T(sR \oplus sM)$, satisfying*

(1) *$D_n^M(x_1, \ldots, x_n) = 0$ if at least two of the x_is are in M, $D_n^M(x_1, \ldots, x_n) \in M$ if exactly one of the x_is is in M and $D_n^M(x_1, \ldots, x_n) \in R$ if all x_is are in R;*

(2) *the restriction of D^M to $T(sR)$ is equal to the differential D defining the (strong) C_∞-structure on R.*

In particular, a strong C_∞-bimodule structure on M is uniquely determined by maps $D_{p,q}^M : R^{\otimes p} \otimes M \otimes R^{\otimes q} \to M$. Furthermore, these defining maps $D_{p,q}^M$ satisfy the relation (2.9), thus M is a C_∞-bimodule. A strong C_∞-algebra is a strong C_∞-bimodule over itself (with defining map as in Example 1.7).

REMARK 2.15. *If k contains $\mathbb{Q}$, any C_∞-bimodule is strong. This follows from Remark 2.3, Proposition 2.9 and the proof of Proposition 2.16 below.*

Moreover when M is free over k, one has

PROPOSITION 2.16. *Let M and R be free over k, and (R, D) be a strong C_∞-algebra (see Example 2.4). If M is a strong C_∞-bimodule over (R, D), it is a C_∞^{op}-bimodule and its dual $M^\star$ is C_∞-bimodule.*

Proof: Denote $D : T(sR) \to T(sR)$ the differential defining the A_∞-structure and D^M the bimodule one. By duality and Proposition 2.13, it is sufficient to prove that a strong C_∞-bimodule M is also a C_∞^{op}-bimodule. Let us show that it is enough to prove the result for the canonical bimodule structure over R. Note that there is a splitting

$$(R \oplus M)^{\otimes i} \cong X \oplus R^{\otimes i} \oplus \bigoplus_{j=0}^{i} R^j \otimes M \otimes R^{i-1-j}.$$

Here $X \subset (R \oplus M)^{\otimes i}$ is the submodule generated by tensors with at least two components in M. Consider the maps $B_i : (R \oplus M)^{\otimes i} \to R \oplus M$ defined to be zero on X, D_i on $R^{\otimes i}$ and $D^m_{j,i-1-j}$ on $R^j \otimes M \otimes R^{i-1-j}$. It is straightforward to check that the maps $(B_i)_{i \geq 1}$ give an A_∞-structure on $R \oplus M$ iff M is an A_∞-bimodule. Moreover, if M is a strong C_∞-bimodule, $R \oplus M$ is a strong C_∞-algebra and it remains to prove the statement for a strong C_∞-algebra.

Let R be a strong C_∞-algebra equipped with its canonical (strong) bimodule structure over itself (Example 2.8). We have to prove that R is a C_∞^{op}-bimodule. Denote $\overline{D} : T(sR) \to sR$ the projection of the differential $D : T(sR) \to T(sR)$. Since D defines a C_∞-structure, $\overline{D}$ factors through the free Lie coalgebra $\mathrm{CoLie}(sR) \to sR$. By hypothesis its dual $\overline{D}^\star$ is a map $(sR)^\star \to Lie\big((sR)^\star\big)$ the free Lie algebra on $(sR)^\star$. Since $R \hookrightarrow R^{\star\star}$ is injective, it is enough to prove that for any $F : (sR)^\star \to Lie\big((sR)^\star\big)$, $f \in (sR)^\star$ $a_1, \ldots, a_n \in R$ and $x \otimes m \otimes y \in A_R^\perp(R)$, one has

$$(2.11) \qquad F(f)\big(\mathrm{sh}(x, a_{i+1} \otimes \ldots \otimes a_n), m, \mathrm{sh}(y, a_1 \otimes \ldots \otimes a_i)\big) = 0.$$

We can work with homogeneous component and use induction, the result for order one component of $F(f)$ being trivial. Thus we can assume $F(f) = [G, H]$ and G, H satisfies identity (2.11). Then, writing z for the term to which we apply $F(f)$, we find

$$
\begin{aligned}
F(f)(z) \quad = \quad & \sum G(x^{(1)} \bullet a^{(2)}) H(x^{(2)} \bullet a^{(3)}, m, y \bullet a^{(1)}) \\
& + \sum G(x \bullet a^{(3)}, m, y^{(1)} \bullet a^{(1)}) H(y^{(2)} \bullet a^{(2)}) \\
& - \sum H(x^{(1)} \bullet a^{(2)}) G(x^{(2)} \bullet a^{(3)}, m, y \bullet a^{(1)}) \\
& - \sum H(x \bullet a^{(3)}, m, y^{(1)} \bullet a^{(1)}) G(y^{(2)} \bullet a^{(2)}).
\end{aligned}
$$

By definition G and H vanish on shuffles, thus all the terms of the first line for which $x^{(1)}$ and $a^{(2)}$ are non trivial are zero. Moreover H satisfies identity (2.11). Thus all the terms for which $a^{(2)}$ is trivial also cancel out. The same analysis works for line 4. Thus for lines 1 and 4 we are left to the terms for which $x^{(1)}$ is trivial and $a^{(2)}$ is not. Those terms cancels out each other by commutativity of k. Line 2 and 3 cancels out by a similar argument. ∎

REMARK 2.17. In particular if k is a characteristic zero field, C_∞ and C_∞^{op} bimodules coincide.

REMARK 2.18. The author realized that C_∞ and C^{op}_∞ should coincide under quite general hypothesis while reading [**HL**]. The proof of Proposition 2.16 is taken from Lemma 7.9 in [**HL**].

Proposition 2.9 can be dualized too. A C^{op}_∞-bimodule is a left (and right by commutativity) module over the shuffle bialgebra through the opposite action $\tilde\rho$. The map $\tilde\rho\colon A^\perp_R(M) \otimes A^\perp(R) \to A^\perp_R(M)$ is the composition

$$A^\perp_R(M) \otimes A^\perp(R) \xrightarrow{\mathrm{id}\otimes to\delta} A^\perp_R(M) \otimes A^\perp(R) \otimes A^\perp(R) \xrightarrow{(\mathrm{sh}\otimes\mathrm{id}\otimes\mathrm{sh})\circ\tau} A^\perp_R(M)$$

where the map τ

$$A^\perp(R) \otimes M \otimes A^\perp(R) \otimes A^\perp(R) \otimes A^\perp(R) \xrightarrow{\tau} A^\perp(R) \otimes A^\perp(R) \otimes M \otimes A^\perp(R) \otimes A^\perp(R)$$

is the permutation of the middle $A^\perp(R)$ factors and t the transposition. Now dualizing the argument of Proposition 2.9 yields

PROPOSITION 2.19. *A R-bimodule M is a C^{op}_∞-bimodule if and only if $A^\perp_R(M)$ is a differential module over the shuffle bialgebra $A^\perp(R)$ for the action $\tilde\rho$. That is to say if the following diagram commutes*

$$
\begin{array}{ccc}
A^\perp_R(M) \otimes A^\perp(R) & \xrightarrow{\tilde\rho} & A^\perp_R(M) \\
{\scriptstyle D^M \otimes \mathrm{id} + \mathrm{id} \otimes D}\downarrow & & \downarrow {\scriptstyle D^M} \\
A^\perp_R(M) \otimes A^\perp(R) & \xrightarrow{\tilde\rho} & A^\perp_R(M).
\end{array}
$$

The equation satisfied by the defining maps D^M_{10} and D^M_{01} for being a C_∞ or a C^{op}_∞-bimodule are the same, namely

$$\forall x \in R, m \in M \quad D^M_{10}(x,m) = (-1)^{|x||m|} D^M_{01}(m,x).$$

From this observation follows the obvious

PROPOSITION 2.20. *If M is either a C_∞ or a C^{op}_∞-bimodule over R, then $H^*(M)$ is a symmetric $H^*(R)$-module.*

2.2. Harrison (co)homology with values in bimodules.

In this section we define Harrison (co)homology for a C_∞-algebras. For simplicity, in this section, we restrict attention to the case where k is a field and C_∞-algebras and bimodules are strong. In particular, applying Proposition 2.16 all modules are C_∞ and C^{op}_∞-bimodules.

We first deal with cohomology. Thus let (R, D) be a (strong) C_∞-algebra and let (M, D^M) be a (strong) C_∞-bimodule over R. Recall that a coderivation $f \in \mathrm{CoDer}\big(T(sR), A^\perp_R(M)\big)$ is determined by its projection $f^i : R^{\otimes i \geq 0} \to M$. Denote $\mathrm{BDer}(R, M)$ the subspace of $\mathrm{CoDer}\big(T(sR), A^\perp_R(M)\big)$ of coderivations f such that the f_i vanishes on the module generated by the shuffles i.e.,

$$f_i\big(\mathrm{sh}(x,y)\big) = 0 \quad \text{for} \quad i \geq 2, x \in R^{k\geq 1}, y \in R^{i-k\geq 1}.$$

LEMMA 2.21. *The map $b(f) = D^M \circ f - (-1)^{|f|} f \circ D$ sends $\mathrm{BDer}(R, M)$ to itself and satisfies $b^2 = 0$.*

Proof: We already know that b maps coderivations into coderivations. Let $x \in R^{\otimes k \geq 1}, y \in R^{\otimes l \geq 1}$ and $f \in \mathrm{BDer}(R, M)$.

$$
\begin{aligned}
b(f)_i(x \bullet y) &= \sum \pm D^M_{i^{(1)}i^{(3)}}\left(x^{(1)} \bullet y^{(1)}, f_{i^{(2)}}(x^{(2)} \bullet y^{(2)}), x^{(3)} \bullet y^{(3)}\right) \\
&= \sum \pm D^M_{i^{(1)}i^{(3)}}\left(x^{(1)} \bullet y^{(1)}, f_{i^{(2)}}(x^{(2)}), x^{(3)} \bullet y^{(2)}\right) \\
&\quad + \pm D^M_{i^{(1)}i^{(3)}}\left(x^{(1)} \bullet y^{(1)}, f_{i^{(2)}}(y^{(2)}), x^{(2)} \bullet y^{(3)}\right) \\
&= 0.
\end{aligned}
$$

The first line follows from the definition of $\mathrm{BDer}(R, M)$ and the other because M is a C_∞-bimodule.

The last statement follows from

$$
\begin{aligned}
b^2(f) &= D\left(D^M \circ f - (-1)^{|f|} f \circ D\right) - (-1)^{|f|+1}\left(D^M \circ f - (-1)^{|f|} f \circ D\right)D \\
&= (-1)^{|f|+1} D^M \circ f \circ D + (-1)^{|f|} D^M \circ f \circ D \\
&= 0.
\end{aligned}
$$

∎

DEFINITION 2.22. *Let (R, D) be a strong C_∞-algebra and (M, D^M) be a strong C_∞-bimodule over R, the Harrison cohomology $Har^*(R, M)$ of R with values in M is the cohomology of the complex $CHar^*(R, M) := \mathrm{BDer}(R, M)$ with differential $b(f) = D^M \circ f - (-1)^{|f|} f \circ D$.*

EXAMPLE 2.23. Let R be a non-graded commutative algebra and M a symmetric R-bimodule. Then the space of coderivations $\mathrm{BDer}(R, M)$ is isomorphic to $\mathrm{Hom}(T(R)/\mathrm{sh}, M)$ and is concentrated in positive degrees. Thus, as in Example 1.10, the Harrison cohomology coincides with the usual one for strictly commutative algebras in degree ≥ 1.

REMARK 2.24. The notation BDer is chosen to put emphasis on the fact that the Harrison cochain complex $\mathrm{BDer}(R, M)$ is a space of B_∞-derivation. More precisely: if R has a structure of B_∞-algebra and $A^\perp_R(M)$ is a differential graded module over the bialgebra $(T(sR), \delta, M^B)$, a derivation of B_∞-algebra from R to M is a map $f : T(sR) \to A^\perp_R(M)$ which commutes with δ and M i.e.

$$
\delta^M \circ f = (\mathrm{id} \otimes f + f \otimes \mathrm{id}) \circ \delta \quad \text{and}
$$

$$
f \circ M^B = \rho(\mathrm{id} \otimes f + f \otimes \mathrm{id}).
$$

When R is a C_∞-algebra and M a C_∞-bimodule over R, $\mathrm{BDer}(R, M)$ is the space of B_∞-derivations from R to M for the B_∞-structure given by the shuffle product and the coderivation given by ρ.

We now define Harrison homology. Let R be a C_∞-algebra and M a C_∞-R-bimodule. We denote $T(sR)/\mathrm{sh}$ the quotient of the shifted tensor coalgebra $T(sR)$ by the image of the shuffle product map $A^\perp(sR) \otimes A^\perp(sR) \to T(sR)$. Reasoning as in the first part of the proof of Lemma 2.21 yields

LEMMA 2.25. *Let R be a C_∞-algebra and M a R-bimodule. If M is a C_∞-bimodule, the Hochschild differential $b : M \otimes T(sR) \to M \otimes T(sR)$ passes to the quotient $M \otimes T(sR)/\mathrm{sh}$.*

DEFINITION 2.26. *Let (R, D) be a strong C_∞-algebra and (M, D^M) a strong C_∞-bimodule over R, the Harrison homology $Har_*(R, M)$ of R with values in M is the homology of the complex $(CHar_*(R, M) := M \otimes C^\perp(R), b)$.*

EXAMPLE 2.27. If R, M are respectively a commutative algebra and a symmetric bimodule, the complex $CHar_*(R, M)$ is the usual Harrison chain complex, up to degree 0 terms.

PROPOSITION 2.28. *Let R be a strong C_∞-algebra and M a strong C_∞-bimodule over R, flat as k-modules. There are converging spectral sequences*

$$E_2^{pq} = Har^{p+q}(H^*(R), H^*(M))_q \Longrightarrow Har^{p+q}(R, M),$$

$$E_{pq}^2 = Har_{p+q}(H^*(R), H^*(M))_q \Longrightarrow Har_{p+q}(R, M).$$

Proof: The shuffle product preserves the grading of $T(sR)$ by tensor powers. Thus the filtration $F_p C_*$ of Proposition 1.19 induces a filtration on the Harrison chain complex $CHar_*(R, M)$. Similarly the filtration $F^p C^*$ restricts to the Harrison cochain complex. Now, the proof of Proposition 1.11 applies also in these cases. ∎

REMARK 2.29. It is of course possible to work with more general ground ring k and general C_∞-algebras and bimodules. In that case, we have to assume that M is a C_∞^{op}-bimodule in statements relative to homology (and a C_∞-bimodule in statement relative to cohomology). Henceforth we shall do so without further comments when there is no risk for confusion, for instance in Theorem 3.1.

3. λ-operations and Hodge decomposition

This section is devoted to the definition and study of the Hodge decomposition for Hochschild cohomology of C_∞-algebras. We first recall quickly some basic facts about λ-rings. The λ-*operations* are standard maps that exists on the Hochschild and cyclic (co)homology of a commutative algebra [**GS1, Lo1**]. They yield a Hodge decomposition in characteristic zero and give a structure of γ-ring with trivial multiplication to the (co)homology groups. A γ-*ring with trivial multiplication* $(A, (\lambda^k))$ is a k-module A equipped with linear maps $\lambda^n : A \to A$ ($n \geq 1$) such that λ^1 is the identity map and

$$\lambda^p \circ \lambda^q = \lambda^{pq}.$$

A γ-ring with trivial multiplication $(A, (\lambda^n))$ has a canonical decreasing filtration $F_\bullet^\gamma A$. For $n \geq 1$, denote $\gamma^n = \sum_{i=0}^{n-1} \binom{k-1}{i}(-1)^{k-i-1}\lambda^{k-i}$. It is standard that λ^k acts as multiplication by k^n on each associated graded module $\mathrm{Gr}^{(n)} A = F_n^\gamma A / F_{n+1}^\gamma A$. In many cases this filtration splits A into pieces $A^{(i)}$ which are the n^i-eigenspaces of the maps λ^n, see [**Lo1**] for more details.

The tensor coalgebra $(T(sR), \delta, \bullet)$ is a graded bialgebra, indeed a Hopf algebra, which is commutative as an algebra. Thus there exist maps $\psi^p : T(sR) \to T(sR)$ defined by

$$(3.12) \qquad\qquad \psi^p = \mathrm{sh}^{p-1} \circ (\delta)^{p-1}$$

which yield a γ-ring with trivial multiplication structure on $T(sR)$ [**Lo2, Lo3, Pa**]. These maps induce the γ-ring structure and Hodge structure in Hochschild (co)homology.

When the ground ring k contains the rational numbers $\mathbb{Q}$, there is a family of orthogonal idempotents $e^{(i)} : T(sR) \to T(sR)$ such that the tensor coalgebra $T(sR) = \bigoplus_{n \geq 0} e^{(i)}(T(sR))$ with $e^{(0)}(T(sR)) = k$ and

$$e^{(1)}(T(sR)) = T(sR)/T(sR) \bullet T(sR) \cong T(sR)/\mathrm{sh}$$

is the set of indecomposable of the shuffle bialgebra $T(sR)$. Furthermore, the idempotents $e^{(i)}$ are linear combinations of the maps ψ^n and $e^{(i)}(T(sR))$ is the n^i-eigenspaces of the map ψ^n.

3.1. Hochschild cohomology decomposition.

In this section we study the λ-operations on Hochschild cohomology of a C_∞-algebra R with values in a C_∞^{op}-bimodule M.

A coderivation $f \in \mathrm{CoDer}(R, M)$ is uniquely defined by its components $f_i : R^{\otimes i \geq 0} \to M$. Thus, for $n \geq 1$, we obtain the coderivation

$$\lambda^n(f) := \left(f_i \circ \psi^n /_{R^{\otimes i}} \right)_{i \geq 0}$$

defined by the maps $f_i \circ \psi^n : R^{\otimes i} \to M$.

THEOREM 3.1. *Let (R, D) be a C_∞-algebra and (M, D^M) be a C_∞^{op}-bimodule over R.*

(1) *The maps $(\lambda^i)_{i \geq 0}$ give a γ-ring with trivial multiplication structure to the Hochschild cochain complex $(\mathrm{CoDer}(R, M), b)$ and the Hochschild cohomology $HH^*(R, M)$.*

(2) *If k contains $\mathbb{Q}$, there is a natural Hodge decomposition*

$$HH^*(R, M) = \prod_{n \geq 0} HH^*_{(n)}(R, M)$$

into eigenspaces for the maps λ^n. Moreover $HH^*_{(1)}(R, M) \cong Har^*(R, M)$ and $HH^*_{(0)}(R, M) \cong H^*(M)$.

(3) *If k is a $\mathbb{Z}/p\mathbb{Z}$-algebra, there is a natural Hodge decomposition*

$$HH^*(R, M) = \bigoplus_{0 \leq n \leq p-1} HH^*_{(n)}(R, M)$$

with each λ^n acts by multiplication by n^i on $HH^*_{(i)}(R, M)$. There is a natural linear map $Har^*(R, M) \to HH^*_{(1)}(R, M)$ inducing an isomorphism $HH^*_{(1)}(R, M)^{q \geq *-p+1} \cong Har^*(R, M)^{q \geq *-p+1}$.

Proof: The identity $\lambda^i \circ \lambda^j = \lambda^{ij}$ is immediate from $\psi^j \circ \psi^i = \psi^{ij}$. Moreover $\lambda^1(f) = f$. To prove the first part of the theorem it remains to check that the maps λ^n are chain complex morphisms. By Definition 2.1, the differential D is both a derivation and a coderivation. Thus the differential D commutes with the maps ψ^p. Hence with the Eulerian idempotents when they are defined. By definition of the differential b, it is sufficient to prove, for $p \geq 1$, that

$$\mathrm{pr}\left(D^M(f(\psi^{p+1})) - D^M(\psi^{p+1}(f)) \right) = 0$$

where pr : $T(sR) \to sR$ is the canonical projection. Let x be in $R^{\otimes n}$. Since the shuffle product $\bullet$ is a map of coalgebra, one has

$$\mathrm{pr}\left(D^M \circ f(\psi^{p+1}(x))\right) = \sum D^M_{**}\left(\pm\, x^{(1)} \bullet x^{(4)} \cdots \bullet x^{(3p-2)} \otimes f_*\left(x^{(2)} \bullet x^{(5)} \bullet\right.\right.$$
$$\left.\left. \cdots \bullet x^{(3p-1)}\right) \otimes x^{(3)} \bullet x^{(6)} \bullet \cdots \bullet x^{(3p)}\right)$$

where the sum is over all possible indexes (recall that we are using Sweedler's notation). By definition 2.10, the terms where $x^{(3)}$ or $x^{(4)}$ are not in k cancel out each others (fixing all other components, it follows immediately from the definition). The same argument works for the terms $x^{(3i)}$ or $x^{(3i+1)}$, $i \leq p-1$. Thus we are left with

$$\mathrm{pr}\left(D^M \circ f(\psi^{p+1}(x))\right) = \sum D^M_{**}\left(\pm\, x^{(1)} \otimes f_*\left(x^{(2)} \bullet x^{(3)} \bullet \cdots \bullet x^{(p+1)}\right) \otimes x^{(p+2)}\right)$$
$$= \mathrm{pr}\left(D^M(f \circ \psi^{p+1}(x))\right)$$

and the first part of the theorem follows.

If $k \supset \mathbb{Q}$, the idempotents $e^{(k)}$ are defined. By the first part of the theorem the Hochschild cochain space splits as the product

$$C^*(R, M) = \prod_{i \geq 0} C^*_{(k)}(R, M)$$

where $C^*_{(i)}(R, M)$ is the subspace of coderivation f whose defining maps f_i factors through $e^{(h)}(T(sR))$. We write $C^*_{(i)}(R, M) = \mathrm{CoDer}(e^{(k)}(T(sR)), A^\perp_R(M))$ by abuse of notation. This yields the Hochschild cohomology decomposition. It is standard that $\psi^n = \sum_{i \geq 0} n^i e^{(i)}$. Moreover

$$(C^*_{(0)}(R, M), b) = (\mathrm{CoDer}(k, A^\perp_R(M)), b) \cong \left(\mathrm{Hom}(k, M), D^M_{00}\right).$$

By section 2.2, Harrison cohomology is well defined. As $e^{(1)}(T(sR)) \cong T(sR)/\mathrm{sh}$, one has

$$C^*_{(1)}(R, M) \cong (\mathrm{Hom}(T(sR)/\mathrm{sh}, A^\perp_R(M)), b) \cong (\mathrm{BDer}(R, M), b) = CHar^*(R, M).$$

If k is a $\mathbb{Z}/p\mathbb{Z}$-algebra, the operators $\overline{e_n^{(i)}} = \sum_{m \geq 0} e_n^{(i+(p-1)m)} : R^{\otimes n} \to R^{\otimes n}$ are well defined for $1 \leq i \leq p-1$ and $n \geq 1$, see [GS2, Section 5]. Denote $\overline{e^{(i)}}$ the map induced by the operators $\overline{e_n^{(i)}}$ for n varying. Note that $\overline{e_n^{(i)}} = e_n^{(i)}$ for $n \leq p-1$. As above, the Hochschild differential commutes with the operators $\overline{e^{(i)}}$, thus the Hochschild cochain complex splits as

$$C^*(R, M) = \bigoplus_{0 \leq i \leq p-1} \overline{C^*_{(i)}}(R, M)$$

where $\overline{C^*_{(i)}}(R, M) = \mathrm{CoDer}(\overline{e^{(i)}}(T(sR)), A^\perp_R(M))$. By [GS2], $(\overline{e^{(1)}}(T(sR))$ lies in the quotient of the cotensor coalgebra $T(sR)/\mathrm{sh}$. It follows that we have a canonical map $Har^*(R, M) \to HH^*_{(1)}(R, M)$ (see Definition 2.22) which is an isomorphism when restricted to components of external degree $q \geq * - p + 1$ because $\overline{e_n^{(1)}} = e_n^{(1)}$ for $n \leq p-1$. Since $p \in k$ is null, reasoning as above we get that the complexes $\overline{C^*_{(i \geq 1)}}(R, M)$ are n^i-eigenspaces for λ^i. ∎

REMARK 3.2. The γ-ring structure given by Proposition 3.1.(1) gives rise to the canonical filtration of complexes $F_\bullet^\gamma(\mathrm{CoDer}(R, M), b)$. Hence there is a spectral sequence

$$E\gamma_1^{p,q} = HH^{p+q}(F_q^\gamma F_{q+1}^\gamma) \Longrightarrow HH^{p+q}(R, M).$$

Denote $F_{ind}^{n,(q)}(R, M) := \mathrm{Im}(H^n(F_q^\gamma) \to HH^{p+q}(R, M))$, the induced filtration on $HH^*(R, M)$. The argument of [**Lo1**, Théorème 3.5] shows that $E_1^{p,2} \cong Har^p(R, M)$ and $F_{ind}^{n,(q)}(R, M)^{*\geq n-q+2} \cong 0$, $F_{ind}^{n,(1)}(R, M) \cong HH^n(R, M)$.

EXAMPLE 3.3. By Proposition 2.19, the Hochschild cohomology $HH^*(R, R^*)$ always has a γ-ring structure. When R is free and R is strong, $HH^*(R, R)$ is also a γ-ring according to Proposition 2.16.

REMARK 3.4. The splitting of the differential graded bialgebra $T(sR)$ (with the differential giving the C_∞-structure and the shuffle product) used in the proof of Theorem 3.1 is also the one given in [**WGS**] for the shuffle bialgebra.

Theorem 3.1 applies to strict algebras. For a strict commutative algebra R, we denote by Ω_R^* the graded exterior algebra of the graded Kähler differential R-module Ω_R^1. The decompositions given by Theorem 3.1 agree with the classical ones for strict algebras according to

PROPOSITION 3.5. *Let (R, d) be a differential graded commutative algebra and M a symmetric bimodule. Then there exist λ-operations on $HH^*(R, M)$. If $k \supset \mathbb{Q}$, the λ-operations yield a Hodge decomposition of the Hochschild cohomology of R:*

$$HH^n(R, M) = \prod_{i\geq 0}^{n} HH_{(i)}^n(R, M) \ \text{for} \ n \geq 1$$

Moreover one has:

i) *If R is unital and $k \supset \mathbb{Q}$, $HH_{(j)}^n(R, M) \supset H^{n-j}(\mathrm{Hom}_R(\Omega_R^j, M), d^*)$ for $n \geq 1$, $j \geq 0$, this inclusion being an isomorphism if R is smooth;*

ii) *If R and M are non-graded, then the decomposition coincides with the one of Gerstenhaber and Schack [**GS1**, **GS2**].*

Proof: By Example 2.11, we know that M is a C_∞^{op}-bimodule over R. By Example 1.10, the Hochschild cochain complex of R as a C_∞-algebra is isomorphic to its usual Hochschild complex as an associative algebra. Through this isomorphism the operation λ^i becomes

(3.13) $f \in \mathrm{Hom}(R^{\otimes n}, M) \mapsto f \circ \psi^i.$

Thus when $k \supset \mathbb{Q}$, the induced splitting coincides with the one of [**GS1**] and $ii)$ is proved. Theorem 3.1 implies that the Hochschild cohomology of R (with its canonical C_∞-structure) admits a Hodge type decomposition. We know from Example 1.10 that this cohomology coincides with the usual Hochschild cohomology.

When R is furthermore unital, there is a canonical isomorphism of cochain complexes

$$\mathrm{Hom}(R^{\otimes n}, M) \cong \mathrm{Hom}_R(R \otimes R^{\otimes n}, M)$$

where the differential on the right is dual to the Hochschild differential on the Hochschild complex $C_*(R, R) = R \otimes R^{\otimes *}$. There is the well-known canonical map $R \otimes T(sR) \xrightarrow{\pi} \Omega_R^*$ given by $\pi(a_0 \otimes a_1 \otimes \ldots \otimes a_n) = a_0 \partial a_1 \ldots \partial a_n$ which is a map of complexes. The differential on Ω_R^* is the one induced by the inner differential $d : R \to R$. Hence we get a chain map

$$\pi^* : (\mathrm{Hom}_R(\Omega_R^j, M), d^*) \to (\mathrm{CoDer}(R, M), b).$$

It is known that $\Omega_R^i \cong R \otimes e^{(i)}(R^{\otimes i})$ [**Lo2**]. Thus the chain map π^* splits and identifies $\mathrm{Hom}_R(\Omega_R^j, M), d^*)$ as a subcomplex of $C_{(i)}^*(R, M)$. Also, when R is smooth, the map π_* is a quasi-isomorphism and Ω_R^1 is projective over R, thus π^* is a quasi-isomorphism too. ∎

REMARK 3.6.   • Proposition 3.5 applies to *non unital* algebras.
- If R and M are non-graded and moreover flat over $k \supset \mathbb{Q}$, then assertion $ii)$ implies that

$$HH_{(i)}^n(R, M) \cong AQ_i^{n-i}(R/k, M)$$

where $AQ_k^*(R/k, M)$ is the higher André-Quillen cohomology of R with coefficients in M.

Recall that for any C_∞-algebra (R, D^C), the map $D_1 : R \to R$ is a differential and that we denote $H^*(R)$ the homology of (R, D_1). Similarly, for an R-bimodule M, $H^*(M)$ is the homology of (M, D_{00}^M). According to Proposition 1.11, there is a spectral sequence abutting to $HH^*(R, M)$. It is in fact a spectral sequence of γ-rings.

PROPOSITION 3.7. *Let (R, D) be a C_∞-algebra with R, $H^*(R)$ flat as a k module and M be a C_∞^{op}-bimodule with $M, H^*(M)$ flat.*
- *The spectral sequence $E_2^{*,*} = HH^*(H^*(R), H^*(M)) \Longrightarrow HH^*(R, M)$ is a spectral sequence of γ-rings (with trivial multiplication).*
- *If $k \supset \mathbb{Q}$, then the above spectral sequence splits into pieces*

$$AQ_{(i)}^{n-i}(H^*(R), H^*(M)) \Longrightarrow HH_{(i)}^*(R, M).$$

Proof: The spectral sequence of Proposition 1.11 is induced by the filtration $F^*C^*(R, M)$. The maps λ^n preserves the filtration, thus the γ-ring structure passes to the spectral sequence. The E_1^{**} term corresponds to the Hochschild cochain complex of the commutative algebra $H^*(R)$ with values in the symmetric bimodule $H^*(M)$. Example 1.10 ensures that the induced operations $\lambda^n : E_1^{**} \to E_1^{**}$ corresponds to Gerstenhaber-Schack standard ones on the cochain complex, hence in its cohomology E_2^{**}.

When $k \supset \mathbb{Q}$, the Hochschild cochain complex splits as a direct product of complexes $\prod_{i \geq 0} C_{(k)}^*(R, M)$. Furthermore, this splitting is induced by the Eulerian idempotent, which are linear combination of the maps λ^n. Thus the filtration on $C^*(R, M)$ is identified with a product of filtered complexes $F^\bullet C_{(i)}^*(R, M)$. As above, we find that the level $E_1^{**} C_{(i)}^*(R, M)$ is given by the weight i part $C_{(i)}^*(H^*(R), H^*(M))$ of $C^*(H^*(R), H^*(M))$. The flatness and rational assumptions ensures that the cohomology of $C_{(i)}^*(H^*(R), H^*(M))$ is the André-Quillen cohomology $AQ_{(i)}^{n-i}(H^*(R), H^*(M))$ see [**Lo2**], Chapter 3. ∎

REMARK 3.8. One easily checks that, when R is flat over $k \supset \mathbb{Q}$, the weight 1 part of the spectral sequence coincides with the Harrison cohomology spectral sequence 2.28. Notice that the spectral sequences also splits with respects to the partial Hodge decomposition if $k \supset \mathbb{Z}/p\mathbb{Z}$.

Let $F : (S, B) \to (R, D)$ be a map of A_∞-algebras and (M, D^M) be an R-bimodule. Let B^M be the coderivation of $A_S^\perp(M)$ given by the defining maps

$$\left(B_{pq}^M : S^{\otimes p} \otimes M \otimes S^{\otimes q} \xrightarrow{F \otimes \mathrm{id} \otimes F} A_R^\perp(M) \xrightarrow{D^M} A_R^\perp(M) \xrightarrow{\mathrm{pr}} M \right)_{p,q \geq 0}.$$

LEMMA 3.9. *The coderivation B^M endows M with a structure of S-bimodule. Furthermore, if F is a map of C_∞-algebras and M is a C_∞^{op}-bimodule, then M is also a C_∞^{op}-bimodule over S.*

Proof: We have to check that $(B^M)^2 = 0$. For $x \in A_R^\perp(M)$, the tensor $B^M(B^M(x))$ is a sum of three kinds of elements: the ones which only involve the maps B_n, the ones which only involve the maps B_{nm}^M $(m, n \geq 0)$ and the ones which involve one B_n and one B_{pq}^M $(n, p, q \geq 0)$. Since $B \circ B = 0$ the sum of elements of the first kind vanishes. Since the degrees $|B| = |B^M| = 1$ are odd, the sum of elements of the third kind is also null. The sum of terms of the second kind vanishes because the projections on M satisfies

$$\mathrm{pr} \left(\sum B_{nm}^M \circ (F \otimes \mathrm{id} \otimes F) \left(B^M(x) \right) \right) = \mathrm{pr} \left(D^M \circ D^M (F \otimes \mathrm{id} \otimes F) \right) = 0.$$

When F is a C_∞-map, each of its components F_i vanish on shuffles. It follows that the coalgebra map $F : A^\perp(S) \to A^\perp(R)$ is also a map of algebras (for the shuffle product). Furthermore, according to Proposition 2.19, $A_R^\perp(M)$ is a differential module over the shuffle algebra $A^\perp(R)$. Thus $A_S^\perp(M)$ is a differential module over the shuffle algebra $A^\perp(S)$. Hence M is a C_∞-bimodule over S. ∎

PROPOSITION 3.10. *Let (R, D) be an A_∞-algebra, (M, D^M) an R-bimodule and (S, B) be a A_∞-algebras.*

- *If there is an A_∞-map $F : (S, B) \to (R, D)$, then there is a linear map*

$$F^* : HH^*(R, M) \to HH^*(S, M)$$

  which is an isomorphism if $F_1 : (S, d_S) \to (R, d_R)$ is a quasi-isomorphism.
- *If (N, D^N) is an R-bimodule and $\phi : (M, D^M) \to (N, D^N)$ is a A-bimodule map, then there is a linear map*

$$\phi_* : HH^*(R, M) \to HH^*(R, N)$$

  which is an isomorphism when $\phi_1 : (M, d^M) \to (N, d^N)$ is a quasi-isomorphism.
- *When R, S are C_∞-algebras, M, N are C_∞-bimodules and F, ϕ C_∞-morphisms, then F^* and ϕ_* are maps of γ-rings.*

Proof: Lemma 3.9 implies that M has an S-bimodule structure given by the map B^M. The morphism of differential coalgebras F induces a morphism of coderivations $F^* : \mathrm{CoDer}(R, M) \to \mathrm{CoDer}(S, M)$. For any x in $C^*(R, M) = \mathrm{CoDer}(R, M)$, one has,

$$\begin{aligned} b(F(x)) &= B^M(x(F)) - x(F(B)) \\ &= B^M(x)(F) - x(D(F)) \end{aligned}$$

hence F is a morphism of complex. If F_1 is a quasi-isomorphism, then $\widetilde{F}_1$ is an isomorphism at the page 1 of the spectral sequences associated to $HH^*(R, M)$, $HH^*(S, M)$ by Proposition 1.11. The first assertion is proved. The second one is analogous using the application $\phi_* : F \in C^*(R, M) \mapsto \phi \circ f \in C^*(R, N)$ instead of F^*.

Moreover when F is a C_∞-map, then M is a C_∞-bimodule by Lemma 3.9 and the λ-operations already commutes with F^* and ϕ_* at the cochain level; the compatibility is proved as in Theorem 3.1. ∎

A C_∞-algebra (R, D) is said to be *formal* if there is a morphism

$$F : (H^*(R), D_2) \to (R, D)$$

of C_∞-algebras with F_1 a quasi-isomorphism.

COROLLARY 3.11. *Let (R, D) be a formal C_∞-algebra, free as a k-module. If there is a C_∞-map $F : (H^*(R), D_2) \to (R, D)$ with F_1 a quasi-isomorphism, then there is a natural isomorphism of γ-rings:*

$$HH^*(R, R) \cong HH^*(H^*(R), H^*(R))$$

and if $k \supset \mathbb{Q}$, an isomorphism of Hodge decomposition

$$HH^*_{(n)}(R, R) \cong HH^*_{(n)}(H^*(R), H^*(R)) \ for \ n \geq 0.$$

Proof: We denote by $\phi : H^*(R) \to R$ the morphism of C_∞-bimodule induced by F. Proposition 3.10 yields a zigzag

$$HH^*(R, R) \xrightarrow{\widetilde{F}^*} HH^*(H^*(R), R) \xleftarrow{\widetilde{\phi}^*} HH^*(H^*(R), H^*(R))$$

where the arrows are isomorphisms of γ-rings. Hence the result. ∎

REMARK 3.12. The definition of *formality* that we use here is quite strong. However, it is enough for our purpose. In the literature, one might find the definition that (R, D) is formal if $(H^*(R), D_2)$ and (R, D) are connected by a chain of C_∞-quasi-isomorphisms. When k is a characteristic zero field, these two definitions agree since one can check that C_∞-quasi-isomorphisms are invertible. This is also the case over any field if one only considers strong C_∞-algebras. Details are left to the reader.

PROPOSITION 3.13. *Let (R, D) be a C_∞-algebra and $F_1 : H^*(R) \to R$ a quasi-isomorphism inducing the product structure on $H^*(R)$. If*

$$Har^n(H^*(R), H^*(R))_{\leq 2-n} = 0 \ for \ n \geq 1$$

then R is formal.

Proof: The techniques of [**GH2**] for homotopy Gerstenhaber algebras apply *mutatis mutandis* to C_∞-algebras as well. Thus, given a quasi-isomorphism $F_1 : H^*(R) \to R$, there is a C_∞-structure $(H^*(R), B)$ and a C_∞-quasi-isomorphism $G : (H^*(R), B) \to (R, D)$ such that $B_1 = 0, B_2 = D_2$ and $G_1 = F_1$. When the Harrison cohomology is concentrated in bidegree $(1, *)$, for the bigrading induced by the tensor power of maps and internal degree of $H^*(R)$, there is a C_∞-morphism $H : (H^*(R), D_2) \to (H^*(R), B)$ with H_1 being the identity map. The composition of this two C_∞-maps gives the formality map. ∎

EXAMPLE 3.14. By a deep result of Tamarkin [**Ta**], it is now well-known that the Hochschild cochain complex of any associative algebra A, over a characteristic zero ring, has a G_∞-structure, hence a C_∞-one, which is (non-canonically) induced by the cup-product and the braces of [**GV**]. For the algebra $A = C^\infty(X)$ of smooth functions on a manifold X, the Hochschild cochain complex $C^*(A, A)$ of multilinear and multidifferential operators on A is a formal C_∞-algebra. Its cohomology is $\Gamma = \Gamma(X, \Lambda TX)$ the polyvector fields on X. We can apply Proposition 3.13 and then Corollary 3.11 (because the Harrison cohomology of Γ vanish) to find

$$HH^*_{(j)}(C^*(A,A), C^*(A,A)) \cong HH^*_{(j)}(\Lambda^*\Gamma(X, \Lambda TX), \Lambda^*\Gamma(X, \Lambda TX))$$

$$\cong \operatorname{Hom}_\Gamma(\Omega^j_\Gamma, \Gamma)$$

$$\cong \Lambda^j s\Omega_\Gamma.$$

The last step follows from the Jacobi-Zariski exact sequence applied to the smooth algebra Γ that leads to $\Omega_\Gamma \cong \Gamma \otimes_R \Omega_R \oplus s(\Omega_R)^*$. Moreover, the Hochschild chain complex $C_*(A, A)$ is a C_∞^{op}-bimodule by Proposition 2.19. From the previous argument one easily gets

$$HH^*_{(j)}(C^*(A,A), C_*(A,A)) \cong \operatorname{Hom}_\Gamma(\Omega^j_\Gamma, \Omega^*_A).$$

EXAMPLE 3.15. When formality does not hold, Proposition 3.7 can be used to study $HH^*(C^*(R,R), C^*(R,R))$. For instance, Let R be a semi-simple separable algebra, then $HH^*(R) = HH^0(R) = Z(R)$ the center of R. It follows that the spectral sequence E_1^{**} is concentrated in bidegree $(*, 0)$ hence collapses. Thus one has

$$HH^*(C^*(R,R), C^*(R,R)) \cong HH^*(Z(R), Z(R))$$

which is an isomorphism of Gerstenhaber algebras and γ-rings on the associated graded to the canonical filtration of $A^\perp(R)$.

PROPOSITION 3.16. *Let k be a characteristic zero field and (R, D) be a C_∞-algebra with $D_1 = 0$. Assume that there is an element $1 \in R^0$ which is a unit for D_2. Let N be a C_∞^{op}-module.*

- *If (R, D_2) is smooth, for any $n \geq 0$, one has*

$$HH^*_{(n)}(R, N) \cong \operatorname{Hom}_R(\Omega^n_R, N).$$

- *If (R, D_2) is not necessarily smooth but satisfies $D_3(1, x, y) = D_3(y, 1, x) = 0$, then*

$$HH^*_{(n)}(R, N) \supset \operatorname{Hom}_R(\Omega^n_R, N).$$

Proof: Let $\pi_n : M \otimes R^{\otimes n} \to M \otimes_{(R,D_2)} \Omega^n_{(R,D_2)}$ be the canonical surjection. It factors through $M \otimes e^{(n)}(R^{\otimes n})$. The C_∞-differential D commutes with $e^{(n)}$. Moreover the map $D_{i \geq 2}(R^{\otimes n}) \subset R^{\otimes \leq n-1}$. Thus, as π_* factors through $M \otimes e^{(n)}(R^{\otimes n})$, the map $\pi_* : (C_*(R, M), b) \to (M \otimes_{(R,D_2)} \Omega^*_{(R,D_2)}, 0)$ is a chain map. This is a generalization of a well-known fact for strict algebras [**Lo2**, chapter 4]. Therefore we obtain a morphism of cochain complexes

$$(\operatorname{Hom}_R(\Omega^n_R, N), 0) \hookrightarrow (C^*(R, N), b).$$

The filtration by the exterior power of $\operatorname{Hom}_R(\Omega^*_{(R,D_2)}, R)$ yields a spectral sequence computing $\operatorname{Hom}_R(\Omega^*_{(R,D_2)}, R)$. The complex map π_* yields a map between this spectral sequence and the one of Proposition 1.11 for $HH^*(R, M)$. When (R, D_2) is smooth, the map π_1 is an isomorphism at page 1 by the Hochschild Kostant Rosenberg theorem, thus on the abutment. If D_3 vanishes when one of the variable is the unit, then the anti-symmetrization map $\varepsilon_n : \Omega^n_{(R,D_2)} \to M \otimes R^{\otimes n}$ is well defined modulo a boundary of $(M \otimes T(sR), b)$, as for strict commutative algebras. Thus we have a well defined map

$$\varepsilon_* : HH^*(R, M) \to \operatorname{Hom}_R(\Omega^n_R, N)$$

which is a section of π_* up to multiplication by non zero scalars [**Lo2**].   ∎

EXAMPLE 3.17. A C_∞-algebra satisfying $D_1 = 0$ is called a minimal C_∞-algebra. Formal Frobenius algebras are a huge class of examples [**Ma**].

3.2. Decomposition in Hochschild homology. In this section we define and study the Hodge decomposition for Hochschild homology of C_∞-algebras. We denote $\overline{\lambda}^p : M \otimes T(sR) \to M \otimes T(sR)$ the map $\mathrm{id} \otimes \psi^p$ for $p \geq 1$, where ψ^p is defined by formula (3.12).

THEOREM 3.18. *Let R be a C_∞-algebra and M an C_∞-bimodule over R.*

(1) *The maps $(\overline{\lambda}^i)_{i \geq 1}$ define a γ-ring (with trivial multiplication) structure on the Hochschild complex $(M \otimes T(sR), b)$ and on Hochschild homology $HH_*(R, M)$.*

(2) *If k contains $\mathbb{Q}$, there is a Hodge decomposition*

$$HH_*(R, M) = \bigoplus_{i \geq 0} HH_*^{(i)}(R, M)$$

into eigenspaces for the maps $\overline{\lambda}^n$. Moreover $HH_*^{(1)}(R, M) \cong Har_*(R, M)$ and $HH_*^{(0)}(R, M) \cong H^*(M, D_{00}^M)$.

(3) *If k is a $\mathbb{Z}/p\mathbb{Z}$-algebra, there is a Hodge decomposition*

$$HH_*(R, M) = \bigoplus_{0 \leq n \leq p-1} HH_*^{(n)}(R, M)$$

with $\overline{\lambda}^n$ acting by multiplication by n^i on $HH_*^{(i)}(R, M)$. Furthermore, there is a natural linear map $HH_*^{(1)}(R, M) \to Har_*(R, M)$ inducing an isomorphism $HH_*^{(1)}(R, M)^{* \leq p-1-n} \cong Har_*(R, M)^{* \leq p-1-n}$.

Proof: The proof is similar to the one of Theorem 3.1. The only slight difference is the compatibility of the maps ψ^{p+1} ($p \geq 1$) with the differential D^M. For $m \in M, x \in T(sR)$ one has

$$D^M\left(\overline{\lambda}^{p+1}(x)\right) = \sum D_{**}^M\left(\pm x^{(3)} \bullet x^{(6)} \cdots \bullet x^{(3p)}\right) \otimes m \otimes x^{(1)} \bullet x^{(4)} \bullet$$
$$\cdots \bullet x^{(3p-2)}\right) \otimes x^{(2)} \bullet x^{(5)} \bullet \cdots \bullet x^{(3p-1)}$$

All the terms for which $x^{(3i+1)}$ or $x^{(3i)}$ is non trivial ($1 \leq i \leq p - 1$) vanish by definition of a C_∞-module. Thus

$$D^M\left(\overline{\lambda}^{p+1}(x)\right) = \pm D_{**}^M\left(x^{(p+2)} \otimes m \otimes x^{(1)}\right) \otimes x^{(2)} \bullet x^{(3)} \bullet \cdots \bullet x^{(p+1)}$$
$$= \overline{\lambda}^{p+1}(D^M(x)).$$

■

EXAMPLE 3.19. When R is a differential graded commutative algebra and M a symmetric bimodule, the γ-ring structures and Hodge decompositions given by Theorem 3.18 coincides with the classical ones [**Lo1, Vi**].

REMARK 3.20. Similarly to Remark 3.2, the γ-ring structure given by Theorem 3.18.(1) gives rise to a canonical filtration of complexes $F_\bullet^\gamma(M \otimes T(sR), b)$ and thus a spectral sequence $E^{\gamma 1}_{p,q} = H_{p+q}(F_q^\gamma F_{q+1}^\gamma) \Longrightarrow HH_{p+q}(R, M)$. The induced filtration $F_{n,(q)}^{ind}(R, M) := \text{Im}(H^n(F_q^\gamma) \to HH^{p+q}(R, M))$ satisfies $E_{p,2}^1 \cong Har_p(R, M)$ and $F_{n,(q)}^{ind}(R, M)^{*\geq q-2-n} \cong 0$, $F_{n,(1)}^{ind}(R, M) \cong HH_n(R, M)$.

PROPOSITION 3.21. *Let (R, D) be a C_∞-algebra with R, $H^*(R)$ flat as a k-module and M be a C_∞^{op}-bimodule such that M and $H^*(M)$ are flat.*

- *The spectral sequence $E_{*,*}^2 = HH_*(H^*(R), H^*(M)) \Longrightarrow HH_*(R, M)$ (see Theorem 1.19) is a spectral sequence of γ-rings (with trivial multiplication).*
- *If $k \supset \mathbb{Q}$, the spectral sequence splits into pieces*
$$AQ_{n-i}^{(i)}(H^*(R), H^*(M)) \Longrightarrow HH_*^{(i)}(R, M).$$

Proof: The proof is dual to the one of Theorem 1.19 and Proposition 3.7 using the dual filtration $F_i C_*(R, M) = M \otimes R^{\otimes * \leq i}$. ∎

REMARK 3.22. One easily checks that when R is flat over $k \supset \mathbb{Q}$, the weight 1 part of the spectral sequence coincides with the Harrison homology spectral sequence of Proposition 2.28.

When the bimodule M is the C_∞-algebra R, the Hochschild complex is a C_∞-algebra. For $i \geq 2$, let $B_i : (R \otimes T(sR))^{\otimes i} \to R \otimes T(sR)$ be the map defined, for $r_k \otimes x_k \in R \otimes T(sR)$, $k = 1 \dots i$, by

$$B_i(r_1 \otimes x_1, \dots, r_i \otimes x_i) = \sum_{j \geq i} \pm D_j\big(x_1^{(3)} \otimes r_1 \otimes x_1^{(1)} \lhd \cdots \lhd x_i^{(3)} \otimes r_i \otimes x_i^{(1)}\big) \otimes x_1^{(2)} \bullet \cdots \bullet x_i^{(2)}$$

where $x \otimes r_1 \otimes y \lhd x' \otimes r_2 \otimes y'$ is obtained from the shuffle product $x \otimes r_1 \otimes y \bullet x' \otimes r_2 \otimes y'$ by taking only shuffles such that r_1 appears before r_2. Take the Hochschild differential b for B_1 and write $B : C_*(R, R) \otimes T(sC_*(R, R)) \to C_*(R, R) \otimes T(sC_*(R, R))$ for the codifferential induced by the maps B_i.

PROPOSITION 3.23.   • *$(C_*(R, R), B)$ is a C_∞-algebra. In particular B_2 induces a structure of commutative algebra on $HH_*(R, R)$.*

- *$B_i(\overline{\lambda^k}^{\otimes i}) = \overline{\lambda^k}(B_i)$ and in particular the operations $\overline{\lambda^k}$ are multiplicative in Hochschild homology.*
- *The spectral sequence $E_{*,*}^2 = HH_*(H^*(R), H^*(R)) \Longrightarrow HH_*(R, R)$ is a spectral sequence of algebras equipped with multiplicative operations $\overline{\lambda^k}$.*

Proof: By commutativity of the shuffle product, the vanishing of the maps $B_{i \geq 2}$ on shuffles amounts to the vanishing of

$$D_{j \geq p+q}(x_1^{(3)} r_1 x_1^{(1)} \lhd \dots \lhd x_p^{(3)} r_p x_p^{(1)} \bullet y_1^{(3)} s_1 y_1^{(1)} \lhd \dots \lhd y_q^{(3)} s_q y_q^{(1)})$$

which follows since R is a C_∞-algebra. Since

$$D(x_1 \bullet \cdots \bullet x_i) = \sum \pm x_1 \bullet \cdots \bullet D(x_j) \bullet \cdots \bullet x_i,$$

the vanishing of B^2 is equivalent to the equation

$$\sum \pm D_* \left(x_1^{(3)} r_1 x_1^{(1)} \vartriangleleft \ldots \vartriangleleft x_i^{(3)} r_i x_i^{(1)} \vartriangleleft x_{i+1}^{(4)} \bullet \cdots \bullet x_j^{(4)} D_* \left(x_{i+1}^{(5)} r_{i+1} x_{i+1}^{(1)} \vartriangleleft \ldots \right.\right.$$
$$\left.\left. \ldots x_j^{(5)} r_j x_j^{(1)} \right) x_{i+1}^{(2)} \bullet \cdots \bullet x_j^{(2)} \vartriangleleft x_{j+1}^{(3)} r_{j+1} x_{j+1}^{(1)} \vartriangleleft \ldots \vartriangleleft x_n^{(3)} r_n x_n^{(1)} \right) = 0.$$

This is the A_∞-equation (1.4) applied to

$$x_1^{(3)} r_1 x_1^{(1)} \vartriangleleft \ldots \vartriangleleft x_n^{(3)} r_n x_n^{(1)}$$

up to terms like

$$D_*(\ldots D_*(x_i^{(1)} \bullet x_{i+1}^{(4)} r_{i+1} x_{i+1}^{(1)} \vartriangleleft \ldots x_j^{(4)} r_j x_j^{(1)}) \ldots)$$

which are trivial by Definition 2.1. Thus the Hochschild complex $(C_*(R,R), B)$ is a C_∞-algebra. In particular its homology for the differential $B_1 = b$ is a commutative algebra.

The maps $\overline{\lambda}^k$ are algebras morphisms (with respect to the shuffle product). As in the proof of Theorem 3.18 (which is the B_1-case), we obtain that $B_i(\overline{\lambda^k}^{\otimes i}) = \overline{\lambda^k}(B_i)$. Furthermore, the map B_i ($i \geq 1$) preserves the filtration $F^*(C_*(R,R))$. It follows that the spectral sequence E^1_{**} is a spectral sequence of commutative algebras. On the page E_1, the product is given by the usual shuffle product on the Hochschild complex of $H^*(R)$. Since the $\overline{\lambda}^k$-operations on page 2 commutes with the shuffle product, the result follows. $\blacksquare$

REMARK 3.24. When R is a strict commutative algebra, one recovers the usual shuffle product of [GJ1].

The functorial properties of Hochschild cohomology holds for homology as well.

PROPOSITION 3.25. *Let (R,D) be an A_∞-algebra, (M, D^M) an R-bimodule and (S,B) be an A_∞-algebra.*

- *An A_∞-morphism $F : (S,B) \to (R,D)$ induces a natural linear map*

$$F_* : HH_*(S,M) \to HH_*(R,M)$$

  which is an isomorphism if $F_1 : (S, B_1) \to (R, D_1)$ is a quasi-isomorphism.
- *Let (N, D^N) be an R-bimodule and let $\phi : (M, D^M) \to (N, D^N)$ be an R-bimodule map. There is a natural linear map $\phi_* : HH_*(R,M) \to HH_*(R,N)$ which is an isomorphism if $\phi_1 : (M, D_{00}^M) \to (N, D_{00}^N)$ is a quasi-isomorphism.*
- *Moreover when R, S are C_∞-algebras, M, N C_∞-bimodules and F, ϕ C_∞-morphisms, then F_* and ϕ_* are maps of γ-rings.*

EXAMPLE 3.26. Recall that, for an associative algebra over a field of characteristic zero, the Hochschild cochain complex $C^*(A,A)$ is a C_∞-algebra (Example 3.14). It is moreover formal when $A = C^\infty(X)$. Thus Proposition 3.25 gives an isomorphism

$$HH_*(C^*(R,R), C_*(R,R)) \cong \Gamma(X, \Lambda^*(TX)^*) \otimes_\Gamma \Omega^*_\Gamma.$$

Recall that if (R,D) is a C_∞-algebra such that $D_1 = 0$, then (R, D_2) is a graded commutative algebra (Example 2.7)

PROPOSITION 3.27. *Let k be a characteristic zero field, (R,D) be a C_∞-algebra such that $D_1 = 0$ and D_2 unital, and M be a C_∞-module.*

- If (R, D_2) is smooth, one has

$$HH_*^{(n)}(R, M) \cong M \otimes_{(R,D_2)} \Omega^n_{(R,D_2)}.$$

- If R is not necessarily smooth but $D_3(1, x, y) = D_3(y, 1, x) = 0$, then

$$HH_*^{(n)}(R, M) \supset M \otimes_{(R,D_2)} \Omega^n_{(R,D_2)}.$$

3.3. The augmentation ideal spectral sequence. In this section, we generalize results of [**WGS**] in the context of C_∞-algebras. In particular, we study the compatibility between the Hodge decomposition and the Gerstenhaber structure, see Theorem 3.31 below.

CONVENTION 3.28. In this section, the ground ring k is either torsion free or a $\mathbb{Z}/p\mathbb{Z}$-algebra. Moreover all k-modules are assumed to be flat.

The (signed) shuffle bialgebra $T(sR)$ has a canonical augmentation $T(sR) \to k \oplus sR$. We wrote $I(sR)$ for its augmentation ideal. There is a decreasing filtration

$$\cdots \subset I(sR)^n \subset I(sR)^{n-1} \subset \cdots \subset I(sR)^1 \subset I(sR)^0 = T(sR).$$

This filtration induces a filtration of Hochschild (co)chain spaces

$$\cdots \subset M \otimes I(sR)^n \subset M \otimes I(sR)^{n-1} \subset \cdots \subset M \otimes I(sR)^1 \subset M \otimes I(sR)^0 = C_*(R, M),$$

$$C^*(R, M) = \mathrm{CoDer}(I(sR)^0, A_R^\perp(M)) \to \mathrm{CoDer}(I(sR)^1, A_R^\perp(M)) \to \cdots$$
$$\cdots \to \mathrm{CoDer}(I(sR)^{n-1}, A_R^\perp(M)) \to \mathrm{CoDer}(I(sR)^n, A_R^\perp(M)) \to \cdots.$$

We called these filtration *the augmentation ideal filtration*.

LEMMA 3.29. *Let R be a C_∞-algebra, M and N respectively a C_∞-bimodule and a C_∞^{op}-bimodule over R. The augmentation ideal filtration of $C_*(R, M)$ and $C^*(R, N)$ are filtration of (co)chain complexes.*

Proof: Since the augmentation ideal filtration is induced by the shuffle product, the result follows as in the proofs of Theorems 3.1, 3.18. ∎

By Lemma 3.29, there are *augmentation ideal* spectral sequences

$$(3.14) \qquad I^1_{pq}(R, M) \;=\; H_{p+q}(M \otimes I(sR)^p/I(sR)^{p+1}),$$

$$(3.15) \qquad I_1^{pq}(R, N) \;=\; H^{p+q}(\mathrm{CoDer}(I(sR)^p/I(sR)^{p+1}, A_R^\perp(N))).$$

PROPOSITION 3.30. *Let R be a C_∞-algebra, M and N two R-bimodules which are respectively a C_∞-bimodule and a C_∞^{op}-bimodule.*

(1) *The spectral sequence $I^1_{pq}(R, M)$ converges to $HH_{p+q}(R, M)$ and the spectral sequence $I_1^{pq}(R, N)$ converges to $HH^{p+q}(R, N)$ if R, M are concentrated in non-negative degrees and N in non-positive degrees.*

(2) *Let k be a field. Then $I_1^{1*}(R, N) \cong Har^{*+1}(R, N)$ and $I^1_{1*}(R, M) \cong Har_{*+1}(R, M)$.*

(3) *When R is free, $I_1^{1*}(R, R)$ is a spectral sequence of Gerstenhaber algebras.*

Proof: It follows from the combinatorial observations of [**WGS**], Section 3 and 4. The only difficulty is to check that all the constructions are compatible with the C_∞-differential. This is straightforward since the differentials D, D^M, D^N (defining the algebra and bimodules structures) are coderivations for the coproduct and moreover compatible with the filtrations (Lemma 3.29). ∎

THEOREM 3.31. *Let k be a field and R be a strong C_∞-algebra (see Example 2.4).*

- *The Harrison cohomology $Har^*(R, R) = HH^*_{(1)}(R, R)$ is stable by the Gerstenhaber bracket.*
- *If $k \supset \mathbb{Q}$, the cup-product and Gerstenhaber bracket are filtered for the Hodge filtration $\mathcal{F}_p HH^*(R, R) = \bigoplus_{n \leq q} HH^*_{(n)}(R, R)$, in the sense that*

$$\mathcal{F}_p HH^*(R, R) \cup \mathcal{F}_q HH^*(R, R) \subset \mathcal{F}_{p+q} HH^*(R, R) \text{ and}$$

$$[\mathcal{F}_p HH^*(R, R), \mathcal{F}_q HH^*(R, R)] \subset \mathcal{F}_{p+q-1} HH^*(R, R)$$

Proof: Since k is a field, the convention 3.28 is satisfied and furthermore, Theorem 3.1 gives a Hodge decomposition if k is of characteristic zero or a partial Hodge decomposition if k is of positive characteristic. Moreover, the identification of the Harrison cohomology also follows from Theorem 3.1. By Proposition 2.16, R is a C_∞^{op}-bimodule over itself. Let g be in $C^*_{(1)}(R, M)$. Since each component $g_i : R^{\otimes i} \to R$ vanishes on shuffles, we obtain $g(x \bullet y) = g(x) \bullet y + (-1)^{|x||g|} x \bullet g(y)$. Thus, for $f, g \in C^*_{(1)}(R, M)$,

$$\text{pr}([f, g](x \bullet y)) = \text{pr}\big(f(g(x) \bullet y) + \pm f(x \bullet g(y)) - \pm g(f(x) \bullet y) + \pm g(x \bullet f(y))\big)$$
$$= 0.$$

Hence $[f, g] \in C^*_{(1)}(R, M)$.

When $k \supset \mathbb{Q}$ it is well-known that there is an isomorphism of algebras $T(sR) \cong S(e^{(1)}(sR))$ where the product on $T(sR)$ is the shuffle product. Furthermore $e^{(i)}(sR) = S^i(e^{(1)}(sR))$. Hence, the filtration $\mathcal{F}_q HH^*(R, R)$ is the filtration induced by the augmentation ideal filtration in cohomology. Let $f \in \mathcal{F}_p HH^*(R, R)$ and $g \in \mathcal{F}_q HH^*(R, R)$; we have to prove that the defining maps $(f \cup g)_m(x_1 \bullet \cdots \bullet x_n) = 0$ for $n \geq p + q$, $m \geq 1$ and $x_i \in e^{(1)}(sR)$. The argument is similar to the first part of the proof. Indeed,

$$(f \cup g)_m(x_1 \bullet \cdots \bullet x_n) = \sum_{i+j+k=m+2} \pm D_k\big(x_1^{(1)} \bullet \cdots \bullet x_n^{(1)} \otimes f_i(x_1^{(2)} \bullet \cdots \bullet x_n^{(2)})$$
$$\otimes x_1^{(3)} \bullet \cdots \bullet x_n^{(3)} \otimes g_j(x_1^{(4)} \bullet \cdots \bullet x_n^{(4)}) \otimes x_1^{(5)} \bullet \cdots \bullet x_n^{(5)}\big).$$

Since $m \geq p + q$, we can assume that there is an index l such that $x_l^{(2)} = 1 = x_l^{(4)}$ (if not either $f_i(x_1^{(2)} \bullet \cdots \bullet x_n^{(2)})$ or $g_j(x_1^{(4)} \bullet \cdots \bullet x_n^{(4)})$ is zero). It follows that $D_k\big(x_1^{(1)} \bullet \cdots \otimes f_i(x_1^{(2)} \bullet \cdots \bullet x_n^{(2)}) \otimes \cdots \otimes g_j(x_1^{(4)} \bullet \cdots \bullet x_n^{(4)}) \cdots \bullet x_n^{(5)}\big)$ is equal to

$$D_k\big((x_1^{(1)} \bullet \cdots \otimes f_i(x_1^{(2)} \bullet \cdots \bullet x_n^{(2)}) \otimes \cdots \otimes g_j(x_1^{(4)} \bullet \cdots \bullet x_n^{(4)}) \cdots \bullet x_n^{(5)}) \bullet x_l\big)$$

which is zero since D_k vanishes on shuffles. Hence, $f \cup g \in \mathcal{F}_{p+q} HH^*(R, R)$. A similar argument shows that $[f, g] \in \mathcal{F}_{p+q-1} HH^*(R, R)$. ∎

REMARK 3.32. Theorem 3.31 applies in particular to differential graded commutative algebras. For non-graded algebras it was first proved in [**BW**]. A careful analysis of the proof of Theorem 3.31 shows that it holds whenever R is free over a ground ring k which contains either $\mathbb{Q}$ or $\mathbb{Z}/p\mathbb{Z}$ (p a prime).

REMARK 3.33. Note that when the spectral sequence $I_1^{pq}(R, R)$ converges, the second assertion in Theorem 3.31 follows immediately from Proposition 3.30.

3.4. Hodge decomposition and cohomology of homotopy Poisson algebras. In this section, for simplicity, we work over a ground ring containing $\mathbb{Q}$. If P is a Poisson algebra, the Hodge decomposition of the Hochschild (co)homology of the underlying commutative algebra identifies with the first page of a spectral sequence computing its Poisson cohomology [**Fr2**]. We want to prove that this result makes sense for homotopy Poisson algebras (P_∞-algebras for short) as well. We briefly recall the definition of P_∞-algebras and refer to [**Gi**] for more details.

DEFINITION 3.34. *Let R be a k-module and $P^\perp(R) := S\big(\mathrm{coLie}(sR)\big)$ be the symmetric coalgebra on the cofree Lie coalgebra over sR. A P_∞-algebra structure on R is given by a coderivation ∇ of degree 1 on $P^\perp(R)$ such that $\nabla^2 = 0$. A map of P_∞-algebra $(R, \nabla) \to (S, \nabla')$ is a graded differential coalgebra map $P^\perp(R) \to P^\perp(S)$.*

The "coalgebra"-structure of $P^\perp(R)$ is obtained by the sum of the symmetric coproduct (i.e. the free cocommutative one) and the lift as a coderivation of the Lie coalgebra cobracket (see [**Gi**] for an explicit formula). As for A_∞-algebras, a P_∞-structure on R is uniquely defined by maps $\nabla_{p_1,\ldots,p_n} : R^{\otimes p_1} \otimes \ldots \otimes R^{\otimes p_n} \to R$ such that $\nabla(x_1,\ldots,x_n)$ $(x_i \in R^{\otimes p_i})$ is antisymmetric with respect to the coordinates x_i and vanish if one of the x_i is a shuffle. There is a forgetful functor from the category of P_∞-algebras to the one of C_∞-algebras. It is determined by considering only the map $D_n : R^{\otimes n} \to R$ which restricts to $\mathrm{coLie}(sR)$. When (R, ∇) is a P_∞-algebra, we denote $\mathrm{CoDer}(R, R)$ the k-module of coderivations of $P^\perp(R)$.

DEFINITION 3.35. *The cohomology of the P_∞-algebra (R, ∇) is the cohomology $HP^*(R, R)$ of the complex $\mathrm{CoDer}(R, R)$ equipped with the differential $[-, \nabla]$. More precisely, one has*

$$[f, \nabla] = f \circ \nabla - (-1)^{|f|} \nabla \circ f \text{ for } f \in \mathrm{CoDer}(R, R).$$

PROPOSITION 3.36. *Let (P, ∇) be a P_∞-algebra, such that P is a C_∞^{op}-bimodule over itself. There is a converging spectral sequence*

$$(3.16) \qquad\qquad E_1^{pq} = HH_{(p)}^{q+p}(P, P) \Longrightarrow HP^{p+q}(P, P).$$

Proof: Since $k \supset \mathbb{Q}$, the cofree Lie coalgebra $\mathrm{coLie}(sP)$ is isomorphic to the indecomposable space $e^{(1)}(T(sP))$ and there is an isomorphism $S^n(e^{(1)})) \cong e^{(n)}$ induced by the shuffle product. Hence

$$P^\perp(P) = \bigoplus_{n \geq 1} S^n \mathrm{coLie}(sP) \cong \bigoplus S^n(e^{(1)}(A^\perp(P))) \cong \bigoplus e^{(n)}(T(sP))).$$

The space $P^\perp(P)$ is filtered by the symmetric power of $\bigoplus S^n \mathrm{coLie}(sP)$. On the associated graded module E_0, the differential reduces to the coderivation defined by the maps $(\nabla_n : R^{\otimes n} \to R)_{n \geq 1}$. Clearly, this is the coderivation defining the C_∞-structure of P. Since P is assumed to be a C_∞-bimodule over its underlying C_∞-algebra structure, the differential $\big[-, \sum_{n \geq 1} \nabla_n\big]$ preserves the decomposition $P^\perp(P) = \bigoplus e^{(n)}(A^\perp(P)))$. It follows that E_1^{p*} is the cohomology of the complex $\mathrm{CoDer}\big(e^{(p)}(A^\perp(P)), A^\perp(P)\big)$, where the coderivations are taken with respect to the coalgebra structure of $A^\perp(P)$, equipped with the Hochschild differential given by the underlying C_∞-algebra structure of P. Thus $E_1^{p*} \cong HH_{(q)}^*(P, P)$, see the proof of Theorem 3.1. ∎

EXAMPLE 3.37. Let $(P, m, [\ ;\ ])$ be a Poisson algebra. The maps $D_2 = m$ and $D_{1,1} = [\ ;\ ]$ endow P with its canonical P_∞-structure and the cohomology of $\mathrm{CoDer}(P^\perp(P), P^\perp(P))$ is its Poisson cohomology $HP^*(P, P)$. Proposition 3.36 implies that there is a spectral sequence converging to $HP^*(P, P)$ whose E_1 term is the Hochschild cohomology of P. This spectral sequence is the dual of the one found by Fresse [**Fr1, Fr2**].

EXAMPLE 3.38. Let $\mathfrak{g}$ be a Lie algebra. The free Poisson algebra on the Lie algebra $\mathfrak{g}$ is $S^*\mathfrak{g}$. Indeed the Lie bracket of $\mathfrak{g}$ extends uniquely on $S^*\mathfrak{g}$ in such a way that the Leibniz rule is satisfied. Since $S^*\mathfrak{g}$ is free as an algebra (thus smooth), according to Proposition 3.5.i), the term E_1 of spectral sequence (3.16) is equal to

$$HH^*_{(p)}(S^*\mathfrak{g}, S^*\mathfrak{g}) \cong HH^p_{(p)}(S^*\mathfrak{g}, S^*\mathfrak{g}) \;\cong\; \mathrm{Hom}_{S^*\mathfrak{g}}(\Omega^p_{S^*\mathfrak{g}}, S^*\mathfrak{g})$$
$$\cong S^p\mathrm{Hom}(\mathfrak{g}, S^*\mathfrak{g}).$$

It follows that the spectral sequence collapses at level 2 with $E_2^n = H^n_{\mathrm{Lie}}(\mathfrak{g}, S^*\mathfrak{g})$ where H^*_{Lie} stands for Lie algebra cohomology.

EXAMPLE 3.39. A C_∞-algebra is a P_∞-algebra by choosing all other defining maps to be trivial. Let R be a P_∞-algebra with $\nabla_{p_1,\ldots,p_n} = 0$ for $n \geq 2$. Then spectral sequence (3.16) collapses at level 1 since the maps $\nabla_{p_1,\ldots,p_n}$ are null for $n \geq 2$. Hence

$$HP^*(R, R) \cong \bigoplus_{p \geq 0} HH^*_{(p)}(R, R).$$

4. An exact sequence à la Jacobi-Zariski

It is well-known that if $K \to S \to R$ is a sequence of strict commutative rings with unit, there is an exact sequence relating the André Quillen (co)homology groups of R viewed as a K-algebra with the ones of S viewed as a K-algebra and R viewed as a S-algebra. This sequence is called the Jacobi-Zariski exact sequence (or the transitivity exact sequence). Under flatness hypothesis and if the rings contained $\mathbb{Q}$, the André Quillen (co)homology corresponds to Harrison (co)homology with degree shifted by one. In particular the exact sequence holds for the Harrison groups of Definitions 2.22, 2.26. We prove here a similar result for C_∞-algebras with units. We first study the category of A_∞ or C_∞-algebras over a C_∞-one.

4.1. Relative A_∞-algebras. In order to make sense of a Jacobi-Zariski exact sequence, we shall first define the notion of a C_∞-algebra over another one. The theory makes sense for A_∞-algebras over an A_∞-algebra as well so that we start working in this more general context. Let $T^\perp(R/S)$ be the coalgebra

$$(4.17) \quad T^\perp(R/S) \;:=\; T(R \oplus S) = \bigoplus_{n, p_0, \ldots, p_n \geq 0} S^{\otimes p_0} \otimes R \otimes S^{\otimes p_1} \otimes \ldots \otimes R \otimes S^{\otimes p_n}.$$

The coalgebra map is the one on $T(R \oplus S)$, that is to say

$$\delta_{R/S}(s_1^{p_0}, \ldots, s_{k_{p_0}}^{p_0}, a_1, \ldots, a_n, s_1^{p_n}, \ldots, s_{k_{p_n}}^{p_n}) =$$
$$\sum (s_1^{p_0}, \ldots, a_j, s_1^{p_j}, \ldots, s_{\ell_{p_j}}^{p_j}) \otimes (s_{\ell_{p_j}+1}^{p_j}, \ldots, a_n, s_1^{p_n}, \ldots, s_{\ell_{p_n}}^{p_n}).$$

The coalgebra $T^{\perp}(sR/sS)$ is co-augmented

$$k \hookrightarrow T^{\perp}(sR/sS) \twoheadrightarrow A^{\perp}(sR/sS).$$

It is straightforward that $\delta_{R/S}$ restricts to $T(S)$ and $T(R)$, hence the following lemma.

LEMMA 4.1. *Both $A^{\perp}(S)$ and $A^{\perp}(R)$ are subcoalgebras of $A^{\perp}(R/S)$.*

Denote $A^{\perp}_{+}(R/S)$ the subspace of $A^{\perp}(R/S)$ which contains at least one factor R so that

$$T^{\perp}(sR/sS) = T(sS) \oplus A^{\perp}_{+}(R/S), \quad A^{\perp}(R/S) = A^{\perp}(S) \oplus A^{\perp}_{+}(R/S).$$

DEFINITION 4.2. *Let (S, B) be an A_{∞}-algebra and R a k-module.*

- *An S-algebra structure on R is an A_{∞}-algebra structure on $R \oplus S$ such that the natural inclusion $S \to S \oplus R$ and the natural projection $S \oplus R \to S$ are maps of A_{∞}-algebras.*
- *A C_{∞}-algebra over S structure on R is a C_{∞}-algebra structure on $R \oplus S$ such that the natural inclusion $S \to S \oplus R$ and the natural projection $S \oplus R \to S$ are maps of C_{∞}-algebras.*

The natural inclusion $S \to S \oplus R$ is the coalgebra map $F : T(sS) \to T(s(S \oplus R))$ with defining maps $F_1 = S \hookrightarrow S \oplus R$ and $F_{i \geq 2} = 0$ (see Remark 1.8). The natural projection $S \oplus R \to S$ is the map $G : T(s(S \oplus R)) \to T(sS)$ with defining maps $G_1 = S \oplus R \twoheadrightarrow S$ and $G_{i \geq 2} = 0$.

In terms of coderivations Definition 4.2 means

PROPOSITION 4.3. *Let (S, B) be an A_{∞}-algebra and R a k-module. A structure of S-algebra on R is uniquely determined by a coderivation $D_{R/S}$ on $A^{\perp}(R/S)$ such that*

    i): $D_{R/S}(A^{\perp}(S)) \subset A^{\perp}(S)$ *and* $(D_{R/S})_{/A^{\perp}(S)} = B$;
    ii): $D_{R/S}(A^{\perp}_{+}(R/S)) \subset A^{\perp}_{+}(R/S)$;
    iii): $(D_{R/S})^2 = 0$.

If, in addition, (S, B) is a C_{∞}-algebra, R is a C_{∞}-algebra over S if $(R, D_{R/S})$ is an S-algebra such that the codifferential $D_{R/S}$ on $A^{\perp}(R/S)$ is a derivation for the shuffle product on $T^{\perp}(sR/sS)$.

In plain English condition **i)** means that the codifferential $D_{R/S}$ restricts to $A^{\perp}(S)$ and that this restriction $D_{R/S}/A^{\perp}(S)$ is equal to B.
Proof: We already know that an A_{∞}-structure on $R \oplus S$ is given by a coderivation of square zero. The claim **i)** and **ii)** follows from the fact that the natural inclusion and natural projection are maps of A_{∞}-algebras. The statement for C_{∞}-structure is an immediate consequence of Definition 2.1. ∎

REMARK 4.4. The definition 4.2 put emphasis on homotopy algebras over a fixed A_{∞} or C_{∞}-structure (S, B). However, it makes perfect sense to study coderivation $D_{R/S} : A^{\perp}(R/S) \to A^{\perp}(R/S)$ with $(D_{R/S})^2 = 0$, restricting to $A^{\perp}(S)$ and with $D_{R/S}(A^{\perp}_{+}(R/S)) \in A^{\perp}_{+}(R/S)$. Such a coderivation $D_{R/S}$ restricts into a codifferential on $A^{\perp}(S)$, hence yielding an A_{∞}-structure on S. Moreover $(R, D_{R/S})$ is a $(S, D_{R/S}/A^{\perp}(S))$-algebra in the sense of Definition 4.2.

LEMMA 4.5. *A structure of S-algebra on R is uniquely determined by maps*

$$D_{p_0,\ldots,p_n} : S^{\otimes p_0} \otimes R \otimes S^{\otimes p_1} \otimes \ldots \otimes R \otimes S^{\otimes p_n} \to R \quad (n \geq 1)$$

satisfying, for all $s_k = a_1^k \otimes \cdots a_{p_k}^k \in S^{\otimes p_k}$ $(k = 0 \ldots n)$ and $r_1, \ldots r_n \in R$,

$$\sum_{i+j=n-1} \sum_{q=0}^{n-1-i} \pm \; D_{k_0,\ldots,k_i}\big(s_0, r_1, \ldots, r_i, s_q^{(1)}, D_{\ell_0,\ldots,\ell_j}\big(s_q^{(2)}, r_{q+1}, \ldots$$

$$\ldots, r_{q+j}, s_{q+j}^{(1)}\big), s_{q+j}^{(2)}, r_{q+j+1}, \ldots, r_n, s_n\big) = 0.$$

Note that in the formula above, we use Sweedler's notation $s^{(1)} \otimes s^{(2)}$ for the deconcatenation coproduct of $s \in S^{\otimes m}$. The indexes $k_0, \ldots, k_i$ and $\ell_0, \ldots, \ell_j$ are uniquely unambiguously defined by the sequences of elements to which they apply. *Proof:* According to Remark 1.4, a structure of S-algebra on R is uniquely determined by maps

$$D_{p_0,\ldots,p_n} : S^{\otimes p_0} \otimes R \otimes S^{\otimes p_1} \otimes \ldots \otimes R \otimes S^{\otimes p_n} \to R \oplus S \quad (n \geq 0).$$

The requirement $D_{R/S}(A_+^\perp(R/S)) \subset A_+^\perp(R/S)$ forces the composition of the maps $D_{p_0,\ldots,p_n}$ with the projection on S to be trivial for $n \geq 1$. For $n = 0$, the maps $D_{p_0} = B_{p_0} : S^{\otimes p_0} \to S$ are determined by the A_∞-structure of S (Proposition 4.3). The formula follows from Remark 1.6 ∎

REMARK 4.6. Let (R, D) be an S-algebra. Lemmas 4.5 and 4.1 imply that the maps $D_n = D_{0,\ldots,0} : R^{\otimes n} \to R$ define a coderivation $\widetilde{D}$ of $A^\perp(R)$. Since $D^2 = 0$, it follows that $\widetilde{D}^2 = 0$. Hence R is an A_∞-algebra. Moreover it is a C_∞-algebra if R is a C_∞-algebra over S.

The notion of an A_∞-*bimodule over an S-algebra R (R/S-bimodule for short)* is the same as in Definition 1.1 with $A^\perp(R)$ replaced by $A^\perp(R/S)$. In other words, a structure of R/S-bimodule on M is given by a codifferential on $A^\perp_{R\oplus S}(M)$.

REMARK 4.7. According to Section 1.1, an R/S-bimodule structure on M is determined by maps

$$D^M_{p_0,\ldots,p_n|q_0,\ldots,q_m} : S^{\otimes p_0} \otimes R \otimes S^{\otimes p_1} \otimes \ldots \otimes S^{p_n} \otimes M \otimes S^{\otimes q_0} \otimes R \otimes \ldots \otimes S^{q_m} \to M$$

where $\{p_0, \ldots, p_n\}$, $\{q_0, \ldots, q_m\}$ are allowed to be the empty set $\emptyset$.

Similarly an A_∞-*morphism of A_∞-algebras over S (S-A_∞-morphism for short)* is a map of A_∞-algebra $F : R \oplus S \to R' \oplus S$ such that the composition

$$S \to S \oplus R \xrightarrow{F} R' \oplus S$$

is the natural inclusion and the composition

$$R \oplus S \xrightarrow{F} R' \oplus S \to S$$

is the natural projection. Equivalently, it is a map of A_∞-algebra $F : A^\perp(R/S) \to A^\perp(R'/S)$ such that F restricts to $A^\perp(S)$ as the identity and $F(A_+^\perp(R/S)) \subset A_+^\perp(R'/S)$. A C_∞-*morphism over S* is an S-A_∞-morphism such that its defining maps $F_{p_0,\ldots,p_n}$ satisfies $F_{p_0,\ldots,p_n}(x \bullet y) = 0$, i.e. vanish on shuffles.

LEMMA 4.8. *An A_∞-morphism $A^\perp(R/S) \to A^\perp(R'/S)$ is uniquely determined by maps*

$$F_{p_0,\dots,p_n} : S^{\otimes p_0} \otimes R \otimes \dots \otimes R \otimes S^{\otimes p_n} \to R' \qquad (n \geq 1)$$

such that the unique coalgebra map F defined by the system $(S^{\otimes p_0} \otimes R \otimes \dots \otimes R \otimes S^{\otimes p_n} \overset{F_{p_0,\dots,p_n}}{\to} R' \hookrightarrow R' \oplus S)$ is a map $F : (A^\perp(R/S), D_{R/S}) \to (A^\perp(R'/S), D_{R'/S})$ of differential coalgebras.

Proof: According to Section 1.1, a map of coalgebras $F : A^\perp(R/S) \to A^\perp(R'/S)$ is uniquely determined by maps

$$F_{p_0,\dots,p_n} : S^{\otimes p_0} \otimes R \otimes S^{\otimes p_1} \otimes S^{\otimes p_1} \dots R \otimes S^{\otimes p_n} \to R' \oplus S.$$

The requirement $F/_{C^\perp(S)} = \mathrm{id}$ implies that $F_1 : S \to R \oplus S$ is the canonical inclusion $S \hookrightarrow R' \oplus S$ and that $F_n : S^{\otimes n} \to R \oplus S$ is trivial. Moreover $F(A^\perp_+(R/S)) \subset A^\perp_+(R'/S)$ implies that the other defining maps take values in $R \subset R \oplus S$. ∎

PROPOSITION 4.9. *If R is an S-algebra, then R is canonically an S-bimodule. Moreover, if R is a C_∞-algebra, then R is a C_∞-bimodule over S.*

Proof: We denote $D_{p_0,\dots,p_n}$ the map defining the S-algebra structure. Note that there is an inclusion $T^S(R) \overset{i}{\hookrightarrow} A^\perp(R/S)$ and that $(i \otimes i) \circ \delta^R = \delta(i)$. Thus the restriction $D^R_{p,q} := D_{p,q}$ defines a coderivation from $T(sR)$ to $A^\perp_S(R)$ of square zero; hence a canonical S-bimodule structure. When S is a C_∞-algebra, and R a C_∞-algebra over R, the vanishing of $D_{R/S} : A^\perp(R \oplus S) \supset T^S(R) \to A^\perp(R \oplus S)$ on shuffles is equivalent to Definition 2.5. ∎

REMARK 4.10. Later on, we also will have to deal with different "ground" homotopy structures on S at the same time. Thus, for two C_∞-algebras (S, B), (S', B'), we define **an A_∞-morphism** $C^\perp(R/S) \to C^\perp(R'/S')$ to be a map of differential coalgebras such that F restricts to $A^\perp(S)$ yielding an A_∞-morphism $(A^\perp(S), B) \to (A^\perp(S'), B')$. We further require that $F(A^\perp_+(R/S)) \subset A^\perp_+(R'/S')$. Such a map is uniquely determined by the maps of Lemma 4.8 together with maps $F_n : S^{\otimes n} \to S'$ (the proof is the same).

In terms of Definition 4.2, such a map is an A_∞-algebra morphism $(R \oplus S, D) \to (R' \oplus S', D')$ such that the composition

$$S \to R \oplus S \to R' \oplus S' \to S'$$

is a prescribed A_∞-map $F : S \to S'$ and moreover F commutes with natural inclusions and projections i.e. the following diagrams commutes

$$
\begin{array}{ccc}
(S, B) & \longrightarrow & (R \oplus S, D) \\
\downarrow & & \downarrow \\
(S', B') & \longrightarrow & (R' \oplus S', D'),
\end{array}
\qquad
\begin{array}{ccc}
(R \oplus S, D) & \longrightarrow & (S, B) \\
\downarrow & & \downarrow \\
(R' \oplus S', D') & \longrightarrow & (S', B').
\end{array}
$$

Low degrees identities satisfied by an A_∞-algebra over a C_∞-algebra :
Let (S, B) be a C_∞-algebra and (R, D) an A_∞-algebra over S.

- The condition $(D)^2 = 0$ implies that the degree one map $D_{0,0} : R \to R$ is a differential that we denote d_R. We also denote $d_S = B_1 : S \to S$.

- The maps $D_{0,1} : R \otimes S \to R$, $D_{1,0} . S \otimes R \to R$ and $D_{0,0,0} : R \otimes R \to R$ are degree 0 maps. Moreover $D_{0,0,0}$ is graded commutative if and only if R is a C_∞-algebra and $D_{1,0}(s,a) = (-1)^{|a| \cdot |s|} D_{0,1}(a,s)$.
- Restricted to $S \otimes R$, the condition $(D)^2 = 0$ implies that $D_{1,0} : S \otimes R \to R$ is a map of differential graded modules.
- Denote by the single letter d the differential induced by d_R and d_S on $A^\perp(R \oplus S)$. The identities satisfied by $D_{p_0,\dots,p_n}$ on $A^\perp(R \oplus S)^{\leq 3}$ are

$$
\begin{aligned}
d_R(D_{0,0,0,0}(a,b,c)) + D_{0,0,0,0}(d_R(a,b,c)) &= D_{0,0}(D_{0,0}(a,b),c) \\
&\quad -D_{0,0}(a,D_{0,0}(b,c)) \\
d_R(D_{0,1,0})(a,s,b) + D_{0,1,0}(d(s,a,b)) &= D_{0,0,0}(D_{0,1}(a,s),b) \\
&\quad +D_{0,0,0}(a,D_{1,0}(s,b)) \\
d_R(D_{1,0,0})(s,a,b) + D_{1,0,0}(d(s,a,b)) &= D_{1,0}(s,D_{0,0}(a,b)) \\
&\quad +D_{0,0}(D_{1,0}(s,a),b) \\
d_R(D_{2,0}(s,t,a)) + D_{2,0}(d(s,t,a)) &= D_{1,0}(D_{S2}(s,t),a) \\
&\quad +D_{1,0}(s,D_{1,0}(t,a))
\end{aligned}
$$

plus the equations similar to the last two ones involving $D_{0,1,0}$, $D_{0,0,1}$ and $D_{0,2}$ instead of $D_{1,0,0}$ and $D_{2,0}$.

These identities imply the following Proposition.

PROPOSITION 4.11. *Let R be an algebra over the C_∞-algebra S and M an R/S-bimodule. Then $H^*(R)$ is an associative $H^*(S)$-algebra, which is graded commutative if R is a C_∞-algebra. Moreover $H^*(M) := H^*(M, D_{\emptyset|\emptyset}^M)$ is a bimodule over the $H^*(S)$-algebra $H^*(R)$.*

4.2. Relative A_∞-algebras over strict C_∞-algebras. When S is a commutative algebra, one can take $k = S$ as ground ring. In particular, Definition 1.1 gives the notion of S-*linear* A_∞-algebra (we also say A_∞-algebra in the category of S-modules); such a structure is a codifferential on $A^{S\perp}(R) := \bigoplus_{n \geq 0} R^{\otimes_S n}$, see Section 1.1, Definition 1.1. We have to make sure that this definition is equivalent to Definition 4.2, where S is equipped with its canonical A_∞-algebra structure. This is the aim of the next Proposition and of Proposition 4.20 below.

PROPOSITION 4.12. *Let (S,d,m) be a strict C_∞-algebra and (R,D) be an S-linear A_∞-algebra.*

i): *R has a natural structure of A_∞-algebra over S (in the sense of Definition 4.2) given by*

$$D_{0,0} = d_R, \qquad D_{1,0}(s,a) = s.a = \pm D_{0,1}(a,s),$$

$$D_{0,\dots,0}(a_1,\dots,a_n) = D_n(a_1,\dots,a_n).$$

ii): *If (M, D_M^R) is an S-linear R-bimodule, the maps*

$$D_{\emptyset|\emptyset}^M = D_{0,0}^M, \qquad D_{\emptyset|1}^M(m,s) = m.s, \qquad D_{1|\emptyset}^M(s,m) = s.m,$$

$$D_{0,\dots,0|0,\dots,0}^M(r_1,\dots,r_p,m,r_1',\dots,r_q') = D_{p,q}^M(r_1,\dots,r_p,m,r_1',\dots,r_q')$$

give M the structure of an R/S-bimodule.

iii): *Let (S, d, m) be a strict C_∞-algebra and R be an S-module. Assume that (R, D) is a C_∞-algebra over S such that*

$$D_{1,0}(s, a) = s.a = (-1)^{|a|.|s|} D_{0,1}(a, s).$$

Then the defining maps $D_{0,\dots,0}$ are S-multilinear; hence defined a structure of S-linear algebra on R.

Proof: For i), we need to prove that $(D_{R/S})^2 = 0$, which reduces to the identities,

$$
\begin{aligned}
D_n^R(a_1, \dots, a_i.s, a_{i+1}, \dots, a_n) &= D_n^R(a_1, \dots, a_i, s.a_{i+1}, \dots, a_n) \quad 1 \le i \le n-1 \\
D_n^R(a_1, \dots, a_n.s) &= D_n^R(a_1, \dots, a_n).s \\
D_n^R(s.a_1, \dots, a_n) &= s.D_n^R(a_1, \dots, a_n)
\end{aligned}
$$

These identities follows by S-linearity.

For ii), the fact that $(D^M)^2 = 0$ reduces to the S-linearity of the maps $D_{p,q}^M$ and the vanishing of $(D^M)^2$ as in i).

The low degrees identities of A_∞-algebras over a C_∞-algebra of Section 4.1 imply that the maps $D_{0,\dots,0}$ are S-linear. Then iii) follows easily. ∎

EXAMPLE 4.13. It follows from Proposition 4.12 that if (S, d, m) is a strict C_∞-algebra and (R, d_R, m_R) is a strict commutative S-algebra, then R is a C_∞-algebra over S with structure maps $D_{p_0, \dots, p_n} = 0$ except for

$$D_0 = d_R, \quad D_{0,0} = m_R, \text{ and } D_{0,1}(r, s) = r.s, \; D_{1,0}(s, r) = s.r$$

for all $(r, s) \in R \otimes S$. Reciprocally, if R is an S-linear C_∞-algebra whose only nontrivial structure maps are $D_0, D_{0,1}, D_{1,0}, D_{0,0}$ then R is a strict S-algebra. This follows easily from the low degrees relations satisfied by a C_∞-algebra over S, see Section 4.1.

REMARK 4.14. Let S be a strict A_∞-algebra and R a strict S-bimodule together with a pairing of differential graded module $\nu : R \otimes R \to R$ left linear in the first variable, right linear in the second and satisfying $\nu(r.s, r') = \nu(r, s.r')$. Then there is an S-linear A_∞-structure on R given by

$$D_{0,0} = d_R, \quad D_{1,0}(s, a) = s.a, \quad D_{0,1}(a, s) = a.s, \quad D_{0,0,0} = \nu.$$

Proposition 4.12.i) also holds in the case where S is a strict A_∞-algebra by requiring that R is an A_∞-algebra in the category of S-bimodules.

4.3. Weakly unital homotopy algebras. The standard Jacobi-Zariski exact sequence holds for unital algebras. Its C_∞-analogue in Section 4.5 also requires unitality assumption. Details on unital A_∞ and C_∞-algebras can be found in [**Tr2, HL2, HL3**]. In fact, we only need weaker unitality assumptions. A **weakly unital** A_∞-algebra (R, D) is an A_∞-algebra equipped with a distinguished element $1 \in R^0$ that satisfies $D_2(1, a) = D_2(a, 1) = a$ for any $a \in R$. Thus unital A_∞-algebras (in the sense of [**Tr2, HL2, HL3**]) are in particular weakly unital. A weakly unital C_∞-algebra is a C_∞-algebra which is weakly unital as an A_∞-algebra.

CONVENTION 4.15. Henceforth, when we write R has a unit, we mean R is weakly unital.

REMARK 4.16. The fact that D_1 is a derivation for D_2 implies that $D_1(1) = 0$. in other words, a weak unit is necessarily a cocycle (for D_1).

REMARK 4.17. If the A_∞-algebra R has unit, then $H^*(R)$ is a unital algebra.

EXAMPLE 4.18. A strict A_∞-algebra is weakly unital if and only if it is a differential graded associative algebra with unit (in the usual sense). For instance if A is a unital associative algebra, then its Hochschild cochain complex $C^*(A, A)$ is weakly unital with unit given by the unit of A viewed as an element of $C^0(A, A)$. Also the cochain complex $C^*(X)$ of a topological or simplicial set X is weakly unital.

We need to extend the definition of weak unitality to the relative setting. Let S be a weakly unital C_∞-algebra, with (weak) unit 1_S. Let R be an S-algebra. The **S-algebra R is said to be weakly unital** if the element $0 \oplus 1_S \in R \oplus S$ is weak unit for the A_∞-algebra $R \oplus S$.

Assume S is a strict unital C_∞-algebra and (R, D^R) is an S-linear A_∞-algebra. The action of the unit $1_S \in S$ is trivial on R, thus $0 \oplus 1_S$ is a weak unit for $R \oplus S$. Therefore we obtain

PROPOSITION 4.19. *Let (S, d, m) be a strict unital C_∞-algebra and (R, D^R) be an S-linear A_∞-algebra. Then R, equipped with the A_∞-algebra structure over S given by Proposition 4.12, is weakly unital if and only if (R, D^R) is weakly unital as an S-linear A_∞-algebra.*

4.4. (Co)homology groups for relative C_∞-algebras. Let M be an R/S-bimodule and let $T^!(sR/oS)$ be the coalgebra defined by Equation (4.17). The *Hochschild (co)homology groups* of the S-algebra R with values in M are the (co)homology groups of the (co)chain complexes

(4.18) $\qquad (C^*(R/S, M), b) := \mathrm{CoDer}(T^\perp(sR/sS), A_{R\oplus S}^\perp(M)), b),$

(4.19) $\qquad (C_*(R/S, M), b) := M \otimes T^\perp(sR/sS), b).$

The differential on the complex $C^*(R/S, M)$ is the Hochschild differential on $C^*(R \oplus S, M) \cong C^*(R/S, M)$ corresponding to the A_∞-algebra structure of $R \oplus S$ (see Definition 4.2). The differential on $C_*(R/S, M)$ is defined similarly. When R is a C_∞-algebra and M a $C_\infty^{(op)}$-bimodule, $R \oplus S$ is automatically a C_∞-algebra and we can define the Harrison (co)chain complexes

(4.20) $\qquad (CHar^*(R/S, M), b) := \mathrm{BDer}(R \oplus S, M), b)$

(4.21) $\qquad (CHar_*(R/S, M), b) := M \otimes C^\perp(sR \oplus sS), b).$

The (co)homology groups of the complexes (4.18), (4.19),(4.20) and (4.21) are denoted $HH^*(R/S, M)$, $HH_*(R/S, M)$, $Har^*(R/S, M)$ and $Har_*(R/S, M)$, respectively.

When S is a strict C_∞-algebra, and R is an S-linear A_∞-algebra, we denote $HH_S^*(R, M)$ and $HH_*^S(R, M)$ the Hochschild (co)homology groups of R over the ground ring S, i.e., those given by Definitions 1.9 and 1.14. Similarly we will denote $Har_S^*(R, M)$, $Har_*^S(R, M)$ the Harrison (co)homology groups.

PROPOSITION 4.20. *Let S be a strict (unital) commutative algebra, R be a A_∞-algebra and M, N R-bimodules which are S-linear and flat over S. There are natural isomorphisms*

$$HH^*(R/S, M) \xleftarrow{\sim} HH_S^*(R, M) : h^*, \quad h_* : HH_*(R/S, M) \xrightarrow{\sim} HH_*^S(R, M).$$

If R is a C_∞-algebra, M a C_∞^{op}-bimodule, N a C_∞-bimodule, then h_* and h^* are isomorphisms of γ-rings. Furthermore, there are natural isomorphisms

$$Har^*(R/S, M) \cong Har_S^*(R, M), \quad Har_*(R/S, N) \cong Har_*^S(R, N).$$

Proof: Let D be the codifferential on $A^{S\perp}(R)$ defining the S-linear A_∞-structure on R. Let $D_{R/S}$ be the codifferential on $A^\perp(R/S)$ defining the A_∞-algebra over S structure on R. There is an obvious projection

$$h : A^\perp(R/S) \longrightarrow \bigoplus_{n\geq 0} R^{\otimes n} \longrightarrow \bigoplus_{n\geq 0} R^{\otimes_S n} = A^{S\perp}(R).$$

This map induces a complex morphism $id \otimes h : M \otimes A^\perp(R/S) \to M \otimes A^{S\perp}(R)$ by S-linearity of the structure morphisms $D_{0,\dots,0}$.

Filtrating $M \otimes A^\perp(R/S)$ by the powers of R, we get a spectral sequence converging to $HH_*(R/S, M)$ whose E^1 term is the homology of $A^\perp(R/S)$ for the differential given by $D_0 = d_R$, $D_{1,0} = l$, $D_{0,1} = r$ and the multiplication $S \otimes S \to S$. In particular the differential restricted to $\bigoplus_{n\geq 0} R \otimes S^{\otimes n} \otimes R$ coincides with the one in the double Bar construction $B(R, S, R)$. Since R is S-flat, the Bar construction $B(R, S, R)$ is quasi-isomorphic to $R \otimes_S R$. Hence

$$E_{**}^1 \cong H^*(M) \otimes_S H^*(R) \otimes_S \dots \otimes_S H^*(R).$$

The filtration by the powers of R of $M \otimes A^{S\perp}(R)$ yields also a spectral sequence (see Proposition 3.21) with isomorphic E_1-term. Moreover, the map h_1 is an isomorphism at page 1 hence is an isomorphism. The cohomology statement is analogous.

Clearly h is a map of coalgebra. Moreover, when R is a C_∞-algebra, it is a map of algebras (with respect to the shuffle product). Thus h commutes with the maps ψ^k inducing the γ-ring structures in (co)homology. It also implies that h factors through the quotient by the shuffles hence the result for Harrison (co)homology. ∎

A homomorphism $F : S \to R$ of commutative algebras induces a canonical structure of commutative S-algebra on R. The following Proposition is the up to homotopy analogue. First we fix some notation:

Notation: If $F : (S, D^S) \to (R, D)$ is a morphism of C_∞-algebras, we denote $F^{[i]}$ the composition $A^\perp(S) \xrightarrow{F} A^\perp(R) \xrightarrow{pr} R^{\otimes i}$, that is to say the component of F which lies in the i-th power of R.

PROPOSITION 4.21. Let $F : (S, D^S) \to (R, D^R)$ be a C_∞-map. Then R has a structure of a C_∞-algebra over S given by the maps

$$D_{p_0,\dots,p_n}(x_0, r_1, \dots, x_n) = \sum_{i=i_0+\dots+i_n+n} D_i^R(F^{[i_0]}(x_0), r_1, \dots, r_n, F^{[i_n]}(x_n)).$$

Proof: According to Lemma 4.5, we have to prove that the coderivation $D_{R/S}$, induced by $D_{p_0,\dots,p_n}$, is of square 0. Since D^S is of square 0 and the degree of $D_{R/S}$ is 1 we find that $(D_{R/S})^2(x_0, r_1, \dots, x_n)$ is equal to

$$\sum \pm D_i^R \left(F^{[i_0]}(x_0), r_1, \dots, D_j^R(F^{[j_k]}(x_k), \dots, F^{[j_k+l]}(x_{k+l})), \dots, r_n, F^{[i_n]}(x_n) \right)$$

$$+ \sum \pm D_i^R \left(F^{[i_0]}(x_0), r_1, \dots, F^{[i_p]}(D^S(x_p)), r_p, \dots, r_n, F^{[i_n]}(x_n) \right)$$

$$= (D^R)^2 \left(\sum F^{[i_0]}(x_0) \otimes r_1 \otimes \dots \otimes r_n \otimes F^{[i_n]}(x_n) \right) = 0$$

The last step follows from $D^R \circ F = F \circ D^S$. ∎

EXAMPLE 4.22. Let S be strict and R be a strict unital associative S-algebra. If R is unital, then there is a ring map $F : S \to R$ and we have $\nu(s, r) = F(s).r$ where ν denotes the S-action. We also denote $F : (S, D^S) \to (R, D)$ the associated A_∞-morphism. The structure of A_∞-algebra over S given by Proposition 4.21 is the same than the one given by Proposition 4.12.i) applied to R viewed as an A_∞-algebra in the category of S-modules.

EXAMPLE 4.23. Let S be any (weakly unital) C_∞-algebra. There is a C_∞-map $F : (S, D^S) \to (S, D^S)$ given by $F_1 = \mathrm{id}$, $F_{n>1} = 0$. In particular S has a canonical C_∞-structure over S (which is the canonical one if S is strict by the previous example). Corollary 4.25 below states that these structure is (co)homologically trivial as expected.

PROPOSITION 4.24. Let M be an R/S-bimodule. Assume R, S, M and their cohomology groups are k-flat. There are converging spectral sequences
$$E_2^{**} = HH^*(H^*(R)/H^*(S), H^*(M)) \Rightarrow HH^*(R/S, M)$$
and
$$E_{**}^2 = HH_*(H^*(R)/H^*(S), H^*(M)) \Rightarrow HH_*(R/S, M).$$

Proof: The spectral sequences are given by the filtration by the power of S. ∎

COROLLARY 4.25. Let S be a weakly unital C_∞-algebra and let M be an S/S-bimodule. There are isomorphisms
$$HH_*(S/S, M) = H^*(M), \qquad HH^*(S/S, M) = H^*(M).$$

Proof: Applying Proposition 4.24, it is sufficient to consider the case of $H^*(S)$, that is of a strict algebra. According to Proposition 4.20, the later case is the well-known computation of Hochschild (co)homology of the ground algebra [Lo2]. ∎

4.5. The Jacobi-Zariski exact sequence. In this section all C_∞-algebras are supposed to be weakly unital.

THEOREM 4.26. Let $K \to S \to R$ be a sequence of weakly unital C_∞-maps, with K, S, R and their cohomology k-flat. Then there is a long exact sequence
$$\cdots \to Har_*(S/K, M) \to Har_*(R/K, M) \to Har_*(R/S, M)$$
$$\to Har_{*-1}(S/K, M) \to Har_{*-1}(R/K, M) \to Har_{*-1}(R/S, M) \to \cdots$$
and also a long exact sequence in cohomology
$$\cdots \to Har^*(R/S, M) \to Har^*(R/K, M) \to Har^*(S/K, M)$$
$$\to Har^{*+1}(R/S, M) \to Har^{*+1}(R/K, M) \to Har^{*+1}(S/K, M) \to \cdots$$
where M is a C_∞-bimodule over R in homology, respectively a C_∞^{op}-bimodule over R in cohomology.

To prove Theorem 4.26, we use the following lemma.

LEMMA 4.27. Let $K \xrightarrow{F} S \xrightarrow{G} R$ be a sequence of C_∞-maps.

i): *There is a C_∞-morphism $C^\perp(S/K) \xrightarrow{\overline{G}} C^\perp(R/K)$ given by the defining maps $\overline{G}_1 = id$, $\overline{G}_{p_0 \geq 2} = 0$ and for $n \geq 2$*

$$\overline{G}_{p_0,\ldots,p_n}(x_0, s_1, \ldots, x_n) = \sum_{i=i_0+\cdots+i_n+n} G_i(F^{[i_0]}(x_0), s_1, \ldots, s_n, F^{[i_n]}(x_n)).$$

ii): *There is a C_∞-morphism $C^\perp(R/K) \xrightarrow{\overline{F}} C^\perp(R/S)$ given by*

$$\overline{F}_{p_0} = F_{p_0}, \quad \overline{F}_{00} = id \text{ and the other maps } \overline{F}_{p_0,\ldots,p_n} = 0.$$

Proof: One has

$$\begin{aligned}
D_{R/K}(\overline{G}(x_0, s_1, \ldots, x_n)) &= \sum D\Big(G \circ F(x_0^{(1)}), G(F(x_0^{(2)}), s_1, \ldots, x_k^{(1)}), \\
&\qquad F(x_k^{(2)}), \ldots, G(F(x_l^{(2)}), s_l, \ldots, x_n^{(1)}), F(x_n^{(2)})\Big) \\
&= \sum G\Big(D^S(F(x_0^{(1)}), F(x_0^{(2)}), s_1, \ldots, x_k^{(1)}, \ldots, F(x_n^{(2)}))\Big) \\
&= \overline{G}(D_{S/K}(x_0, s_1, \ldots, x_n)).
\end{aligned}$$

It proves i). The proof of ii) is similar.   ∎

Proof of Theorem 4.26: Let $F : C^\perp(K) \to C^\perp(S)$, $G : C^\perp(S) \to C^\perp(R)$ be two C_∞-maps. By Lemma 4.27, they induce chain maps

$$M \otimes C^\perp(S/K) \xrightarrow{G} M \otimes C^\perp(R/K) \text{ and } M \otimes C^\perp(R/K) \xrightarrow{F} M \otimes A^\perp(R/S).$$

Let $c(G)$ be the cone of the chain map G. That is to say

$$c(G) := M \otimes C^\perp(R/K) \oplus M \otimes C^\perp(S/K)[1].$$

In particular we have an exact sequence

$$\begin{aligned}
\cdots &\to Har_*(S/K, M) \to Har_*(R/K, M) \to H_*(c(G)) \to \\
&\quad Har_{*-1}(S/K, M) \to Har_{*-1}(R/K, M) \to \cdots
\end{aligned}$$

The homology spectral sequence will follow once we prove that there is a natural isomorphism $H_*(c(G)) \cong Har_*(R/S, M)$. The morphism F induces a chain map

$$M \otimes C^\perp(R/K) \oplus M \otimes C^\perp(S/K)[1] \xrightarrow{i} M \otimes C^\perp(R/S) \oplus M \otimes C^\perp(S/S)[1].$$

The target of i is isomorphic to the cone $c(F)$ of $M \otimes C^\perp(S/S) \to M \otimes C^\perp(R/S)$. There is also the inclusion of chain complexes

$$M \otimes C^\perp(R/S) \xrightarrow{j} M \otimes C^\perp(R/S) \oplus M \otimes C^\perp(S/S)[1].$$

Corollary 4.25 implies that $Har_*(S/S)$ is trivial, thus $H_*(c(F)) \cong Har_*(R/S, M)$ The spectral sequences of Proposition 4.24 also yield converging spectral sequences for $c(G)$ and $c(F)$. Applying the Jacobi Zariski exact sequence for strict commutative unital rings, we get that, at page 1 of the spectral sequences, the map i^1 is a quasi-isomorphism. Similarly the map j^1 is an isomorphism at page 1. It follows that i and j are quasi-isomorphisms, hence $H_*(c(G)) \cong Har_*(R/S, M)$ as claimed.

The existence of the cohomology exact sequence is proved in the same way. ∎

5. Applications to string topology

In this section we apply the machinery of previous sections to string topology. We assume that our ground ring k is a field of characteristic different from 2.

Let X be a topological space, the singular cochain $C^*(X)$ is an associative differential graded algebra (thus an A_∞-algebra) and the singular chains $C_*(X)$ forms a differential graded coalgebra. String topology is concerned about algebraic structures on Hochschild (co)homology of singular cochains because of

THEOREM 5.1 (Jones [**Jo**]). *If X is simply connected, then there are isomorphisms*

$$HH^*(C^*(X), C_*(X)) \cong H_*(LX),$$
$$HH_*(C^*(X), C^*(X)) \cong H^*(LX).$$

Degree issues: one has to be careful that the isomorphisms in Theorem 5.1 above are isomorphisms preserving the *cohomological degree*. As $x \in H_i(LX)$ has cohomological degree $-i$, the isomorphism reads as $HH^{-i}(C^*(X), C_*(X)) \cong H_i(LX)$ and similarly in Hochschild homology. Note that our convention for the degree of Hochschild cohomology is the opposite of the one in [**FTV**].

5.1. C_∞-structures on cochain algebras. The chain coalgebra $C_*(X)$ and cochain algebra $C^*(X)$ are not (co)commutative. Nevertheless the existence of Steenrod's $\smile_1$-product leads to the existence of natural C_∞-(co)algebras structures. The definition of C_∞-coalgebras is dual to C_∞-algebras. More precisely

- A A_∞-*coalgebra structure* on a k-module R is given by a square zero derivation ∂ of degree -1 on $A_\perp(R) := \prod_{i \geq 1}(sR)^{\otimes n}$, the completed tensor algebra equipped with the (continuous) concatenation

$$\mu(sx_1 \ldots sx_p, sy_1, \ldots sy_q) = sx_1 \otimes sx_2 \otimes \ldots sy_{q-1} \otimes sy_q.$$

  Coderivations on $A_\perp(R)$ are in one-to-one correspondence with family of maps $\partial^i : R \to R^{\otimes i}$ by dualizing the argument of Remark 1.5.
- The shuffle coproduct is defined by

$$\Delta^{sh}(sx_1 \ldots sx_n) = \sum \pm \left(sx_{\sigma^{-1}(1)} \otimes \cdots \otimes sx_{\sigma^{-1}(p)}\right) \otimes \left(sx_{\sigma^{-1}(p+1)} \otimes \cdots \right.$$
$$\left. \cdots \otimes sx_{\sigma^{-1}(n)}\right)$$

  where the sum is over shuffles $\sigma \in S_n$, making $A_\perp(R)$ a commutative bialgebra. A C_∞-*coalgebra* is an A_∞-coalgebra (R, ∂) such that $(R, \Delta^{sh}, \mu, \partial)$ is a differential graded bialgebra (in other words a B_∞-coalgebra).

It is easy to define A_∞-coalgebras maps, A_∞-comodules and their C_∞-analogs in the same way [**TZ**].

PROPOSITION 5.2. *Let k be a field of characteristic zero. There exists a natural C_∞-coalgebra structure on $C_*(X)$ and C_∞-algebra structure on $C^*(X)$, with $C_*(X)$ being a C_∞^{op}-module over $C^*(X)$, such that ∂^1 and D_1 are the singular differentials and, furthermore, the induced (co)algebras structures on $H_*(X)$, $H^*(X)$ are the usual ones.*

Proof: The singular cochains $C^*(X)$ are equipped with a brace algebra structure [**GV**] and thus a B_∞-structure. By a fundamental result of Tamarkin [**Ta**, **GH1**], a B_∞-structure yields a C_∞-structure, (which is the restriction of a G_∞-structure), with defining maps $D_i : C_*(X)^{\otimes i} \to C_*(X)$. Furthermore, D_1 is the usual differential on singular cochains and D_2 induces the cup-product in cohomology. The dual of the defining maps $D_i : C_*(X)^{\otimes i} \to C_*(X)$ yield a C_∞-coalgebra structure on $C_*(X)$. Moreover $C_*(X)$ inherits a C_∞^{op}-comodule structure by Proposition 2.19. Alternatively, one can use an acyclic models argument as in [**Sm**].

∎

REMARK 5.3. The Proposition above holds for non-simply connected spaces. For simply connected X, Rational homotopy theory gives strict C_∞-structures equivalent to the differential graded algebra $C^*(X)$.

For string topology applications, one needs a Poincaré duality between chains and cochain. We use Tradler's terminology [**Tr1**, **Tr2**]. Given any A_∞-algebra (R, D_R), a A_∞-**inner product** on R is a bimodule map $G : R \to R^*$ (where $R^* = \mathrm{Hom}(R, k)$ is the dual of R). We denote D_R, D_{R^*}, the codifferentials defining the canonical R-module structure of R and R^*. An A_∞-algebra R is said to have a **Poincaré duality structure** if R has an A_∞-inner product together with a bimodule map $F : R^* \to R$ such that $G : \left(A_R^{\perp}(R), D_R\right) \leftrightarrows \left(A_R^{\perp}(R^*), D_{R^*}\right) : F$ are quasi-isomorphisms which are quasi-inverse of each others (morphisms are not assumed to be of degree 0).

For *finely triangulated* oriented spaces, one can find C_∞-structures on chains and cochains together with a Poincaré duality. By finely triangulated we mean that the closure of every simplex has the homology of a point. The following Lemma is taken from an appendix of Sullivan [**Su**] together with an application of Tradler and Zeinalian [**TZ**]. We write $C^*(X), C_*(X)$ for the simplicial complexes associated to the triangulation of a space. Hopefully, the context should always makes clear if we are working with singular chains or the ones from a triangulation. We denote by $d : C_*(X) \to C_{*-1}(X)$ the differential and by $\Delta : C_*(X) \to C_*(X) \otimes C_*(X)$ the diagonal. We also write respectively d, $\cup$ for the differential and the cup-product on $C^*(X)$ (induced by Δ).

LEMMA 5.4. *Let k be a field of characteristic different from 2 and 3 and X be a triangulated oriented closed space with Poincaré duality such that the closure of every simplex has the homology (with coefficient in k) of a point. There exists a counital C_∞-coalgebra structure on $C_*(X)$ with structure maps $\delta^i : C_*(X) \to C_*(X)^{\otimes i}$ such that*

   i): *δ^1 is the simplicial differential and $\delta^2 = \dfrac{1}{2}(\Delta + \Delta^{op})$;*

   ii): *there exists a quasi-isomorphism of A_∞-coalgebras $F : (C_*(X), \delta) \to (C_*(X), d + \Delta)$;*

   iii): *the cochains $C^*(X)$ inherits a unital C_∞-structure by duality and there is an A_∞-algebra quasi-isomorphism $F : (C^*(X), D) \to (C^*(X), d + \cup)$;*

   iv): *there is a Poincaré duality $C_*(X) \overset{\equiv}{\to} C^*(X)$ of A_∞-modules inducing the Poincaré duality isomorphism in (co)homology.*

Proof: The triangulation of X yields a simplicial complex K^X and a homeomorphism $|K^X| \cong X$. The complex $C_*(X)$ is the simplicial space $C_*(K^X)$. By assumption, the closure of a q-cell (aka q-simplex) of K^X has the homology of a point. Statement *iv)* is in [**TZ**] as well as A_∞-analogs of *i)*, *iii)*. As in [**TZ**], a map between simplicial complexes is said to be *local* if all simplexes $c \in C_*(X)$ are mapped to $\prod_{i \geq 1} C_*(\bar{c})^{\otimes i}$, where $C_*(\bar{c})$ is the subcomplex generated by the closure $\bar{c}$

of c. By assumption $C_*(\bar{c})$ is *contractible*, *i.e.*, is quasi-isomorphic to k concentrated in degree 0. Let $\Delta : C_*(X) \to C_*(X) \otimes C_*(X)$ be a cell approximation to the diagonal. For instance one can take the Alexander-Whitney diagonal. Assertion *iii*) is obvious consequence of *i*) and *ii*). The proof of *i*) and *ii*) is essentially contained in [**Su**]. Here we only assume that our field is of characteristic different from 2 and 3. Let us outlined the argument:

Similarly to Example 2.4, a strong C_∞-coalgebra structure on $C_*(X)$ is given by a structure of differential graded Lie algebra on the free Lie algebra $L(X) := \mathrm{Lie}(C_*(X)[1])$ generated by the vector space $C_*(X)[1]$. We denote $\delta : L(X) \to L(X)$ the differential. A strong C_∞-coalgebra is a C_∞-coalgebra. Clearly δ is uniquely determined by its restrictions $\delta^i : C_*(X) \to C_*(X)^{\otimes i}$. Note that, since k is of characteristic different from 2 and 3, the identity $\delta^2 = 0$ is equivalent to $[\delta, \delta] = 0$ and the Jacobi identity for δ is equivalent to $[\delta, [\delta, \delta]] = 0$. We proceed by induction to construct both δ and the quasi-isomorphism $F : (C_*(X), \delta) \to (C_*(X), d + \Delta)$. We define $F_1 = \mathrm{id}$ and $\delta^1 = d$, which are local maps. Thus

$$(F \otimes F) \circ \delta = (d + \Delta) \circ F + O(2)$$

where $O(i)$ means that we restrict to components of $L(X)$ lying in the subspace $\bigoplus_{j \leq i-1} C_*(X)^{\otimes j}$. By *i*) we have to take $\delta_2 = \frac{1}{2}(\Delta + \Delta^{\mathrm{op}})$ which is local and cocommutative, hence with values in $L(X)$. The identity $\delta^2 = 0 + O(3)$ boils down to the fact that Δ is a map of chain complexes. We have to find F_2. We only have to do so locally. The compatibility between F and δ in $O(3)$ is equivalent to

$$[F_2, d] = \frac{1}{2}(\Delta - \Delta^{\mathrm{op}}).$$

The right part is a cocycle in the complex of endomorphisms $(\mathrm{End}(C_*(\bar{\sigma})), [-, d])$ for every simplex σ. Since $C_*(\bar{\sigma})$ is contractible, the complex $(\mathrm{End}(C_*(\bar{\sigma}))), [-, d])$ has trivial homology and the existence of F_2 follows. Assume by induction that $\delta_1, \ldots, \delta_n$, $F_1, \ldots, F_n$ have already been chosen and satisfy *i*), *ii*) and *iii*) up to $O(n + 1)$.

We first define $\delta_{n+1} : C_*(X) \to L^n(X)$. By hypothesis we have $[\delta, \delta] = E_{n+1} + O(n + 1)$ with $E_{n+1} \subset L^{n+1}(X)$. Since $\delta^2 = \frac{1}{2}[\delta, \delta]$, the Jacobi identity gives $[\delta, [\delta, \delta]] = 0$ and thus

$$[d, [\delta, \delta]] + O(n + 2) = [d, E_{n+1}] + O(n + 1) = 0 + O(n + 2).$$

Thus $[d, E_{n+1}] \subset L^{n+1}(X)$ is equal to zero. Again, the contractibility of each $C(\bar{\sigma}))$ implies that we can find a local map δ_{n+1} such that $E_{n+1} = [d, \delta_{n+1}]$. By definition of E_{n+1}, we have

$$[\delta + \delta_{n+1}, \delta + \delta_{n+1}] = O(n + 2)$$

that is *i*) up to $O(n + 2)$.

The induction hypothesis ensures that

$$F(\delta) + (d + \Delta)(F) = G_{n+1} + O(n + 2)$$

with $G_{n+1} \subset F^n(X)$ equal to

$$\sum_{2 \leq k \leq n} F_k(\delta_{n+2-k}) - \Delta(F_n).$$

A straightforward computation of $\rho^2(F) = 0$, where $\rho(F) = F \circ \delta + (d + \Delta) \circ F$, using that $d + \Delta$ gives an A_∞-coalgebra structure on $C_*(X)$, shows that

$$[d, G_{n+1}] + E_{n+1} = 0.$$

Now a map $F_{n+1} : C_*(X) \to C_*(X)^{\otimes n+1}$ makes $F + F_{n+1}$ satisfies $ii)$ up to $O(n+2)$ if and only if

(5.22) $[d, F_{n+1}] + \delta_{n+1} + G_{n+1} = 0.$

The map $\delta_{n+1} + G_{n+1}$ is a local cycle by above, hence a local map F_{n+1} could be chosen to satisfy (5.22). This concludes the induction.   ∎

REMARK 5.5. Lemma 5.4 actually holds when $C_*(X)$ is replaced by any simplicial complex in which the closure of any q-cell ($q \geq 0$) is contractible. It seems reasonable that it also holds if X is an oriented regular CW-complex. Note that cellular approximation to the diagonal can be constructed using the same ideas, see [**Su**, Remark A.3]. Also note that the C_∞-structures given by Lemma 5.4 are *not* canonical. Furthermore, the C_∞-structure given by Lemma 5.4 is strong.

5.2. Hodge decomposition for string topology. Hochschild cohomology of singular chains of any space X has a Hodge decomposition according to Proposition 5.2.

PROPOSITION 5.6. *Let k be a characteristic zero field. There exists Hodge decompositions*

$$HH^*(C^*(X), C^*(X)) = \prod_{i \geq 0} HH^*_{(i)}(C^*(X), C^*(X)),$$

$$HH^*(C^*(X), C_*(X)) = \prod_{i \geq 0} HH^*_{(i)}(C^*(X), C_*(X)),$$

$$HH_*(C^*(X), C^*(X)) = \bigoplus_{i \geq 0} HH^{(i)}_*(C^*(X), C^*(X)),$$

$$HH_*(C^*(X), C_*(X)) = \bigoplus_{i \geq 0} HH^{(i)}_*(C^*(X), C_*(X)).$$

The Hodge filtration $\mathcal{F}_i HH^*(C^*(X), C^*(X)) = \bigoplus_{n \leq i} HH^*_{(n)}(C^*(X), C^*(X))$ is a filtration of Gerstenhaber algebras. Moreover

$$HH^*_{(0)}(C^*(X), C^*(X)) = H^*(X) = HH^{(0)}_*(C^*(X), C^*(X)),$$

$$HH^*_{(0)}(C^*(X), C_*(X)) = H_*(X) = HH^{(0)}_*(C^*(X), C_*(X)).$$

For $i \geq 1$ there are spectral sequences

$$HH^{p+q}_{(i)}(H^*(X), H^*(X))_p \implies HH^{p+q}_{(i)}(C^*(X), C^*(X))$$
$$HH^{p+q}_{(i)}(H^*(X), H_*(X))_p \implies HH^{p+q}_{(i)}(C^*(X), C_*(X))$$
$$HH^{(i)}_{p+q}(H^*(X), H^*(X))_p \implies HH^{(i)}_{p+q}(C^*(X), C^*(X))$$
$$HH^{(i)}_{p+q}(H^*(X), H_*(X))_p \implies HH^{(i)}_{p+q}(C^*(X), C_*(X)).$$

Proof: According to Proposition 5.2, $C^*(X)$ is a C_∞-algebra and $C_*(X)$ is a C_∞^{op}-bimodule. Propositions 2.16, 2.13 ensure that $C_*(X)$ and $C^*(X)$ are both C_∞ and C_∞^{op}-bimodules. Now, the Hodge decompositions follow from Theorems 3.1, 3.18. Since $D_1 : C^*(X) \to C^*(X)$ and $\partial^1 : C_*(X) \to C_*(X)$ are the singular differential, the identification of the weight 0-part is immediate. There is a filtration of Gerstenhaber algebras according to Theorem 3.31. The spectral sequences are given by Propositions 3.7, 3.21. ∎

In presence of Poincaré duality for chains, the Hochschild cohomology of the cochain algebra lies in the realm of "string topology". Indeed, there is an isomorphism

$$H_*(LX) \cong HH^*(C^*(X), C_*(X)) \cong HH^*(C^*(X), C^*(X))[d]$$

if X is an oriented manifold of dimension d [**CJ, Mer, FTV2**]. The isomorphism $H_*(LX) \cong HH^*(C^*(X), C^*(X))$ is an isomorphism of algebras with respect to Chas-Sullivan product [**CS**] on the left and the cup product on the right, see [**CJ, Co, Mer**]. When X is a triangulated oriented Poincaré duality space, applying Sullivan's techniques as in Lemma 5.4, Tradler and Zeinalian proved that the Hochschild cohomology

$$HH^*(C^*(X), C^*(X)) \cong HH^*(C^*(X), C_*(X))[d]$$

is a BV-algebra (whose underlying Gerstenhaber algebra is the usual one) [**TZ**]. The intrinsic reason for the existence of this BV structure is that a Poincaré duality is a up to homotopy version of a Frobenius structure and that for Frobenius algebras, the Gerstenhaber structure in Hochschild cohomology is always BV [**Me**]. This result and our preliminary work leads to

THEOREM 5.7. *Let k be a field of characteristic different from 2 and 3 and X be a triangulated oriented closed space with Poincaré duality (of dimension d), such that the closure of every simplex has the homology of a point.*

- *There is a BV-structure on $HH^*(C^*(X), C^*(X))$ and a compatible γ-ring structure.*
- *If X is simply connected, there is a BV-algebra structure on $\mathbb{H}_*(LX) := H_{*+d}(LX)$ and a compatible γ-ring structure. When X is a manifold the underlying product of the BV-structure is the Chas-Sullivan loop product.*

By a BV-structure on a graded space H^* and compatible γ-ring structure we mean the following:

(1) H^* is both a BV-algebra and a γ-ring.
(2) The *BV*-operator Δ and the γ-ring maps λ^k satisfy

$$\lambda^k(\Delta) = k\Delta(\lambda^k).$$

(3) There is an "ideal augmentation" spectral sequence $J_1^{pq} \Rightarrow H^{p+q}$ of BV algebras.
(4) On the induced filtration J_∞^{p*} of the abutment H^*, one has, for any $x \in J_\infty^{p*}$ and $k \geq 1$,

$$\lambda^k(x) = k^p x \bmod J_\infty^{p+1*}.$$

(5) If $k \supset \mathbb{Q}$, there is a Hodge decomposition $H^* = \prod_{i \geq 0} H_{(i)}^*$ (given by the associated graded of the filtration J_∞^{**}) such that the filtered space $\mathcal{F}_p H^* := \bigoplus H_{(n \leq p)}^*$ is a filtered BV-algebra.

As a consequence of Theorem 5.7, $Har^*(C^*(X), C^*(X))$ has an induced Lie algebra structure. Moreover $J_\infty^{0*}/J_\infty^{1*} \cong H_*(X)$ always splits.

Proof: We apply Lemma 5.4 to get a C_∞-algebra structure (given by a differential D) on $C^*(X)$. Assertion *iii*) of this lemma ensures that there is a quasi-isomorphism of A_∞-algebras $F : (C^*(X), D) \to (C^*(X), d + \cup)$. Proposition 3.10 implies that

$$(5.23) \quad HH^*((C^*(X), D), (C^*(X), D)) \cong HH^*((C^*(X), d + \cup), (C^*(X), d + \cup)).$$

Thus we only need to prove the theorem for $C^*(X)$ endowed with its C^∞-structure. The proof of 3.10 shows that the isomorphism (5.23) is the composition of the following isomorphisms:

$$F_* : HH^*((C^*(X), D), (C^*(X), D)) \to HH^*((C^*(X), D), (C^*(X), d + \cup)) \text{ and}$$

$$HH^*((C^*(X), D), (C^*(X), d + \cup)) \leftarrow HH^*((C^*(X), d + \cup), (C^*(X), d + \cup)) : F^*.$$

Since $(C^*(X), d + \cup)$ is an A_∞-algebra, formula (1.7) yields a ring structure on $HH^*((C^*(X), D), (C^*(X), d + \cup))$ and F_* and F^* are rings morphisms. Thus the cohomology $HH^*((C^*(X), D), (C^*(X), D))$ and $HH^*((C^*(X), d + \cup), (C^*(X), d + \cup))$ are isomorphic as rings.

By Theorem 3.1 there is a γ-ring structure on $HH^*(C^*(X), C_*(X))$. The Poincaré duality structure quasi-isomorphism $\Xi : C_*(X) \to C^*(X)$ and Proposition 3.10 implies that there is an isomorphism of γ-rings

$$HH^*(C^*(X), C_*(X)) \cong HH^*(C^*(X), C^*(X)).$$

The compatibility between the γ-ring structure and the Gerstenhaber structure follows from Proposition 3.30. The existence of the BV-structure is asserted by Tradler-Zeinalian [**TZ**] as stated above. Note that the BV-structure identifies with Connes's operator $B^* : HH^*(C^*(X), C^*(X)) \to HH^*(C^*(X), C^*(X))$ through the isomorphism $HH^*(C^*(X), C_*(X)) \cong HH^*(C^*(X), C^*(X))$ [**Tr2**]. It is proved in [**Lo1**] that $kB(\lambda^k) = \lambda^k(B)$ on $T(sR)$. Thus by duality we get the BV-compatibility.

If X is simply connected, Theorem 5.1 ensures that

$$H_*(LX) \cong HH^*((C^*(X), d + \cup), (C_*(X), d + \cup)) \cong HH^*(C^*(X), C_*(X))$$
$$\cong HH^{*-d}(C^*(X), C^*(X))$$

where the last isomorphism is induced by naturality and the Poincaré duality quasi-isomorphism Ξ. Thus the BV-structure is transferred to $H_*(LX)$. ∎

EXAMPLE 5.8. Let $X = S^3$ with its usual simplicial structure and the associated triangulation and k be a field of characteristic different from 2. Its cochain complex is a C_∞-algebra. The term E_2^{pq} of the spectral sequence 3.7 is

$$HH^{p+q}(H^*(S^3), H^*(S^3)) = HH^{p+q}(k[y], k[y]) \text{ where } |y| = 3.$$

It is a spectral sequence of γ-rings. An easy computation yields that this page of the spectral sequence has a Hodge decomposition where the only non trivial terms are

$$HH_{(p)}^{-2p}(H^*(S^3), H^*(S^3)) = k, \qquad HH_{(p)}^{-2p+3}(H^*(S^3), H^*(S^3)) = k \quad (p \geq 0).$$

The Hodge decomposition above holds even if char$(k) > 0$. In that case, the computation yields a partial Hodge decomposition with the same terms but with the subscript (p) being taken modulo char$(k) - 1$ for $p > 0$, i.e.

$$HH^*_{(p)}(H^*(S^3), H^*(S^3)) = HH^*_{(p+n(\text{char}(k)-1))}(H^*(S^3), H^*(S^3))$$

for $1 \le p \le \text{char}(k) - 1$. The total degree of an element in $HH^*(H^*(S^3), H^*(S^3))$ enables to split off the various terms of the partial decomposition, thus giving the claimed Hodge decomposition above. The higher differentials necessarily vanish and one finds that $HH^*(C^*(S^3), C^*(S^3))$ has a decomposition

$$HH^{-2p+3}(C^*(S^3), C^*(S^3)) = HH^{-2p+3}_{(p)}(C^*(S^3), C^*(S^3)) = k,$$

$$HH^{-2p}(C^*(S^3), C^*(S^3)) = HH^{-2p}_{(p)}(C^*(S^3), C^*(S^3)) = k$$

where $p \ge 0$. By Theorem 5.7, the BV-operator commutes with the λ-operations and the ring structure is the same as the one of the Hochschild cohomology of its singular cochains (viewed as an associative differential graded algebra). Thus we have an isomorphism of rings

$$\mathbb{H}_*(LS^3) \cong HH^*(C^*(S^3), C^*(S^3)) \cong k[u, v] \text{ with } |u| = 3, |v| = -2,$$

see [**FTV**] for example (the degrees are cohomological ones). The weight p-piece of the cohomology is the component $k[u]v^p$. In particular the λ-operations also commute with the loop product and the Hodge decomposition is graded for the BV-structure. An analogous computation using spectral sequence 3.21 gives

$$HH_{-2p}(C^*(S^3), C^*(S^3)) = HH^{(p)}_{-2p}(H^*(S^3), H^*(S^3)) = k \text{ and}$$

$$HH_{-2p-3}(C^*(S^3), C^*(S^3)) = HH^{(p)}_{-2p-3}(H^*(S^3), H^*(S^3)) = k$$

for $p \ge 0$ and other terms are null.

The computation for S^3 are straightforwardly generalized to all spheres. For odd dimensional simply connected spheres one has isomorphism of rings ($n \ge 1$)

$$\mathbb{H}_*(LS^{2n+1}) = HH^*(C^*(S^{2n+1}), C^*(S^{2n+1})) = k[u, v]$$

with $|u| = 2n + 1$, $|v| = -2n$ and the weight p-component of the Hodge decomposition is

$$\mathbb{H}^{(p)}_*(LS^{2n+1}) = HH^*_{(p)}(C^*(S^{2n+1}), C^*(S^{2n+1})) = kv^p \oplus kuv^p.$$

For even dimensional (simply connected) spheres, one has an isomorphism of rings ($n \ge 1$)

$$\mathbb{H}_*(LS^{2n}) = HH^*(C^*(S^{2n}), C^*(S^{2n})) = k[v, w] \oplus k[u]/(u^2)$$

with $|u| = 2n$, $|v| = 2 - 4n$ and $|w| = 1$. The weight p-component of the Hodge decomposition is

$$\mathbb{H}^{(p \ge 1)}_*(LS^{2n}) = kv^p \oplus kwv^{p-1}, \quad \mathbb{H}^{(0)}_*(LS^{2n}) = k[u]/(u^2).$$

In particular the BV-structure is graded with respect to the Hodge decomposition. Furthermore, denoting $s^{-i}k =: k[i]$ the module k concentrated in cohomological degree i (hence homological degree $-i$), the homology spectral sequence yields that the groups

$$H^k(LS^{2n+1}) \cong HH_{-k}(C^*(S^{2n+1}), C^*(S^{2n+1})) \text{ and}$$

$$H^k(LS^{2n}) \cong HH_{-k}(C^*(S^{2n}), C^*(S^{2n}))$$

have Hodge decomposition where the weight p-pieces are

$$HH_*^{(p\geq0)}(C^*(S^{2n+1}), C^*(S^{2n+1})) = k[2p + 2n + 1] \oplus k[2p] \text{ and}$$

$$HH_*^{(p\geq1)}(C^*(S^{2n}), C^*(S^{2n})) = k[p(4n-2) + 2n] \oplus k[p(4n-2) - 2n + 1].$$

Of course $HH_*^{(0)}(C^*(S^{2n}), C^*(S^{2n})) = H^*(S^{2n}) = k[2n] \oplus k$.

EXAMPLE 5.9. If $\text{char}(k) = 0$, the Harrison (co)homology groups of $C^*(S^n)$ immediately follow from Theorem 3.1 and Example 5.8:

$$Har^*(C^*(S^{2n+1}), C^*(S^{2n+1})) = k[-2n] \oplus k[1],$$

$$Har^*(C^*(S^{2n}), C^*(S^{2n})) = k[2 - 4n] \oplus k[1],$$

$$Har_*(C^*(S^{2n+1}), C^*(S^{2n+1})) = k[2n + 3] \oplus k[2],$$

$$Har_*(C^*(S^{2n}), C^*(S^{2n})) = k[6n - 2] \oplus k[2n - 1]$$

where $k[i]$ still means k concentrated in cohomological degree i. If $\text{char}(k) = p > 0$ then

$$Har^*(C^*(S^{2n+1}), C^*(S^{2n+1})) = \prod_{i\geq0} k[-2n(pi - i + 1)] \oplus k[1 - 2n(pi - i)],$$

$$Har^*(C^*(S^{2n}), C^*(S^{2n})) = \prod_{i\geq0} k[(2 - 4n)(pi - i + 1)] \oplus k[1 + i(2 - 4n)(p - 1)].$$

EXAMPLE 5.10. For $X = \mathbb{CP}_n$, the loop homology ring is $H^*(X) = k[x]/(x^{n+1})$ (where $|x| = 2$), see [**CJY**]. When k is of characteristic different from $n + 1$, the spectral sequence 3.7 also collapses at page 2 and a straightforward computation yields an isomorphism of rings

$$H_*(L\mathbb{CP}_n) \cong HH^*(C^*(\mathbb{CP}_n), C^*(\mathbb{CP}_n)) \cong k[u, v, w]/(u^{n+1}, u^n v, u^n w)$$

where $|u| = 2$, $|v| = -2n$ and $|w| = 1$. Furthermore the Hodge decomposition is given by

$$H_*^{(p\geq1)}(L\mathbb{CP}_n) = (v^p k[u] \oplus w v^{p-1} k[u])/(u^n v, u^n w)$$

and $H_*^{(0)}(L\mathbb{CP}_n) = k[u]/(u^{n+1})$. As in Example 5.8, we get a Hodge decomposition even if $\text{char}(k) > 0$. In particular, the Harrison cohomology groups are

$$Har^*(C^*(\mathbb{CP}_n), C^*(\mathbb{CP}_n)) = (k[u]/(u^n))[-2n] \oplus (k[u]/(u^n))[1]$$

if $\text{char}(k) = 0$ and

$$Har^*(C^*(\mathbb{CP}_n), C^*(\mathbb{CP}_n)) = (k[u]/(u^n)) \prod_{i\geq0} k[-2n(pi - i + 1)] \oplus k[1 - 2in(p - 1)]$$

if $\text{char}(k) = p$. The Hodge decomposition in Hochschild homology is given by

$$HH_*^{(p)}(C^*(\mathbb{CP}_n), C^*(\mathbb{CP}_n)) = k[x]/(x^n)[2np - 2n + 1] \oplus k[x]/(x^n)[2np + 2],$$

$$HH_*^{(0)}(C^*(\mathbb{CP}_n), C^*(\mathbb{CP}_n)) = k[x]/(x^{n+1})$$

where the degrees are cohomological degrees. In particular, if $\text{char}(k) = 0$, the only non trivial Harrison homology groups are

$$Har_{2i-1}(C^*(\mathbb{CP}_n), C^*(\mathbb{CP}_n)) = k \quad \text{for} \quad 1 \leq i \leq n,$$

$$Har_{2i}(C^*(\mathbb{CP}_n), C^*(\mathbb{CP}_n)) = k \quad \text{for} \quad n + 1 \leq i \leq 2n.$$

6. Concluding remarks

- At the same time as a first draft of this paper, Hamilton and Lazarev [**HL**] (also see the recent updated versions [**HL2, HL3, HL4**]) wrote a paper about cohomology of homotopy algebras, using Kontsevich framework of formal non-commutative geometry. In particular they study Harrison and Hochschild cohomology of C_∞-algebras over a field of characteristic zero and give a Hodge decomposition of $HH^*(R,R)$ and $HH^*(R,R^*)$. Using that the C_∞-structure is determined by maps $D_i : R^{\otimes i} \to R$, it is easy to check that their definitions are dual and equivalent to ours in this special cases. They also prove that the above cohomology theories yields the good obstruction theory. They finally apply it to (a different from our) issue in string topology, namely the homotopy invariance of the Gerstenhaber algebra structure.

- The Connes operator $B : C^*(R,M) \to C^{*-1}(R,M)$ is well defined for C_∞-algebras and commutes with the Hochschild differential, thus one can define cyclic (co)homology of a C_∞-algebra R see [**GJ1, Tr2, HL**]. Furthermore, it is easy to check that the λ-operations and Hodge decomposition passes to the various cyclic homology theories in characteristic zero [**HL**]. In positive characteristic, the λ-operations passes to cyclic (co)homology but not to negative cyclic (co)homology.

- Besides the BV algebra structure, there are other string topology operations on $\mathbb{H}_*(LM)$ as well as in equivariant homology $H_*^{S^1}(LM)$, which come from an action of Sullivan chord diagram on LM. It seems interesting to obtain compatibility conditions between the λ-operation/Hodge decomposition and the full scope of string topology operation. It might be achieved by combining the techniques of this paper and [**TZ2**].

- There are power maps $\gamma^k : LM \to LM$ which sends a loop $f : S^1 \to M$ to the loop $u \mapsto f(ku)$. It seems reasonable to expect that these power maps coincides with our λ-operation for simply connected spaces.

References

[Ba] J. H. Baues, *The double bar and cobar constructions*, Compos. Math. 43 (1981), 331-341

[BW] N. Bergeron, L. Wolfgang *The decomposition of Hochschild cohomology and Gerstenhaber operations*, J. Pure Appl. Algebra 79 (1995) 109–129

[CS] M. Chas, D. Sullivan *String Topology*, preprint GT/9911159 (1999)

[Co] R. Cohen *Multiplicative properties of Atiyah duality*, Homology, Homotopy, and its Applications, vol 6 no. 1 (2004), 269–281

[CG] R. Cohen, V. Godin *A polarized view to string topology*, Topology, Geometry, and Quantum Field theory, Lond. Math. Soc. lecture notes vol. 308 (2004), 127–154

[CJ] R. Cohen, J.D.S. Jones *A homotopic realization of string topology*, Math. Annalen, vol 324. 773-798 (2002)

[CJY] R. Cohen, J.D.S. Jone, J. Yan *The loop homology algebra of spheres and projective spaces*, in Categorical Decomposition Techniques in Algebraic Topology, Prog. Math. 215 (2004), 77–92

[EKMM] A. Elmendorf, I. Kriz, M. Mandell, J.P.May, *Rings, modules, and algebras in stable homotopy theory*, Mathematical Surveys and Monographs, 47. American Mathematical Society, Providence, RI, 1997.

[FTV] Y. Félix, Y. Thomas, M. Vigué *The Hochschild cohomology of a closed manifold*, Publ. IHES, 99, (2004), 235-252

[FTV2] Y. Félix, Y. Thomas, M. Vigué *Rational string topology*, J. Eur. Math. Soc. (JEMS) 9 (2007), no. 1, 123–156.

[Fr1] B. Fresse, *Homologie de Quillen pour les algèbres de Poisson*, C.R. Acad. Sci. Paris Sér. I Math. 326(9) (1998), 1053–1058

[Fr2] B. Fresse, *Théorie des opérades de Koszul et homologie des algèbres de Poisson*, preprint

[Ge] M. Gerstenhaber, *The Cohomology Structure Of An Associative ring* Ann. Maths. 78(2) (1963)

[GS1] M. Gerstenhaber, S. Schack, *A Hodge-type decomposition for commutative algebra cohomology* J. Pure Appl. Algebra 48 (1987), no. 3, 229–247

[GS2] M. Gerstenhaber, S. Schack, *The shuffle bialgebra and the cohomology of commutative algebras* J. Pure Appl. Algebra 70 (1991), 263–272

[GV] M. Gerstenhaber, A. Voronov, *Homotopy G-algebras and moduli space operad*, Internat. Math. Res. Notices (1995), no. 3, 141–153

[GJ1] E. Getzler, J.D.S. Jones, A_∞-*algebras and the cyclic bar complex*, Illinois J. Math. 34 (1990) 12–159

[GJ2] E. Getzler, J.D.S. Jones *Operads, homotopy algebra and iterated integrals for double loop spaces*, preprint hep-th/9403055 (1994)

[Gi] G. Ginot, *Homologie et modèle minimal des algèbres de Gerstenhaber*, Ann. Math. Blaise Pascal 11 (2004), no. 1, 95–127

[GH1] G. Ginot, G. Halbout *A formality theorem for Poisson manifold*, Let. Math. Phys. 66 (2003) 37–64

[GH2] G. Ginot, G. Halbout *Lifts of G_∞-morphism to C_∞ and L_∞-morphisms*, Proc. Amer. Math. Soc. 134 (2006) 621–630.

[GK] V. Ginzburg, M. Kapranov, *Koszul duality for operads*, Duke Math. J. 76 (1994), No 1, 203–272

[HL] A. Hamilton, A. Lazarev *Homotopy algebras and noncommutative geometry*, preprint, math.QA/0410621

[HL2] A. Hamilton, A. Lazarev *Cohomology theories for homotopy algebras and noncommutative geometry*, preprint, math.QA/0707.2311

[HL3] A. Hamilton, A. Lazarev *Symplectic C_∞-algebras*, preprint, math.QA/0707.3951.

[HL4] A. Hamilton, A. Lazarev *Symplectic A_∞-algebras and string topology operations*, preprint, math.QA/0707.4003.

[Hi] H. Hiller, *λ-rings and algebraic K-theory*. J. Pure Appl. Algebra 20 (1981), no. 3, 241–266.

[HS] J. Huebschmann, J. Stasheff *Formal solution of the master equation via HPT and deformation theory*, Forum. Math. 14 (2002), no. 6, 847–868

[Jo] J.D.S. Jones *Cyclic homology and equivariant homology*, Inv. Math. 87, no.2 (1987), 403–423

[Ka] T. Kadeishvili *On the homology theory of fiber spaces*, Russian Mathematics Surveys 6 (1980), 231–238.

[KST] T. Kimura, J. Stasheff, A. Voronov *Homology of moduli spaces of curves and commutative homotopy algebras*, Comm. Math. Phys. 171 (1995), 1–25

[Lo1] J.-L. Loday, *Opérations sur l'homologie cyclique des algèbres commutatives*, Invent. Math. 96 (1989), No. 1, 205–230

[Lo2] J.-L. Loday, *Cyclic homology*, Springer Verlag (1993)

[Lo3] J.-L. Loday, *Série de Hausdorff, idempotents Eulériens et algèbres de Hopf*, Expo. Math. 12 (1994), 165–178

[Ma] Y. Manin, *Frobenius Manifolds, Quantum Cohomology and Moduli Spaces*, Colloquium publications 47 (1991), American Mathematical Society

[MMSS] M. Mandell, J.P. May, S. Schwede, B. Shipley, *Model categories of diagram spectra*, Proc. London Math. Soc. (3) 82 (2001), no. 2, 441–512.

[Me] L. Menichi *Batalin-Vilkovisky algebras and cyclic cohomology of Hopf algebras*, K-Theory 32 (2004), 231–251.

[Mer] S. Merkulov, *De Rham model for string topology*, Int. Math. Res. Not. 2004, no. 55, 2955–2981.

[Pa] F. Patras, *La décomposition en poids des algèbres de Hopf*, Ann. Inst. Fourier 43 (1993), No. 4, 1067–1087

[St] J.D. Stasheff, *Homotopy associativity of H-spaces I, II*, Trans. Amer. Math. Soc. 108 (1963), 275–292

[Sm] V. Smirnov, *Simplicial and operad methods in algebraic topology*, Transl. Math. Monographs 198, American Mathematical Society (2001)

[Su] D. Sullivan, *Appendix to Infinity structure of Poincaré duality spaces*, Algebr. Geom. Topol.
7 (2007), 233–260.

[Ta] D. Tamarkin, *Another proof of M. Kontsevich's formality theorem*, math.QA/9803025

[Tr1] T. Tradler, *Infinity-inner-products on a A-infinity-algebras*, preprint arXiv AT:0108027

[Tr2] T. Tradler, *The BV Algebra on Hochschild Cohomology Induced by Infinity Inner Products*,
preprint arXiv QA:0210150

[TZ] T. Tradler, M. Zeinalian *Infinity structure of Poincaré duality spaces*, Algebr. Geom. Topol.
7 (2007), 233–260.

[TZ2] T. Tradler, M. Zeinalian *On the cyclic Deligne conjecture*, J. Pure Appl. Algebra 204 (2006),
no. 2, 280–299

[Vi] M. Vigué *Décompositions de l'homologie cyclique des algèbres différentielles graduées*, K-
theory 4 (1991) -399–410

[WGS] J. Wu, M. Gerstenhaber, J. Stasheff, *On the Hodge decomposition of differential graded
bi-algebras* J. Pure Appl. Algebra 162 (2001), no. 1, 103–125

UPMC Paris 6, Institut de Mathématiques de Jussieu - Equipe Analyse Algébrique,
Case 82, 4 place Jussieu F-75252 Paris Cedex 05, France

E-mail address: ginot@math.jussieu.fr

URL: http://www.math.jussieu.fr/~ginot

What is the Jacobian of a Riemann Surface with Boundary?

Thomas M. Fiore and Igor Kriz

ABSTRACT. We define the Jacobian of a Riemann surface with analytically parametrized boundary components. These Jacobians belong to a moduli space of "open abelian varieties" which satisfies gluing axioms similar to those of Riemann surfaces, and therefore allows a notion of "conformal field theory" to be defined on this space. We further prove that chiral conformal field theories corresponding to even lattices factor through this moduli space of open abelian varieties.

1. Introduction

The main purpose of the present note is to generalize the notion of the Jacobian of a Riemann surface to Riemann surfaces with real-analytically parametrized boundary (or, in other words, conformal field theory worldsheets). The Jacobian of a closed surface is an abelian variety. What structure of "open abelian variety" captures the relevant data in the "Jacobian" of a CFT worldsheet? If we considered Riemann surfaces with punctures instead of parametrized boundary components, the right answer could be easily phrased in terms of mixed Hodge structures.

But in worldsheets, we see more structure, and some of it is infinite-dimensional. For example, even to a disk with analytically parametrized boundary, one naturally assigns an infinite-dimensional symplectic form and a restricted maximal isotropic space (cf. [7]). Any structure we propose should certainly include such data. Additionally, in worldsheets, boundary components can have inbound or outbound orientation, and an inbound and outbound boundary component can be glued to produce another worldsheet. So another test of having the right notion of "open abelian variety" is that it should enjoy a similar gluing structure.

We should point out that it is actually a remarkably strong requirement that a structure such as a (closed) abelian variety could somehow be "glued together" from "genus 0" data similar to the situation we described above for a disk. One quickly convinces oneself that naive approaches based on modelling somehow the 1-forms

2000 *Mathematics Subject Classification.* Primary 14H40, 81T40; Secondary 18C10, 32G15.

The first author was supported at the University of Chicago by NSF Grant DMS-0501208. At the Universitat Autònoma de Barcelona he was supported by Grant SB2006-0085 of the Programa Nacional de ayudas para la movilidad de profesores de universidad e investigadores españoles y extranjeros. The second author was supported in part by NSF Grant DMS-0305583.

on a Riemann surface, together with mixed Hodge-type integral structure data, fail to produce the required gluing. In fact, in some sense, the desired structure must be "pure" rather than "mixed". Note that there is no way of "gluing" a pure Hodge structure out of a mixed Hodge structure which does not already contain it: in the case of a closed Riemann surface with punctures, the mixed Hodge structure on its first cohomogy contains the pure Hodge structure of the original closed surface, so no gluing is involved. Clearly, the situation is different when we are gluing a non-zero genus surface from a genus 0 surface with parametrized boundary.

There is, however, a yet stronger test. When L is an even lattice (together with a $\mathbb{Z}/2$-valued bilinear form b satisfying a suitable condition), one has a notion of conformal field theory associated with L ([**9, 4**]). It could be argued that the definition only uses additive data, so the lattice conformal theories should "factor through open abelian varieties". In some sense, if one considers the conjectured space of open abelian varieties to be the "Jacobian" of the moduli space of world-sheets (with all its structure), then one could interpret this as a sort of "Abelian Langlands correspondence" for that space. This test is also severe, as lattice conformal field theories are known to be unexpectedly tricky. For example, the definition of operator assigned to a worldsheet appears to depend on the order of boundary components, and a subtle discussion is needed to remove this (unacceptable) dependency. This will be clarified in Section 5 below.

In this paper, we indeed propose a notion of an open abelian variety and answer both test questions in the affirmative. Of course, one has to start out by being precise about what exact abstract structure captures the notion of gluing, and then generalize the notion of conformal field theory to be defined on such abstract structures. Following ideas of Segal [**9**], this was done in [**1, 4, 5**], with a correction in [**3**]. The desired structure is called *stack of pseudo commutative monoids with cancellation* (SPCMC - see [**3**] for a correct definition) and a CFT is a pseudo morphism of certain SPCMC's

$$(1) \qquad\qquad \mathcal{C} \to C(\mathcal{M}, \mathcal{H}).$$

(The papers [**4, 5**] used the word "lax" instead of "pseudo", but the first author [**1**] discovered that "pseudo" conforms more with existing terminology of higher category theory.)[1]

In the present paper, the meaning of the target of the map (1), which is defined in [**5**], plays only a marginal role. The source of the map (1), however, is important: it is the SPCMC of Segal's worldsheets. Those are 2-dimensional real-analytic manifolds with boundary which have a complex structure and real-analytically parametrized boundary components. The notion of SPCMC, which is defined in [**4, 5**], is designed to capture the operations of disjoint union and gluing in $\mathcal{C}$, along with the fact that $\mathcal{C}$ is a groupoid (under holomorphic maps compatible with the boundary parametrizations), and in fact a stack over the Grothendieck topology of complex-analytic manifolds and open covers. In particular, gluing in

---

[1]It should be pointed out that instead of SPCMC's, we could use other known structures present on worldsheets which can be used for axiomatizing CFT, for example the 'cobordism approach' based on PROPs; our structure satisfies those axioms as well. As shown in [**2**], however, when one carefully treats the cobordism approach so no relevant axioms are omitted, the discussion is comparable to SPCMC's.

$\mathcal{C}$ is defined by noticing that the parametrized boundary components of a worldsheet can have two possible orientations with respect to the complex structure - one usually calls them inbound and outbound. Now from a worldsheet X, another worldsheet, usually denoted by $X^\triangledown$ (despite of the ambiguity of the symbol), can be obtained by gluing an inbound boundary component of X to an outbound, using the parametrizations. The notion of SPCMC is designed to capture all the algebraic properties of these operations.

The definition (1) may seem mysterious, but roughly speaking, we can imagine we have a certain finite set of labels A, Hilbert spaces H_a for $a \in A$, and for every worldsheet X with a map ϕ assigning to each boundary component c of X a label $\phi(c) \in A$, a finite-dimensional vector space $M_{X,\phi}$ and a trace class element

$$(2) \qquad U_{X,\phi} \in M_{X,\phi} \otimes \bigotimes_c H^*_{\phi(c)} \hat{\otimes} \bigotimes_d H_{\phi(d)}$$

where the tensor products are over inbound boundary components c and outbound boundary components d of X. The symbol $\hat{\otimes}$ means Hilbert tensor product, and H^* means the Hilbert dual of H. These elements (called vacuum elements) are required to satisfy certain properties which we will not list here. However, one important example is in order. When X has no boundary components (is a closed surface), (2) becomes simply an element of M_X (ϕ is dummy), and it follows from the structure that M_X is a representation of the mapping class group $Mod(X)$.

However, physicists noticed that in some cases (e.g. the lattice theories) more is true, namely that the representation of the mapping class group $Mod(X)$ on M_X extends to the Siegel modular group $Sp(2g, \mathbb{Z})$ where g is the genus of X (there is a natural map $Mod(X) \to Sp(2g, \mathbb{Z})$ by taking 1st cohomology). The question therefore arises: what does it mean for a CFT to be "Siegel-modular", or, in other words, to depend only on the cohomology of the worldsheet X?

It is the main purpose of this note to provide one possible answer to this question. Our approach is to define a pseudo morphism of SPCMC's

$$(3) \qquad\qquad \mathcal{C} \to \mathcal{J}$$

where $\mathcal{J}$ is, roughly speaking, the SPCMC of all possible 'structures that look like cohomologies of worldsheets'. We define precisely what this means, and call such structures 'open abelian varieties'.

Defining the SPCMC of open abelian varieties is our main result. We also show that the (chiral) lattice CFT corresponding to an even lattice indeed factors through a CFT on $\mathcal{J}$ by the map (3), which explains its Siegel modularity. The reader is invited to notice that such a discussion would be very difficult, if not impossible, if the notion of SPCMC were not developed.

The present paper is organized as follows. In Section 2, we define open abelian varieties, and discuss their moduli stack. In Section 3, we discuss gluing of open abelian varieties, and their SPCMC structure. In Section 4, we discuss the Jacobian map from the SPCMC of worldsheets to the SPCMC of open abelian varieties. In

Section 5, we shall discuss the lattice conformal field theory on the SPCMC of open abelian varieties.

2. Open abelian varieties

In this section, we will introduce a generalization of the concept of a (principally polarized) abelian variety to a notion which contains "Jacobians" of Riemann surfaces with parametrized boundary. Again, the main motivation is to a purely algebraic notion of "gluing". While the concept we develop is far from obvious, one can actually find in it a very pretty "open" version of principally polarized weight 1 pure Hodge structure with non-negative Hodge-degrees.

CONSTRUCTION 2.1. Let us start with the space V_1 of real-analytic functions

$$f : \mathbb{R} \to \mathbb{R}$$

for which there exists a number Δ_f such that

$$f(x + 2\pi) = f(x) + \Delta_f.$$

We may then alternately think of V_1 as a space of "branched" functions on S^1 by applying the map e^{iz}. There is an antisymmetric form S on V_1 given by

$$(4) \qquad S(f, g) = \int_{S^1} f \, dg - \Delta_f g(0) - \frac{1}{2} \Delta_f \Delta_g$$

(the integral over S^1 is interpreted as the integral from 0 to 2π). Note that in (4), the term $\Delta_f g(0)$ could have been equally well replaced by $\Delta_g f(0)$. The point is to choose the terms so that $S(f, g) = -S(g, f)$.

Given a pair of disjoint finite sets A^+ and A^-, we set

$$A = A^+ \amalg A^-.$$

(We think of A^+ as the set of *outbound* and A^- as the set of *inbound* boundary components within a connected component.) Define

$$V_A = \{ f = (f_i)_i \in \prod_{i \in A^+ \amalg A^-} V_1 \mid \sum_{i \in A^+} \Delta_{f_i} - \sum_{i \in A^-} \Delta_{f_i} = 0 \} / \langle (1)_i \rangle.$$

Now choose a linear ordering on the set A. For $i \in A$, define $\epsilon_i = 1$ if $i \in A^+$ and $\epsilon_i = -1$ if $i \in A^-$. Define for $f, g \in V_A$,

$$(5) \qquad S_<(f, g) = \sum_{i \in A} \epsilon_i S(f_i, g_i) - \frac{1}{2} \sum_{i < j \in A} \epsilon_i \epsilon_j (\Delta_{f_i} \Delta_{g_j} - \Delta_{g_i} \Delta_{f_j}).$$

Note that since the space V_A is fixed, we can also give it an integral structure, i.e. choose once and for all a topological basis $B_<$ on which $S_<$ is hyperbolic.

The exact choice does not matter. Note also that although the form (5) depends on the ordering of A, the antisymmetric forms $S_<$ for different orderings $<$ are easily calculated from each other, by adding differences of the corresponding terms

$$(6) \qquad \frac{1}{2} \sum_{i < j} \epsilon_i \epsilon_j (\Delta_{f_i} \Delta_{g_j} - \Delta_{g_i} \Delta_{f_j}).$$

Instead of speaking of an ordering and an antisymmetric form, it will be more useful for us to speak of a collection of antisymmetric forms $S_<$ related to each other by the said formulas. We shall speak of antisymmetric forms $S_<$ related *in the standard way*.

REMARK 2.2. It will be important in the sequel to note that the form $S_<$ in fact only depends on the *cyclic* ordering, i.e. if we take the smallest element 1 of A and make it the greatest element without changing the order of the other elements, then the form S does not change. To see this, note that the operation just described results in adding to S the term

$$(7) \qquad \sum_{1 \neq j \in A} \epsilon_1 \epsilon_j (\Delta_{f_1} \Delta_{g_j} - \Delta_{g_1} \Delta_{f_j}).$$

But we are also assuming

$$(8) \qquad \sum_{j \in A} \epsilon_j \Delta_{f_j} = 0 = \sum_{j \in A} \epsilon_j \Delta_{g_j},$$

so (7) is equal to

$$\epsilon_1^2 (\Delta_{f_1} \Delta_{g_1} - \Delta_{g_1} \Delta_{f_1}) = 0.$$

REMARK 2.3. There is another way of relating the forms $S_<$, $S_{<'}$ for different orders $<$, $<'$ which will be of importance to us. Consider functions $f = (f_i)_i$ and $g = (g_i)_i$ as above. Then define

$$(9) \qquad f_i' = f_i - \sum \{\epsilon_j \Delta_{f_j} | j < i \text{ and } i <' j\}.$$

We will refer to the map $f \mapsto f'$ given by (9) as the *standard transformation*

$$V_A \xrightarrow{\cong} V_A$$

corresponding to the change of the order $<$ to $<'$. The relation we have in mind is established by the following result.

LEMMA 2.4. *We have*

$$S_<(f, g) = S_{<'}(f', g').$$

PROOF. We have

$$S_<(f, g) = \sum_{i \in A} \epsilon_i \left(\int_{S^1} f_i \, dg_i - \Delta_{f_i} g_i(0) - \tfrac{1}{2} \Delta_{f_i} \Delta_{g_i} \right)$$

$$- \tfrac{1}{2} \sum_{i < j} \epsilon_i \epsilon_j (\Delta_{f_i} \Delta_{g_j} - \Delta_{g_i} \Delta_{f_j})$$

$$= \sum_{i \in A} \epsilon_i \left(\int_{S^1} f_i' \, dg_i' + \sum_{j < i, i <' j} \int_{S^1} \epsilon_j \Delta_{f_j} dg_i' - \Delta_{f_i'} (g_i'(0) + \sum_{j < i, i <' j} \epsilon_j \Delta_{g_j}) \right)$$

$$- \tfrac{1}{2} \Delta_{f_i'} \Delta_{g_i'}) - \tfrac{1}{2} \sum_{i < j} \epsilon_i \epsilon_j (\Delta_{f_i'} \Delta_{g_j'} - \Delta_{g_i'} \Delta_{f_j'})$$

$$= \sum_{i \in A} \epsilon_i \left(\int_{S^1} f'_i dg'_i - \Delta_{f'_i} g'_i(0) - \tfrac{1}{2} \Delta_{f'_i} \Delta_{g'_i} \right)$$

$$+ \sum_{i \in A} \epsilon_i \left(\sum_{j < i, i <' j} \epsilon_j \Delta_{f'_j} \Delta_{g'_i} \right) - \sum_{i \in A} \epsilon_i \Delta_{f'_i} \left(\sum_{j < i, i <' j} \epsilon_j \Delta_{g'_j} \right)$$

$$- \tfrac{1}{2} \sum_{i < j} \epsilon_i \epsilon_j \left(\Delta_{f'_i} \Delta_{g'_j} - \Delta_{g'_i} \Delta_{f'_j} \right)$$

$$= \sum_{i \in A} \epsilon_i S(f'_i, g'_i)$$

$$+ \sum_{j < i, i <' j} \epsilon_i \epsilon_j \left(\Delta_{f'_j} \Delta_{g'_i} - \Delta_{g'_j} \Delta_{f'_i} \right)$$

$$- \tfrac{1}{2} \sum_{i < j} \epsilon_i \epsilon_j \left(\Delta_{f'_i} \Delta_{g'_j} - \Delta_{g'_i} \Delta_{f'_j} \right)$$

$$= \sum_{i \in A} \epsilon_i S(f'_i, g'_i)$$

$$- \tfrac{1}{2} \sum_{j < i, i <' j} \epsilon_i \epsilon_j \left(\Delta_{f'_i} \Delta_{g'_j} - \Delta_{g'_i} \Delta_{f'_j} \right)$$

$$+ \tfrac{1}{2} \sum_{i < j, j <' i} \epsilon_i \epsilon_j \left(\Delta_{f'_i} \Delta_{g'_j} - \Delta_{g'_i} \Delta_{f'_j} \right)$$

$$- \tfrac{1}{2} \sum_{i < j} \left(\Delta_{f'_i} \Delta_{g'_j} - \Delta_{g'_i} \Delta_{f'_j} \right)$$

$$= \sum_{i \in A} \epsilon_i S(f'_i, g'_i) - \tfrac{1}{2} \sum_{i <' j} \epsilon_i \epsilon_j \left(\Delta_{f'_i} \Delta_{g'_j} - \Delta_{g'_i} \Delta_{f'_j} \right)$$
$$= S_{<'}(f', g').$$

$\square$

It is also of interest to us that when $<'$ is obtained from $<$ by moving the greatest element i to the lowest, then $f'_j = f_j$ for $j \neq i$, and f'_i is obtained from f_i by adding the constant function equal to $\epsilon_i \Delta_{f_i}$. This means that when $<$ and $<'$ correspond to the same cyclic order, the standard transformation is not necessarily the identity, but is given by adding to each f_i a constant function which is a fixed integral multiple of Δ_{f_i}.

DEFINITION 2.5. An *open abelian variety* $(C, U, S, W, \iota, V_{\mathbb{Z}}^{\perp})$ consists of a (possibly empty) set C of finite sets (called open connected components) $A = A^+ \amalg A^-$ (whose elements are called *outbound* and *inbound boundary components* respectively), a real vector space U with, for each system of linear orders $<$ of each

$A \in C$, an embedding

(10)
$$V := \bigoplus_{A \in C} V_A \xrightarrow[\subseteq]{\iota_<} U$$

such that the image $\iota_< V$ is of finite codimension. (Note that the image $\iota_< V$ does not depend on $<$.) Further, for different choices of orders $<$ and $<'$, the embeddings $\iota_<$ and $\iota_{<'}$ are related by composing with the standard transformation (see Remark 2.3). Further, a nondegenerate real symplectic form S is given on U, and (10) maps the form

$$\bigoplus_{A \in C} S_< \text{ on } \bigoplus_{A \in C} V_A.$$

to S. (Note that by Remark 2.3, it suffices to verify this assumption for one $\iota_<$.)

Next, there is given a smooth (in the standard sense, see below) complex isotropic subspace $W \subset U_{\mathbb{C}}$ such that

(11)
$$W \oplus \overline{W} = U_{\mathbb{C}},$$

(12)
$$2iS(x, \overline{x}) > 0 \text{ for all } x \in W$$

(here $\overline{W}$ denotes the complex conjugate of W).

Additionally, there is an *integral structure*, which is the following subtle data. First, there is an integral structure on the S-complement $V^\perp$ (=annihilator) of $\iota_< V$, which means there is a subgroup $V_{\mathbb{Z}}^\perp$ of $V^\perp$ on which S is isomorphic (but not by a given isomorphism) to a hyperbolic antisymmetric form.

Next, we impose an identification on open abelian varieties according to the following rule. Denote by $V_{const,\mathbb{Z}}$ the subgroup of V consisting of functions which are constant, and have integral value, on every boundary component. Similarly, let $V_{deg,\mathbb{Z}}$ denote the subspace of V of functions which have integral degree on each boundary component, i.e. $\Delta_{f_j} \in \mathbb{Z}$ for all $j \in A \in C$. Fix a system of linear orders $<$ on each $A \in C$. Then two open abelian varieties $(C_1, U_1, S_1, W_1, \iota_1, V_{\mathbb{Z},1}^\perp)$ and $(C_2, U_2, S_2, W_2, \iota_2, V_{\mathbb{Z},2}^\perp)$ are identified if $C_1 = C_2$, $U_1 = U_2$, $S_1 = S_2$, $W_1 = W_2$ and the selection of the map $\iota_<$ and $V_{\mathbb{Z}}^\perp$ is subject to the following rules: We require

(13)
$$V_{\mathbb{Z},1}^\perp \subseteq V_{\mathbb{Z},2}^\perp \oplus \iota_2 V_{const,\mathbb{Z}},$$

(14)
$$(\iota_1 - \iota_2)(V_{deg,\mathbb{Z}}) \subseteq V_{\mathbb{Z},2}^\perp \oplus \iota_2 V_{const,\mathbb{Z}}.$$

Note that $\iota_1 - \iota_2$ is a homomorphism. (Note that condition (14) implies that $\iota_1 - \iota_2$ on V only depends on the degree, as it is determined by its restriction to $V_{deg,\mathbb{Z}}$, and the target of that map is discrete. It then follows that on elements of V of constant degree, in particular on $V_{const,\mathbb{Z}}$, $\iota_1 = \iota_2$. Because of this, one can replace ι_2 by ι_1, and/or $V_{\mathbb{Z},1}^\perp$ by $V_{\mathbb{Z},2}^\perp$ in (13). Also, because of this and (13), we may replace ι_2 by ι_1 and/or $V_{\mathbb{Z},2}^\perp$ by $V_{\mathbb{Z},1}^\perp$ on the right hand side of (14).)

Note also that by Remark 2.3, the choice of $<$ does not matter in this identification, since the identification is invariant under standard transformation. Note also that by the same remark, fixing $<$, we may replace $\iota_<$ by $\iota_{<'}$ for any system of orders $<'$ which defines the same cyclic order on each $A \in C$ without changing the open abelian variety.

To define smoothness of a subspace W, recall that we have a standard polarization of $V_{\mathbb{C}}$ given by the isotropic subspaces V^+, V^- of functions on each copy of S^1 which holomorphically (resp. antiholomorphically) extend to the unit disk (recall that polarizations do not depend on adding or subtracting finite-dimensional subspaces). Now by smoothness of W we mean that the projection of W to V^+ is a Fredholm operator and the projection of W to V^- is a smooth operator, i.e. its singular values (considering the Hilbert structures on W, V^- given by (12) and the analogous form on V^-) decrease exponentially.

Deciding which maps to call *morphisms* of open abelian varieties is an interesting problem. For the purpose of the present paper, we will choose morphisms to be only isomorphisms, which is unambiguous.

DEFINITION 2.6. An *isomorphism* of open abelian varieties

$$(C, U, S, W, \iota) \to (C', U', S', W', \iota')$$

consists of a bijection $b : C \to C'$, and for each $A \in C$ a bijection $b_A : A \to b(A)$ preserving inbound and outbound boundary components, an isomorphism $\phi : U \to U'$ such that $\phi(W) = W'$, ϕ carries S to S', and for each system of orders $<$ of all $A \in C$, if we denote by $<_b$ the order induced by the system b_A on $b(A)$, b_A and ϕ conjugate $\iota_<$ to an embedding which defines the same open abelian variety as $\iota'_{<_b}$.

In this paper, the category of open abelian varieties will be chosen to be the category whose objects are open abelian varieties and whose morphisms are isomorphisms.

The identifications imposed in Definition 2.5 can be viewed more systematically in the following way: Consider a particular embedding $\iota_0 : V \to U$, and a particular hyperbolic basis of $V_{\mathbb{Z}}^{\perp}$. Then we can identify U with $V \oplus V_{\mathbb{R}}^{\perp}$ via this embedding. Now consider the group of all linear transformations

$$\phi : V \oplus V^{\perp} \to V \oplus V^{\perp}$$

which can be represented by 2×2 matrices

$$\begin{pmatrix} \phi_{VV} & \phi_{VV^{\perp}} \\ \phi_{V^{\perp}V} & \phi_{V^{\perp}V^{\perp}} \end{pmatrix}$$

such that the map $\phi_{V^{\perp}V^{\perp}}$ is an integral symplectic transformation, $\phi_{VV^{\perp}}(V_{\mathbb{Z}}^{\perp}) \subseteq V_{const,\mathbb{Z}}$, $\phi_{V^{\perp}V}(V_{deg,\mathbb{Z}}) \subseteq V_{\mathbb{Z}}^{\perp}$, $(\phi_{VV} - Id)(V_{deg,\mathbb{Z}}) \subseteq V_{const,\mathbb{Z}}$. It is easy to check that linear transformations of this type form a discrete group, which we denote by $Sp_{open}(V, \mathbb{Z})$. This can be considered the group of identifications of open abelian variety data.

More precisely, let us compute the moduli space of open abelian varieties for a given set of open connected components. From the definition, it follows that the moduli space is

(15) $U(W) \backslash Sp_{sm}(U) / Sp_{open}(V, \mathbb{Z})$.

The group $U(W)$ is the Hilbert unitary group on W, the group $Sp_{sm}(U)$ is the real symplectic group of U which when expressed as 2×2 matrices in the decomposition $W \oplus \overline{W}$, the off-diagonal terms are smooth operators.

It is worth noting that the group $Sp_{sm}(U)$ is in fact also contractible, so the moduli space is a "$K(\pi, 1)$-stack". To prove this, by Kuiper's theorem, it suffices to show that the coset space

$$(16) \qquad\qquad U(W) \backslash Sp_{sm}(U)$$

is contractible. Expressing the form S as a 2×2 matrix as discussed above, it is of the form

$$\begin{pmatrix} 0 & -iI \\ iI & 0 \end{pmatrix},$$

so (16) is isomorphic to the contractible space

$$\left\{ \exp \begin{pmatrix} 0 & A \\ A & 0 \end{pmatrix} \,\middle|\, A \text{ is symmetric smooth} \right\}.$$

REMARK 2.7. An open abelian variety with no open connected (and hence no boundary) components is simply a real symplectic space U with integral structure and decomposition

$$U_0 - W \oplus \overline{W}$$

where W is positive-definite isotropic, in other words, $S_{W \times W} = 0$ and $2iS(x, \overline{x}) >$ 0 for all $x \in W$. This is equivalent data to an abelian variety over $\mathbb{C}$ as in [6].

3. Gluing and SPCMC structure

If $(I, +, 0)$ is a monoidal category, a strict 2-functor X from I^2 into the 2-category of small categories is a *pseudo commutative monoid with cancellation* if it is equipped with the basic operations of

$$(17) \qquad\qquad \text{addition} + : X_{a,b} \times X_{c,d} \to X_{a+c,b+d}$$

$$(18) \qquad\qquad \text{unit } 0 \in X_{0,0}$$

$$(19) \qquad\qquad \text{and gluing } \nabla : X_{a+c,b+c} \to X_{a,b}$$

which satisfy the following axioms up to coherent isomorphisms: commutativity, associativity, unitality, transitivity, distributivity, and trivial cancellation is trivial. In this paper, $(I, +, 0)$ is category of finite sets with disjoint union and 0 is the empty set. We abbreviate the phrase *stack of pseudo commutative monoids with cancellation* by *SPCMC*.

THEOREM 3.1. *There exists an SPCMC structure on the set of open abelian varieties.*

REMARK 3.2. Before embarking on this story, let us briefly note the following curious fact: although open abelian varieties model the notion of open connected components, it does not model the notion of closed connected components. Moreover, for the same reason, while one can define genus as one half of the codimension of V in U (which agrees with the usual genus in the case of a Jacobian of a surface), the structure does not model the genus of an individual open connected component. It is worthwhile pointing out that one can consider a variant of our notion which would keep track of both closed and open connected components, and would be simply a sequence of closed and open abelian varieties with one connected component in our sense. Such structure would also form an SPCMC by our arguments.

The proof of Theorem 3.1 will occupy the remainder of this section. First, note that the stack structure over complex manifolds and coverings follows from the moduli space remarks at the end of the last section. Also, the operation of sum is obvious, realized by direct sum in the obvious sense. So the main point to discuss is gluing.

We have the decomposition

$$(20) \qquad\qquad U \cong V \oplus V^{\perp}$$

where $V^{\perp}$ is the S-annihilator of V in U. We therefore have a canonical projection given by the decomposition (20)

$$(21) \qquad\qquad p : U \to V.$$

Composing with the projection q_A from V to V_A for a connected component A, we get a projection

$$(22) \qquad\qquad p_A : U \to V_A.$$

Composing further, for $j \in A$, with the projection

$$q_{A,j} : V_A \to V_1/\mathbb{R}$$

(where V_1 is the space of real analytic branched functions on S^1 as in Construction 2.1 and $\mathbb{R}$ is generated by the constants), we get a projection

$$(23) \qquad\qquad p_{A,j} : U \to V_1/\mathbb{R}.$$

All these maps of course also have complex forms, which we will denote by the same symbol.

Now the idea of gluing an inbound boundary component $i \in A^-$ to an outbound boundary component $j \in B^+$, $A, B \in C$, is to set

$$(24) \qquad U^{\triangledown} = \{a \in U | p_{A,i}(a) = p_{B,j}(a)\}/Im(V_1).$$

Here by $Im(V_1)$ we denote the image of V_1 in U by sending an element $x \in V_1$ to the sum of $\iota_<(x_i)$ and $\iota_<(x_j)$ where x_i (resp. x_j) is the same function as x on the i'th (resp. j'th) boundary component and zero everywhere else. The order $<$ is selected in such a way that i immediately precedes j (see discussion of Cases 1 and 2 below). Then $Im(V_1)$, by our assumptions, S-annihilates $\{a \in U | p_{A,i}(a) = p_{B,j}(a)\}$, so we can choose $S^{\triangledown}$ as the form induced by S. However, we will need to show that it is a non-degenerate symplectic form. To this end, we will actually first give an independent formula for gluing W, and then show that it is compatible with (24).

To glue W, we simply take

$$(25) \qquad\qquad W^{\triangledown} = \{a \in W | p_{A,i}(a) = p_{B,j}(a)\}.$$

Next, we will define the set of open connected components $C^{\triangledown}$ after gluing, which will give us a space $V^{\triangledown}$ defined the same way as V, with C replaced by $C^{\triangledown}$, and an embedding $\iota_<^{\triangledown}$ after gluing corresponding to a system of orders $<$ before gluing. Of course, one can choose the order $<$, since for different orders the embeddings must be related by composing with a standard transformation. Now there are two principal cases to distinguish:

Case 1: $A = B$. In this case, define $C^{\triangledown}$ as the set of components $E^{\triangledown} = E$ when $E \neq A \in C$, and $A^{\triangledown} = A - \{i, j\}$ provided $A^{\triangledown} \neq \emptyset$. Next, assume $i < j < k$ for all $k \in A - \{i, j\}$. Then let the order $<$ after gluing be given by omitting i, j. Further, assume that the boundary component corresponding to i is inbound. We define for $x \in V^{\triangledown}$, $\iota_{<}^{\triangledown}(x)$ to be the projection of $\iota_{<}(y)$ for any $y = (y_k) \in V$ where $y_k = x_k$ when $k \neq i, j$, and $y_i = y_j$ is arbitrary. By definition, this embedding preserves the symplectic form.

Case 2: $A \neq B$. Then $C^{\triangledown}$ is the set of $E^{\triangledown} = E$ where $A, B \neq E \in C$, and $A^{\triangledown} = B^{\triangledown} = (A \cup B) - \{i, j\}$, provided $A^{\triangledown} \neq \emptyset$. Then assume that i is the greatest element of A and j is the least element of B. Assume again that the i'th boundary component is inbound. Let the ordering on the glued connected component $(A \cup B) - \{i, j\}$ be obtained by juxtaposing the ordering on $A - \{i\}$ before the ordering on $B - \{j\}$. Again, for $x \in V^{\triangledown}$, we define $\iota_{<}^{\triangledown}(x)$ to be the projection of $\iota_{<}(y)$ for any $y = (y_k) \in V$ where $y_k = x_k$ when $k \neq i, j$, and $y_i = y_j$ is arbitrary. (Note that in this case, there is a subtlety due to the fact that x_k is only defined up to adding two different constants for $k \in A, B$; what we mean is that the difference of the constants is fixed by the requirement $y_i = y_j$.) Again, we see that this embedding preserves antisymmetric forms.

REMARK 3.3. In the Cases 1 and 2, when $A^{\triangledown} = \emptyset$, it simply gets deleted from the data (see comments in the paragraph below Theorem 3.1 at the beginning of this section). It does not affect the rest of the gluing procedure.

It remains to relate the formulas (24), (25), and prove that S remains non-degenerate. First, since we have complete control over the structure of $U^{\triangledown}$, it is easy to see that

$$(26) \qquad \begin{aligned} g^{\triangledown} &= g + 1 \quad \text{in Case 1,} \\ g^{\triangledown} &= g \qquad\;\; \text{in Case 2} \end{aligned}$$

where $g^{\triangledown}$ denotes $1/2$ times the codimension of $V^{\triangledown}$ in $U^{\triangledown}$. Additionally, since $S^{\triangledown}$ is induced from S (at least for the particular choice of orderings), we know that $W^{\triangledown} \subset U_{\mathbb{C}}^{\triangledown}$, $\overline{W}^{\triangledown} \subset U_{\mathbb{C}}^{\triangledown}$ are isotropic and $S^{\triangledown}$-dual to each other, so in particular

$$W^{\triangledown} \cap \overline{W}^{\triangledown} = 0$$

and thus that the natural map

$$(27) \qquad W^{\triangledown} \oplus \overline{W}^{\triangledown} \to U_{\mathbb{C}}^{\triangledown}$$

is injective. What remains to be shown is that, viewing (27) as an inclusion,

$$(28) \qquad W^{\triangledown} + \overline{W}^{\triangledown} = U_{\mathbb{C}}^{\triangledown},$$

or in other words that the map (27) is onto. To show this, we will take advantage of Segal's method [9] of relative dimension. Choosing a polarization of

$$(29) \qquad V_{\mathbb{C}} = V^{+} \oplus V^{-}$$

compatible with W (for example as discussed in the last section), let

$$W_0 = Im(p|_W), \quad \overline{W}_0 = Im(p|_{\overline{W}}).$$

Here $p|_W$ respectively $p|_{\overline{W}}$ denotes the restriction to W respectively $\overline{W}$ of the complexification $p_{\mathbb{C}} : U_{\mathbb{C}} \to V_{\mathbb{C}}$ of the projection p in equations (20) and (21). Denoting relative dimension with respect to the positive space V^+ by dim_{V^+}, i.e.

$$dim_{V^+}(Q) = index(\pi_Q)$$

for $Q \subset V_{\mathbb{C}}$ where $\pi_Q : Q \to V^+$ is the projection given by the decomposition, we get

(30) $\quad dim(Ker(p|_W)) + dim_{V^+}W_0 + dim(Ker(p|_{\overline{W}})) + dim_{V^-}\overline{W}_0 = 2g$

(since W_0 and $\overline{W}_0$ generate $V_{\mathbb{C}}$,

$$dim_{V^+}W_0 + dim_{V^-}\overline{W}_0 = dim(W_0 \cap \overline{W}_0)).$$

But now one has

(31) $\quad dim(Ker(p|_{W^\triangledown})) + dim_{V^\triangledown+}W_0^\triangledown \geq dim(Ker(p|_W)) + dim_{V^+}W_0 + \epsilon$

where ϵ is 1 in Case 1 and 0 in Case 2 (this shift arises because of our treatment of the constants on connected components). Equality arises if and only if

(32) $$W_0^\triangledown + \overline{W}_0^\triangledown = V_{\mathbb{C}}^\triangledown.$$

Similarly, we have

(33) $\quad dim(Ker(p|_{\overline{W}^\triangledown})) + dim_{V^\triangledown-}\overline{W}_0^\triangledown \geq dim(Ker(p|_{\overline{W}})) + dim_{V^-}\overline{W}_0 + \epsilon$

and

(34) $\quad dim(Ker(p|_{W^\triangledown})) + dim_{V^+}W_0^\triangledown + dim(Ker(p|_{\overline{W}^\triangledown})) + dim_{V^-}\overline{W}_0^\triangledown \leq 2g + 2\epsilon.$

Comparing (30), (31), (33), (34), we see that equality must arise in (31), (33), so we have (32), which implies (28) by (26) and the comment preceeding (32).

Now integral structure is discussed as follows. First of all, $V_{\mathbb{Z}}^{\triangledown\perp}$ is generated by $V_{\mathbb{Z}}^\perp$ in Case 2, and is generated by $V_{\mathbb{Z}}^\perp$ and elements which have integral degree on i and differ by an integral value on i, j, and have 0 projection to the other boundary components (well defined since we are in the same boundary component) in Case 1. Such elements must generate $V_{\mathbb{C}}^\perp$ by the discussion of the previous paragraph. Additionally, a direct verification proves that equivalence is preserved by gluing.

To define the operations of an SPCMC as defined in [3], we need to soup up our gluing definition to glue simultaneously several pairs of boundary components, each consisting of one inbound and one outbound boundary component.

Regarding the gluing of U and W, there are obvious generalizations of formulas (24) and (25) for multiple pairs of components. The trickiest part is the discussion of the ordering of boundary components, since in the case of multiple boundary components, we can no longer rely on distinguishing two cases as we did above. The procedure for generalizing to the case of gluing several pairs is as follows: First, note that for an open abelian variety $\mathfrak{X}$, we can associate an antisymmetric form $S_<$ with *any* ordering of the entire set of boundary components of $\mathfrak{X}$, regardless of the open connected components. Simply relate the forms corresponding to the orderings in the standard way, and the embeddings $\iota_<$ by composing with the standard transformations. (Note that even though the components of an element in each open connected component are only defined up to a separate additive constant, this does not affect standard transformations.) For the operation of disjoint union, we simply juxtapose the order (this is possible, as permuting cyclically the boundary components of each disjoint summand does not change the form S). The general procedure for gluing is to change the order of boundary components (while relating S in the standard way and $\iota_<$'s by composing with standard transformations) so that all pairs of boundary components to be glued are arranged so that the outbound component immediately follows the inbound, i.e. the inbound is i'th and the outbound is $i+1$'st, if the boundary components are indexed by integers. The key observation is that permuting i and $i+1$ past another boundary component will not change the value of the form S, since the terms of (5) involving i and $i+1$ cancel out, since f_i and f_{i+1} are the same function when gluing. Similarly, the standard transformations corresponding to such permutations are identities on functions where f_i and f_{i+1} coincide. More generally, embeddings with respect to orders of this specified form which are related by composing with standard transformations before gluing remain related by composing with standard transformations after gluing, since terms coming from the glued boundary components cancel out.

After such arrangement we take the induced embedding $\iota_<^\nabla$ to be associated with the order $<$ which omits all the pairs of the glued boundary components, and leaves the order of the others unchanged. For a direct definition of the integral structure, $V_{\mathbb{Z}}^{\nabla \perp}$ is generated by $V_{\mathbb{Z}}^\perp$ and elements which can be lifted to an element f of the sum of copies of V_1 over all the boundary components in such a way that $f_k = 0$ on any boundary component not glued, f_i has integral degree and $f_i - f_j$ is a constant integral function when i, j are glued. We see that this composite gluing produces an open abelian variety, since it will be, for a particular order selected, isomorphic to the open abelian variety obtained by gluing the pairs of boundary components successively.

Next, we must prove that the disjoint union and gluing operations just defined have the coherence isomorphisms and diagrams required in an SPCMC [3].

The coherence isomorphisms correspond simply to the identities required for a commutative monoid with cancellation (Def 3.4 of [3]). The identities are commutativity, associativity, and unitality of sum, unitality and transitivity of cancellation, and distributivity of cancellation under addition. The isomorphisms are by definition determined by what they do on W, where sum corresponds to direct sum, and gluing is given by the generalization of (25) to multiple pairs. This is coherent with respect to the obvious maps. It is also easy to see that the corresponding maps are compatible with the $\iota_<$'s and the integral structure.

Having defined the coherence isomorphisms, we need to consider the commutativity of coherence diagrams. Those diagrams are defined in [3]. All the diagrams required are of the following form. Denote by $X_{a,b}$ the set of open abelian varieties with inbound (resp. outbound) boundary components indexed by the finite set a (resp. b). Recall the notation for the basic operations in equations (17), (18), and (19). We consider all *words* $\mathfrak{W}$ which can be written using n distinct variables $x_1, ..., x_n$, each x_i representing an open abelian variety with inbound (resp. outbound) boundary components indexed by v_i (resp. w_i). The v_i's and w_i's are in turn words in m variables $a_1, ..., a_m$ (representing finite sets), using the finite set-level operations $+$, 0. No variable a_i is allowed to occur more than once among the v_i's, or among the w_i's. However, a variable occuring among the v_i's may also occur among the w_i's (note that otherwise, the operation ∇ could not be applied).

Now coherence diagrams [3] are obtained by the following procedure: Alter a word $\mathfrak{W}$ repeatedly by applying one of the identities (commutativity, associativity, unitality of $+$, unitality and transitivity of $\triangledown$, and distributivity). Denote the word obtained by the end result of this sequence of alterations by $\mathfrak{W}'$. Then it is possible that the same word $\mathfrak{W}'$ could also be obtained from $\mathfrak{W}$ by a different sequence of alterations. Any time this occurs, we have an obvious corresponding coherence diagram. Our task is to show that all such diagrams commute.

However, this is quite easy, since an isomorphism between open abelian varieties is determined by the isomorphism of the W's. Now we have a canonical injection

$$(35) \qquad\qquad W_{\mathfrak{X}^\triangledown} \to W_{\mathfrak{X}},$$

and also canonical projections

$$(36) \qquad\qquad W_{\mathfrak{X}_1 + \mathfrak{X}_2} \to W_{\mathfrak{X}_i}.$$

Therefore, by induction, we obtain a map

$$(37) \qquad\qquad W_{\mathfrak{W}} \xrightarrow{\phi^i_{\mathfrak{W}}} W_{\mathfrak{X}_i},$$

$i = 1, ..., n$, whose product is injective. By considering all types of coherence isomorphisms again (units, $\triangledown$-transitivity, $+$-commutativity and associativity), we see that the maps (35) and (36) commute with the maps induced by the coherence isomorphisms. Consequently, the two paths p_1 and p_2 from the word $\mathfrak{W}$ to the word $\mathfrak{W}'$ induce a commutative diagram

$$(38)$$

$j = 1, 2$, $i = 1, ..., n$. Since however the product of the maps $\phi^i_{\mathfrak{W}'}$ is injective, we conclude that $p_{1*} = p_{2*}$, as required.

4. The Jacobian of a worldsheet with boundary

In this paper, a *worldsheet* Σ is a Riemann surface whose boundary components $c_1, ..., c_n$ are parametrized by analytic diffeomorphisms

$$\phi_i : S^1 \to c_i.$$

Taking a chart of Σ (and thus identifying with a subset of $\mathbb{C}$), boundary components oriented counterclockwise (resp. clockwise) are called *inbound* (resp. *outbound*). Worldsheets form an SPCMC $\mathcal{C}$, as proved in [3].

THEOREM 4.1. *There exists a morphism of SPCMC's*

$$(39) \qquad\qquad T : \mathcal{C} \to \mathcal{J}.$$

extending the Torelli map on the moduli stack of closed Riemann surfaces.

We will also call the map T the Torelli map, by extension of the closed case. The proof of Theorem 4.1 will occupy the remainder of this section.

DEFINITION 4.2. A *cut worldsheet* is a pair (Σ, Γ) where Σ is a worldsheet and

$$\Gamma \subset \Sigma$$

is a graph, i.e. a 1-dimensional CW complex whose edges are piecewise analytic, subject to the two conditions. First, the boundary components c_i are required to be edges of Γ and the points $\phi_i(1)$ are the vertices (in particular, the boundary components are not subdivided). Recall that the ϕ_i are the boundary parametrizations from above. Second, the connected components of $\Sigma - \Gamma$ must be surfaces of genus 0 and their number must be equal to the number of the connected components of Σ.

Thus, Γ basically cuts each connected component of Σ into a surface of genus 0 without disconnecting it.

LEMMA 4.3. *A structure of a cut worldsheet (we will say simply cut structure) exists on every worldsheet.*

PROOF. Without loss of generality, we can assume Σ is connected. To construct Γ, we can first choose a set of disjoint collectively non-separating curves in Σ which cut it to a surface Σ' of genus 0, and let the vertices of Γ be the images of 0 under the parametrizations. Then connect the vertices by disjoint open edges which cut Σ into a disk.   □

It will be convenient to be a little more specific about the choice of cut structure constructed in the proof of Lemma 4.3. Note that a cut structure on a connected worldsheet specifies a cyclic order of boundary components: changing for the moment the orientation of the boundary components to outbound if necessary, this is simply the order in which the boundary components appear if we travel the boundary of the disk obtained by cutting the worldsheet along Γ. Now, if Σ is connected, we will call (Σ, Γ) a *standard cut structure* on Σ if the cyclic order of the boundary components of the genus 0 worldsheet Σ' defined in the proof of Lemma 4.3 is of the form

(40) $$c_1, ..., c_n, d_1, ..., d_{2g},$$

where $c_1, ..., c_n$ are the boundary components of Σ, and Σ is obtained from Σ' by gluing d_{2i-1} with d_{2i}, $i = 1, ..., g$. We may refer to the pairs d_{2i-1}, d_{2i} as *pairs of hidden boundary components* of Σ'. A cut structure on a general worldsheet Σ will be called standard if its restriction to every connected component of Σ is standard.

Now for a Riemann surface with standard cut structure (Σ, Γ), we define an open abelian variety $T(\Sigma, \Gamma)$ as follows:

Without loss of generality, we may assume that Σ is not closed, for in the closed case we just take the ordinary Jacobian. We may further assume that Σ is connected, as there is an obvious operation of direct sum on open abelian varieties (as already remarked). Under the assumption, then, there is only one open connected component A, and its elements are the boundary components of Σ. Let, then, W be the space of holomorphic functions

$$f : \Sigma - \Gamma \to \mathbb{C}$$

which extend to holomorphic functions on the universal cover $\tilde{\Sigma}$

$$\tilde{f} : \tilde{\Sigma} \to \mathbb{C}$$

such that for every deck transformation

$$\sigma : \tilde{\Sigma} \to \tilde{\Sigma}$$

there exists a number $n_{\sigma,f} \in \mathbb{C}$ such that

$$\tilde{f}(\sigma z) - \tilde{f}(z) = n_{\sigma,f} \text{ for all } z \in \tilde{\Sigma},$$

factored out by the space of functions constant on each connected component. The space $\overline{W}$ is defined analogously with the word "holomorphic" replaced by "antiholomorphic". Then we must define

$$U_{\mathbb{C}} = W \oplus \overline{W}.$$

To define the form S on U, first define, for $f \in U$, a 1-form ω_f on Σ by

$$\omega_f = d\tilde{f}.$$

Then define the ordering $<$ of boundary components as the order in which the boundary components occur on the boundary of $\Sigma - \Gamma$ in the counterclockwise direction. (Recall that only the cyclic order matters.) Then define

(41) $$S(f, g) = \int_\Sigma \omega_f \omega_g.$$

LEMMA 4.4. *The restriction*

(42) $$U \to V_A$$

is onto. More precisely, (42) has a splitting which is canonical on functions of degree 0 on each boundary component, and canonical in the general case subject to selecting a standard cut structure on Σ.

PROOF. Assume without loss of generality that Σ is connected and not closed. Recall that by the Dirichlet principle, for a (single-valued) real-analytic function ϕ_0 on $\partial\Sigma$, there exists a unique harmonic function ϕ on Σ such that

$$\phi|_{\partial\Sigma} = \phi_0.$$

We can then represent uniquely

$$\phi \in W \oplus \overline{W},$$

which gives a canonical splitting of (42) on functions of degree 0. To find a splitting on functions of non-zero degrees, note that, using the notation (40), $c_1, ..., c_n, d_2,$ $d_4, ..., d_{2g}$ and the paths $p_1, ..., p_g$ on the boundary of the disk D from the vertex v_i of Γ on d_{2i-1} and the corresponding point on d_{2i} form a basis of $H_1(\Sigma, \mathbb{Z})$. Therefore, there exists a harmonic form with any given residues along $c_1, ..., c_n$ with sum 0, and residues 0 along $d_2, ..., d_{2g}, p_1, ..., p_g$. Integrating the form we obtain a function ϕ, and subtracting ϕ from the original function reduces the general case to the degree 0 case in terms of existence and uniqueness.   □

LEMMA 4.5. *Let (Σ, Γ) be a genus 0 cut worldsheet and let $<$ be an order of boundary components compatible with the cyclic order specified by the cut. Then, for real analytic functions f, g on $\partial\Sigma$,*

$$(43) \qquad\qquad S(\tilde{f}, \tilde{g}) = S_<(f, g)$$

where $S_<$ is the form defined in Construction 2.1, S is (41), and $\tilde{f}, \tilde{g}$ are the harmonic continuations of f, g to the disk obtained by cutting Σ along Γ.

PROOF. Let D be the disk obtained from Σ by cutting along Γ. By Stokes' theorem, we have

$$(44) \qquad\qquad S(\tilde{f}, \tilde{g}) = \int_D \omega_{\tilde{f}}\omega_{\tilde{g}} = \int_{\partial D} \tilde{f}d\tilde{g}.$$

We claim that the right hand side is equal to (5) in the order specified. To see this, we can assume that all the boundary components are outbound, and the graph Γ has no vertices except the vertices $v_1, ..., v_n$ on the boundary components $c_1, ..., c_n$, and edges connecting v_i, v_{i+1}, $i = 1, ..., n-1$ (since we can always reach such case by continuous deformation which does not change the value of (44)).

In this case, denoting by f_i, g_i the restrictions of f, g to c_i, the contribution to the right hand side of (44) other than from the boundary components $c_1, ..., c_n$ is

$$(g_2(0) - g_1(0) - \Delta_{g_1})\Delta_{f_1} + (g_3(0) - g_2(0) - \Delta_{g_2})(\Delta_{f_1} + \Delta_{f_2}) + \cdots$$
$$\cdots + (g_n(0) - g_{n-1}(0) - \Delta_{g_{n-1}})(\Delta_{f_1} + \cdots + \Delta_{f_{n-1}})$$

$$= -\sum_{i=1}^{n} g_i(0)\Delta_{f_i} - \sum_{i \leq j} \Delta_{f_i}\Delta_{g_j}$$

$$= -\sum_{i=1}^{n} g_i(0)\Delta_{f_i} - \frac{1}{2}\sum_{i=1}^{n} \Delta_{f_i}\Delta_{g_i} - \frac{1}{2}\sum_{i<j}(\Delta_{f_i}\Delta_{g_j} - \Delta_{g_i}\Delta_{f_j}).$$

$\square$

LEMMA 4.6. *The conclusion of Lemma (4.5) extends to all worldsheets with standard cut structure, provided*

$$\tilde{f}|_{d_{2i-1}} = \tilde{f}|_{d_{2i}} \text{ of degree } 0$$

and

$$\tilde{g}|_{d_{2i-1}} = \tilde{g}|_{d_{2i}} \text{ of degree } 0.$$

PROOF. It suffices to assume, without loss of generality, that Σ is connected. Then simply apply Lemma (4.5) to Σ'. The additional terms related to d_{2i-1}, d_{2i} cancel out. $\square$

Note that the function $\tilde{f}$ in Lemma (4.6) is determined uniquely by f and Γ. Thus, fixing Γ, we can now define an open abelian variety $T(\Sigma, \Gamma)$ by choosing W as above, and letting the map (10) be defined by the correspondence $f \mapsto \tilde{f}$. Regarding the integral structure, a function $f \in V^\perp$ is integral if all the numbers $n_{\sigma,f}$ are integers. By the proof of Lemma (4.4), this is equivalent to putting

$$V_{\mathbb{Z}}^\perp = \{f \in U|\ f|_{\partial\Sigma} = 0,\ \deg(f|_{d_{2i}}) \in \mathbb{Z},\ f|_{d_{2i-1}} - f|_{d_{2i}} \in \mathbb{Z}\}.$$

To show correctness of our definition, it remains to show that $T(\Sigma, \Gamma)$ does not depend on the choice of standard cut structure Γ. In other words, we need to show that the open abelian varieties constructed by two different choices Γ_1, Γ_2 of Γ are related by conditions (13) and (14). Let us use the same notation as in (13) and (14), with ι_i, $V_{\mathbb{Z},i}^{\perp}$ constructed from Γ_i. Assume again, without loss of generality, that Σ is connected. Looking first at (13), we see from the above comments that for $f \in V_{\mathbb{Z},1}^{\perp}$, df has integral periods with respect to $H_1(\Sigma, \mathbb{Z})$ and f has 0 degrees on the boundary components. These conditions do not depend on Γ_i. However, there is an additional condition that the branch of the function f on the disk D obtained by cutting Σ along Γ has 0 restriction to the boundary components of Σ. We see that changing the fundamental domain D results in possibly selecting different branches of the function on the boundary components of Σ, which results in adding an integral linear combination of the periods of df, which are integral constant functions, as claimed.

Regarding (14), we have already shown that the selection of $\tilde{f}$ is canonical in case of f having 0 degrees, so we know (14) in this case. In the general case, again, if $f \in V_{deg,\mathbb{Z}}$, then $d\iota_i f$ have integral periods with respect to generators of $H_1(\Sigma, \mathbb{Z})$. In addition, the restrictions of f_1 and f_2 to the boundary component c_j differ at most by selection of a branch (since we use different fundamental domains for calculating the restriction), i.e. by an integral constant function. This proves (14).

To complete the proof of Theorem 4.1, it remains to show that the map T is compatible with gluing. We follow again the two cases of the definition of gluing in the previous section.

Case 1: $A = B$. Assume, without loss of generality, that Σ is connected, Γ is a standard cut structure on Σ, and the boundary components are $c_1, ..., c_n$, as in (40). Without loss of generality, then, Σ^{∇} is obtained from Σ by gluing c_{n-1} and c_n. Then the projection Γ^{∇} of Γ onto Σ^{∇} defines a standard cut structure on Σ^{∇}, and

$$(45) \qquad\qquad T(\Sigma, \Gamma)^{\nabla} = T(\Sigma^{\nabla}, \Gamma^{\nabla})$$

by definition.

Case 2: $A \neq B$. Without loss of generality, $\Sigma = \Sigma_1 \amalg \Sigma_2$ and we have standard cut structures Γ_i on Σ_i such that

$$\Gamma = \Gamma_1 \amalg \Gamma_2,$$

and the boundary components of Σ_i' are

$$c_{i,1}, ..., c_{i,n_i}, d_{i,1}, ..., d_{i,2g_i}.$$

Without loss of generality, further, we are gluing c_{1,n_1} to $c_{2,1}$. Then we obtain a standard cut structure Γ^{∇} on Σ^{∇} by taking the projection of $\Gamma_1 \cup \Gamma_2$ and omitting the edge corresponding to c_{1,n_1} (or equivalently, $c_{2,1}$). Again, by definition, we then have (45).

The compatibility of T with disjoint union is obvious, as is compatibility with coherence isomorphisms (the point here, again, being that isomorphisms of open abelian varieties are determined by the isomorphisms of the W's, so the more subtle structure does not need to be discussed to prove commutativity of diagrams).

5. The lattice conformal field theory on the SPCMC of open abelian varieties

We begin by the same considerations as in [4], starting on p. 351. Consider an even lattice L and a bilinear form

$$b : L \times L \to \mathbb{Z}/2$$

which satisfies

$$b(x, x) \equiv \frac{1}{2}\langle x, x\rangle \mod 2.$$

Let $T = L_{\mathbb{C}}/L$. We let T_{S^1} denote the space of all real analytic maps $S^1 \to T$. We choose a universal cover T'_{S^1} of T_{S^1}, which can be considered as a space of maps $[0, 1] \to L_{\mathbb{C}}$. On T'_{S^1}, we have a cocycle

$$c(\tilde{f}, \tilde{g}) = \exp \frac{2\pi i}{2} \oint_{S^1} (\tilde{f}d\tilde{g} - \Delta_{\tilde{f}}g(0) + b(\Delta_{\tilde{f}}, \Delta_{\tilde{g}}))$$

but L is canonically a normal subgroup of the resulting $\mathbb{C}^\times$-central extension $\tilde{T}'_{S^1}$, so we obtain a canonical $\mathbb{C}^\times$-central extension $\tilde{T}_{S^1} = \tilde{T}'_{S^1}/L$,

$$1 \to \mathbb{C}^\times \to \tilde{T}_{S^1} \to T_{S^1} \to 1.$$

For $\lambda \in L'/L$ where L' is the dual of L, there is now a level 1 Hilbert representation $\mathcal{H}_\lambda$ of $\tilde{T}_{S^1}$ (the real subgroup acts by unitary bounded operators) distinguished by the fact that the constant subgroup $T \subset \tilde{T}_{S^1}$ acts by $e^{2\pi i\langle ?, \lambda\rangle}$. Our conformal field theory associated with L, b has L'/L as its set of labels and $\mathcal{H}_\lambda$ as its Hilbert spaces.

Now consider an open abelian variety $Y = (C, U, S, W, \iota)$. Assume without loss of generality that there is only one open connected component A. Consider the pullback

(46)
$$\begin{array}{ccc} \tilde{W} & \longrightarrow & W \\ \downarrow & & \downarrow \\ \bigoplus_{j \in A} V_1 & \longrightarrow & V_A \end{array}$$

("putting back the constants"). Assuming there is only one connected component, (46) gives a short exact sequence

(47) $$0 \to \mathbb{C} \to \tilde{W} \to W \to 0.$$

Now let $U_{\mathbb{Z}}^0 \subset U_{\mathbb{Z}}$ be the sum of $V_{\mathbb{Z}}^\perp$ and the lattice spanned by $1_j \in V_0 \cdot j$, $j \in A$. Then

$$W_L = \{w \in \tilde{W} \otimes L \mid S(w, u) \in L \text{ for every } u \in U_{\mathbb{Z}}^0\}/L$$

($L \subset L_{\mathbb{C}} \subset \tilde{W} \otimes L$ is embedded by the first map (47) tensored with L). We note that when $Y = T(\Sigma)$ for a worldsheet Σ, then W_L is canonically identified with the space of holomorphic functions $\Sigma \to T = L_{\mathbb{C}}/L$. Next, we construct a restriction homomorphism

(48) $$r : W_L \to \prod_{j \in A} T_{S^1}.$$

In fact, this map is induced simply by tensoring with L the pullback to $\tilde{W}$ of the projection

(49) $$r' : W \to V_{\mathbb{C}}.$$

In fact, let us note that we can assume without loss of generality that

(50)   (49) is injective.

Otherwise, Y is a direct sum of $Ker(r') \oplus \overline{Ker(r')}$ (a closed abelian variety) and its S-complement.

Next, note that

(51)   The canonical central extension $\widetilde{\prod\limits_{j \in A} T_{S^1}}$ canonically splits

when pulled back to W_L.

But in fact, this is completely analogous to the case of surfaces (since the data used there depend only on the Jacobian), which is treated in [4], formulas (58)-(61). Then in the present case, the conformal field theory data is given by the space of fixed points

(52) $$\left(\bigotimes_{j \in A} \hat{\mathcal{H}}^{(*)}_{\lambda_j} \right)^{W_L}$$

for labels λ_j, $j \in A$ (to simplify notation, the superscript $(*)$ stands for the dual when $j \in A^-$ and is void when $j \in A^+$). Here $\mathcal{H}_\lambda$, $\lambda \in L'/L$ are the level 1 irreducible representations of $\tilde{T}_{S^1}$. In the case of a closed abelian variety Y, the data required are given simply by the space of theta functions on $Y \otimes L$ (see formula (98) of [4]).

The main statement to prove is that the dimension of the space (52) is equal to

(53) $$|L'/L|^g$$

where g is the genus of Y when we have the condition

$$\sum_{j \in A} \epsilon_j \lambda_j = 0 \in L'/L$$

where $\epsilon_j = 1$ resp. -1 when j is outbound resp. inbound, and the dimension of the space (52) is 0 otherwise. To this end, choose a "reference" surface Σ of genus 0 (i.e. a disk in $\mathbb{C}$ with a collection of disjoint open disks inside it removed) which has boundary components which match those of Y, with opposite orientation. Now the beginning point is that

(54) $$\bigoplus \{ \bigotimes_{j \in A} \hat{\mathcal{H}}_{\lambda_j} \mid \sum_j \epsilon_j \lambda_j = 0 \}$$

is contained in the space of sections of the line bundle associated with the principal bundle

(55) $$\widetilde{\prod_{j \in A} T_{S^1}} / Hol(\Sigma, T)$$

over

$$\prod_{j\in A} T_{S^1}/Hol(\Sigma, T)$$

(In fact, the only reason equality does not occur is convergence issues; a proof follows from the theory of loop groups [7], we do not give the details.)

So this shows that the sum of (52) over $\sum_j \epsilon_j \lambda_j = 0$ is contained in (and equal to if we can prove a certain convergence condition) the space of sections of the line bundle associated with the principal bundle

(56) $$W_L \backslash \widetilde{\prod_{j\in A} T_j}/Hol(\Sigma, T)$$

over

(57) $$W_L \backslash \prod_{j\in A} T_j/Hol(\Sigma, T).$$

But (57) is the closed abelian variety A obtained by gluing $T\Sigma$ to Y tensored with L, and (56) is the θ-bundle.

So it remains to show the convergence condition. Again, the method is analogous to [4], Lemma 3. One first uses the boson-fermion correspondence to show the convergence of the "tower modes" of the vacuum operator, i.e. the summand of momentum 0. Lemma 5 then deals with sum over different momenta. The sum over momenta is treated exactly in the same way in the present case. To discuss the tower modes, there is also boson-fermion correspondence in the category of open abelian varieties. It suffices to discuss the genus 0 case, where on the fermionic side, the vacuum is represented simply by the space W (or more precisely its image in the appropriate Grassmanian). But that element is smooth because we are working in the smooth moduli space.

References

[1] T.M. Fiore: Pseudo limits, biadjoints, and pseudo algebras: categorical foundations of conformal field theory, *Mem. Amer. Math. Soc.* 182 (2006).

[2] T.M. Fiore: On the cobordism and commutative monoid with cancellation approaches to conformal field theory, *J. Pure Appl. Algebra* 209 (2007) 583-620.

[3] T.M. Fiore, P. Hu, I. Kriz: Laplaza sets, or how to select coherence diagrams for pseudo algebras, to appear in the Advances of Mathematics.

[4] P. Hu, I. Kriz: Conformal field theory and elliptic cohomology, *Adv. Math.* 189 (2004) 325-412.

[5] P. Hu, I. Kriz: Closed and open conformal field theories and their anomalies, *Comm. Math. Phys.* 254 (2005) 221-253.

[6] D. Mumford: *Abelian varieties*, Tata Inst. of Fund. Res. Studies in Math. 5, London 1970.

[7] A. Pressley, G. Segal: *Loop groups*, Oxford Math. Monographs, Oxford University Press 1986.

[8] B. Riemann: Theorie der Abelschen Functionen, *J. Reine und Angew. Mathematik* 54 (1857) 101-155.

[9] G. Segal: The definition of conformal field theory, *Topology, geometry and quantum field theory*, London Math. Soc. Lecture Note Ser. 308, Cambridge University Press, 2004, 421-577.

[10] G. Segal: Unitary representations of some infinite-dimensional groups, *Comm. Math. Phys.* 80 (1981). 301-342.

THOMAS M. FIORE, DEPARTMENT OF MATHEMATICS, UNIVERSITY OF CHICAGO, 5734 S. UNIVERSITY, CHICAGO, IL 60637, USA, AND, DEPARTAMENT DE MATEMÀTIQUES, UNIVERSITAT AUTÒNOMA DE BARCELONA, 08193 BELLATERRA (BARCELONA), SPAIN
E-mail address: fiore@math.uchicago.edu

IGOR KRIZ, DEPARTMENT OF MATHEMATICS, UNIVERSITY OF MICHIGAN, 2074 EAST HALL, 530 CHURCH STREET, ANN ARBOR, MI 48109-1043, USA
E-mail address: ikriz@umich.edu

Pure weight perfect Modules on divisorial schemes

Toshiro Hiranouchi and Satoshi Mochizuki

ABSTRACT. We introduce the notion of weight for pseudo-coherent Modules on a scheme. For a divisorial scheme X and a regular closed immersion $i : Y \hookrightarrow X$ of codimension r, We show that there is a canonical derived Morita equivalence between the DG-category of perfect complexes on X whose cohomological supports are in Y and the DG-category of bounded complexes of weight r pseudo-coherent $\mathcal{O}_X$-Modules supported on Y. This implies that there is a canonical isomorphism between their K-groups (resp. cyclic homology groups). As an application, we decide a generator of the topological filtration on non-connected K-theory (resp. cyclic homology theory) for affine Cohen-Macaulay schemes.

1. Introduction

The aim of this paper is to introduce the notion of *weight* on a class of pseudo-coherent Modules on a scheme. Let X be a divisorial scheme (in the sense of [14], cf. Def. 2.12) and $i : Y \hookrightarrow X$ a regular closed immersion of codimension r. A pseudo-coherent $\mathcal{O}_X$-Module supported on Y is said to be *of (Thomason-Trobaugh) weight r* if it is of Tor-dimension $\leq r$. Here the word "weight" is coming from the weight of the Adams operations in [15]. We denote by $\mathbf{Wt}^r(X \text{ on } Y)$ the exact category of pseudo-coherent $\mathcal{O}_X$-Modules of weight r supported on the subspace Y and $\mathbf{Perf}(X \text{ on } Y)$ the exact category of perfect complexes on X whose cohomological supports are in Y. We shall prove the following theorem:

THEOREM (Th. 3.3). *There is an equivalence of categories from the derived category of the exact category of bounded complexes in $\mathbf{Wt}^r(X \text{ on } Y)$ to the derived category of $\mathbf{Perf}(X \text{ on } Y)$.*

From this theorem we have an isomorphism between the Bass-Thomason-Trobaugh non-connected K-theory $K^B(X \text{ on } Y)$ [25](resp. the Keller-Weibel cyclic homology $HC(X \text{ on } Y)$ [16], [29]) and the Schlichting non-connected K-theory $K^S(\mathbf{Wt}^r(X \text{ on } Y))$ [20] (resp. $HC(\mathbf{Wt}^r(X \text{ on } Y))$ [17]). That is, we have

$$K_q^B(X \text{ on } Y) \simeq K_q^S(\mathbf{Wt}^r(X \text{ on } Y))$$
$$(\text{resp. } HC_q(X \text{ on } Y) \simeq HC_q(\mathbf{Wt}^r(X \text{ on } Y))),$$

1991 *Mathematics Subject Classification.* Primary 19D10; 19D35 Secondary 19D55.
The first author is partially supported by GCOE, Kyoto University.
This research is supported by JSPS core-to-core program 18005.

for *all* $q \in \mathbb{Z}$. For the connected K-theory, this result is nothing other than Exercise 5.7 in [25]. However, there are only hints in loc. cit. We could not find a reference where the exercise was carried out in detail[1].

As mentioned in Exercise 5.7 of [25], our theorem implies, for divisorial schemes, the conjecture of Gersten [6] on describing the homotopy fiber of $K^B(X) \to K^B(X \smallsetminus Y)$ by the K-theory of a certain exact category (see also [7] and [18]). For a general closed immersion, there is an example due to Deligne ([6]) which suggests difficulty of the conjecture. However, the example indicates that for an appropriate scheme X, we have a good class of pseudo-coherent $\mathcal{O}_X$-Modules, that is, Modules of *pure weight*. This notion is closely related to Weibel's K-dimensional conjecture [28] (see Conj. 5.3), Gersten's conjecture [5] and its consequences. These subjects will be treated in a sequel to this paper. Notice that there are different notions of pure weight by Grayson [8] and Walker [27] and these two notions are compatible in a particular situation [26]. The authors hope to compare the Thomason-Trobaugh weight with Walker's.

In §2, we recall some fundamental facts on Tor-dimension of $\mathcal{O}_X$-Modules, Verdier's coherator theory and so on. In §3, we define the notion of weight and state our main theorem again. In §4 we prove our main theorem. Finally, we give some applications of the theorem in §5.

Convention. A *complex* means a chain complex whose boundary morphism is increase level of term by one. For an additive category $\mathcal{A}$, we denote by $\mathbf{Ch}(\mathcal{A})$ the category of chain complexes in $\mathcal{A}$. The word "$\mathcal{O}_X$-Module" means a sheaf on a scheme X which is a sheaf of modules over the sheaf of rings $\mathcal{O}_X$. We denote by $\mathbf{Mod}(X)$ the abelian category of $\mathcal{O}_X$-Modules and $\mathbf{Qcoh}(X)$ the category of quasi-coherent $\mathcal{O}_X$-Modules. An *algebraic vector bundle* over the scheme X is a locally free $\mathcal{O}_X$-Module of finite rank. In particular a *line bundle* is an algebraic vector bundle of rank one (= an invertible sheaf). For the terminologies of algebraic K-theory, we follow [22]. For example, for a complicial biWaldhausen category $\mathcal{C}$ we denote its derived category by $\mathcal{T}(\mathcal{C})$. In particular, for an exact category $\mathcal{E}$ we define $\mathcal{D}(\mathcal{E}) := \mathcal{T}(\mathbf{Ch}(\mathcal{E}))$.

Acknowledgments. The second author is thankful to Masana Harada, Charles A. Weibel for giving several comments to Exercise 5.7 in [25], Marco Schlichting for teaching about elementary questions of negative K-theory via e-mail, Paul Balmer for bringing him the preprint [1] and Mark E. Walker for sending the thesis [26] to him.

2. Preliminary

2.1. Tor-dimension. We briefly review the definition and fundamental properties of *Tor-dimension* of $\mathcal{O}_X$-Modules on a scheme X.

DEFINITION 2.1. Let $\mathcal{M}$ be an $\mathcal{O}_X$-Module.

(i) $\mathcal{M}$ is *flat* if the functor $? \otimes_{\mathcal{O}_X} \mathcal{M} : \mathbf{Mod}(X) \to \mathbf{Mod}(X)$ defined by $\mathcal{N} \mapsto \mathcal{N} \otimes_{\mathcal{O}_X} \mathcal{M}$ is exact.

---

[1]In Proposition 2 of [19], there is a proof of the exercise for Grothendieck groups ($q = 0$) when X is the spectrum of a Cohen-Macaulay local ring and Y is the closed point of X. Furthermore, in Theorem 3.1 of [30], we have a proof of the theorem in the case of $\mathrm{Codim}_X Y = 1$ and $q \geq 0$.

(ii) A *Tor-dimension* of $\mathcal{M}$ is the minimal integer n such that there is a resolution of $\mathcal{M}$,

$$0 \to \mathcal{F}_n \to \mathcal{F}_{n-1} \to \cdots \to \mathcal{F}_0 \to \mathcal{M} \to 0,$$

where all $\mathcal{F}_i$ are flat. We write as $\mathrm{Td}(\mathcal{M}) = n$.

Now we list some well-known facts on Tor-dimension.

LEMMA 2.2 ([**14**], Exp. I, 5.8.3; [**11**], 6.5.7.1). *Let $\mathcal{M}$ be an $\mathcal{O}_X$-Module.*

(i) *If $\mathcal{M}$ is a flat and finitely presented $\mathcal{O}_X$-Module, then $\mathcal{M}$ is an algebraic vector bundle.*

(ii) *The following conditions are equivalent:*
(a) $\mathrm{Td}(\mathcal{M}) \leqq d$.
(b) *For any $\mathcal{O}_X$-Module $\mathcal{N}$ and any $n > d$, we have $\mathrm{Tor}_n^{\mathcal{O}_X}(\mathcal{M}, \mathcal{N}) = 0$.*
(c) *For any $\mathcal{O}_X$-Module $\mathcal{N}$, we have $\mathrm{Tor}_{d+1}^{\mathcal{O}_X}(\mathcal{M}, \mathcal{N}) = 0$.*
(d) *If there is an exact sequence*

$$0 \to \mathcal{N}_d \to \mathcal{F}_{d-1} \to \mathcal{F}_{d-2} \to \cdots \to \mathcal{F}_0 \to \mathcal{M} \to 0,$$

where all $\mathcal{F}_i$ are flat, then $\mathcal{N}_d$ is also flat.

(iii) *For any short exact sequence of $\mathcal{O}_X$-Modules*

$$0 \to \mathcal{M} \to \mathcal{M}' \to \mathcal{M}'' \to 0,$$

we have $\mathrm{Td}(\mathcal{M}') \leqq \max\{\mathrm{Td}(\mathcal{M}), \mathrm{Td}(\mathcal{M}'')\}$.

(iv) *For any $x \in X$ and quasi-coherent $\mathcal{O}_X$-Modules $\mathcal{M}, \mathcal{N}$, we have an isomorphism*

$$\mathcal{TOR}_n^{\mathcal{O}_X}(\mathcal{M}, \mathcal{N})_x \xrightarrow{\sim} \mathrm{Tor}_n^{\mathcal{O}_{X,x}}(\mathcal{M}_x, \mathcal{N}_x).$$

As its consequence, we have the following formula.

$$\mathrm{Td}(\mathcal{M}) \leqq \sup_{x \in X} \mathrm{Td}_{\mathcal{O}_{X,x}}(\mathcal{M}_x).$$

We define a notion similar to Tor-dimension for unbounded complexes.

DEFINITION 2.3 ([**25**], Def. 2.2.11). Let $E^\bullet$ be a complex of $\mathcal{O}_X$-Modules.

(i) $E^\bullet$ has *(globally) finite Tor-amplitude* if there are integers $a \leqq b$ and for all $\mathcal{O}_X$-Module $\mathcal{F}$, $\mathrm{H}^k(E^\bullet \otimes_{\mathcal{O}_X}^L \mathcal{F}) = 0$ unless $a \leqq k \leqq b$. (In the situation, we say that $E^\bullet$ *has Tor-amplitude contained in $[a, b]$*).

(ii) $E^\bullet$ has *locally finite Tor-amplitude* if X is covered by opens U such that $E^\bullet|_U$ has finite Tor-amplitude.

REMARK 2.4. (i) If the scheme X is quasi-compact, then every locally finite Tor-amplitude complex $E^\bullet$ of $\mathcal{O}_X$-Modules is globally finite Tor-amplitude.

(ii) For three vertexes of a distinguished triangle in the derived category of $\mathbf{Mod}(X)$, if two of these three vertexes are globally finite Tor-amplitude then the third vertex is also.

2.2. The coherator. There are two abelian categories $\mathbf{Qcoh}(X)$ and $\mathbf{Mod}(X)$ and the canonical inclusion functor $\phi_X : \mathbf{Qcoh}(X) \hookrightarrow \mathbf{Mod}(X)$ which is exact, closed under extensions, reflects exactness, preserves and reflects infinite direct sums. In general, the functor ϕ_X does not preserve injective objects in $\mathbf{Qcoh}(X)$. But for coherent schemes, there is the theory of "coherator"(cf. [14], Exp. II; [25], Appendix B).

DEFINITION 2.5 ([12], 1.2.1, 1.2.7; [13], Exp. VI; [25], B.7). A scheme X is said to be *quasi-separated* if the diagonal map $X \to X \times X$ is quasi-compact or equivalently if the intersection of any pair of affine open sets in X is quasi-compact. It is said to be *semi-separated* if there is a basis for the topology of X which is consisting of affine open sets and closed under finite intersection. It is said to be *coherent* (resp. *strictly coherent*)[2]if it is quasi-compact and quasi-separated (resp. quasi-compact and semi-separated).

Note that every separated scheme is semi-separated and every semi-separated scheme is quasi-separated. Therefore every strictly coherent scheme is coherent.

PROPOSITION 2.6 ([14], Exp. II, 3.2; [25], Appendix B). *Let X be a coherent scheme.*

(i) *ϕ_X has a right adjoint functor, the coherator $Q_X : \mathbf{Mod}(X) \to \mathbf{Qcoh}(X)$ and the canonical adjunction map* id $\to Q_X\phi_X$ *is an isomorphism. In particular $\mathbf{Qcoh}(X)$ has enough injective and closed under limits.*

(ii) *The coherator Q_X preserves colimits.*

(iii) *We further assume that X is strictly coherent. For any $E^\bullet \in \mathcal{D}^+(\mathbf{Qcoh}(X))$ and $F^\bullet \in \mathcal{D}^+(\mathbf{Mod}(X))$ with quasi-coherent cohomology, the canonical adjunction maps $E^\bullet \to RQ_X\phi_X E^\bullet$ and $\phi_X RQ_X F^\bullet \to F^\bullet$ are quasi-isomorphisms.*

2.3. Perfect and pseudo-coherent complexes. We review the notion of pseudo-coherent and perfect complexes on a scheme X.

DEFINITION 2.7 ([14], Exp. I; [25], §2.2). Let $E^\bullet$ be a complex of $\mathcal{O}_X$-Modules.

(i) $E^\bullet$ is *strictly perfect* (resp. *strictly pseudo-coherent*) if it is a bounded complex (resp. bounded above complex) of algebraic vector bundles.

(ii) $E^\bullet$ is *perfect* (resp. *n-pseudo-coherent*) if it is locally quasi-isomorphic (resp. n-quasi-isomorphic) to strictly perfect complexes. More precisely, for any point $x \in X$, there is a neighborhood U of x in X, a strictly perfect complex $F^\bullet$, and a quasi-isomorphism (resp. an n-quasi-isomorphism) $F^\bullet \xrightarrow{\sim} E^\bullet|_U$. $E^\bullet$ is said to be *pseudo-coherent* if it is n-pseudo-coherent for all integer n.

LEMMA 2.8 ([25], §2.2). *Let $E^\bullet$ be a complex of $\mathcal{O}_X$-Modules on X.*

(i) *If $E^\bullet$ is strictly pseudo-coherent, then it is pseudo-coherent.*

(ii) *In general, a pseudo-coherent complex may not be locally quasi-isomorphic to a strictly pseudo-coherent complex. But if $E^\bullet$ is pseudo-coherent complex of quasi-coherent $\mathcal{O}_X$-Modules, then $E^\bullet$ is locally quasi-isomorphic to a strictly pseudo-coherent complex.*

---

[2]The notion of coherence is coming from topoi theory but we use this notion for schemes (cf. [13]).

(iii) *If $E^\bullet$ is a pseudo-coherent, then all cohomology sheaf $H^i(E^\bullet)$ is quasi-coherent. In particular, a pseudo-coherent $\mathcal{O}_X$-Module is a quasi-coherent $\mathcal{O}_X$-Module. Moreover if we assume X is quasi-compact and $E^\bullet$ is pseudo-coherent, then $E^\bullet$ is cohomologically bounded above.*

(iv) *Moreover if we assume X is noetherian, we have the following equivalent conditions.*

(a) *$E^\bullet$ is pseudo-coherent.*

(b) *$E^\bullet$ is cohomologically bounded above and all the cohomology sheaf $H^k(E^\bullet)$ are coherent $\mathcal{O}_X$-Modules.*

In particular, a pseudo-coherent $\mathcal{O}_X$-Module is coherent.

(v) *The complex $E^\bullet$ is perfect if and only if $E^\bullet$ is pseudo-coherent and has locally finite Tor-amplitude.*

(vi) *Pseudo-coherence and perfection have 2 out of 3 properties. Namely, let $E^\bullet$, $F^\bullet$ and $G^\bullet$ be the three vertexes of a distinguished triangle in the derived category of $\mathbf{Mod}(X)$ and if two of these three vertexes are pseudo-coherent (resp. perfect) then the third vertex is also.*

(vii) *For any complexes of $\mathcal{O}_X$-Modules $F^\bullet$ and $G^\bullet$, $F^\bullet \oplus G^\bullet$ is pseudo-coherent (resp. perfect) if and only if $F^\bullet$ and $G^\bullet$ are.*

(viii) *A strictly bounded complex of perfect $\mathcal{O}_X$-Modules $E^\bullet$ is perfect.*

DEFINITION 2.9. (1) For any $\mathcal{O}_X$-Module $\mathcal{F}$, we denote its support by

$$\operatorname{Supp}\mathcal{F} := \{x \in X; \mathcal{F}_x \neq 0\}.$$

(ii) ([24], 3.2) For a complex of $\mathcal{O}_X$-Modules $E^\bullet$, the *cohomological support* of $E^\bullet$ is the subspace $\operatorname{Supph}E^\bullet \subset X$ those points $x \in X$ at which the stalk complex of $\mathcal{O}_{X,x}$-module $E_x^\bullet$ is not acyclic.

(iii) For any closed subset Y of X, we denote by $\mathbf{Perf}(X \text{ on } Y)$ (resp. $\mathbf{Perf}_{\mathrm{qc}}(X \text{ on } Y)$, $\mathbf{sPerf}(X \text{ on } Y)$) the complicial biWaldhausen category of globally finite Tor-amplitude perfect complexes (resp. globally finite Tor-amplitude perfect complexes of quasi-coherent $\mathcal{O}_X$-Modules, strictly perfect complexes) whose cohomological support on Y. Here, the cofibrations are the degree-wise split monomorphisms, and the weak equivalences are the quasi-isomorphisms. Put

$$\mathbf{Perf}^r(X) := \bigcup_{\operatorname{Codim}_X Y \geq r} \mathbf{Perf}(X \text{ on } Y).$$

LEMMA 2.10. (i) *For any short exact sequence of $\mathcal{O}_X$-Modules*

$$0 \to \mathcal{F} \to \mathcal{G} \to \mathcal{H} \to 0,$$

we have $\operatorname{Supp}\mathcal{G} = \operatorname{Supp}\mathcal{F} \cup \operatorname{Supp}\mathcal{H}$.

(ii) *For a complex of $\mathcal{O}_X$-Modules $E^\bullet$, we have*

$$\operatorname{Supph}E^\bullet = \bigcup_{n \in \mathbb{Z}} \operatorname{Supp}H^n(E^\bullet).$$

LEMMA 2.11. *For a strictly coherent scheme X and its closed set Y, the canonical inclusion functor $\mathbf{Perf}_{\mathrm{qc}}(X \text{ on } Y) \hookrightarrow \mathbf{Perf}(X \text{ on } Y)$ induces an equivalence of categories between their derived categories.*

PROOF. It is enough to show the inverse functor of $T(\mathbf{Perf}_{qc}(X \text{ on } Y)) \to T(\mathbf{Perf}(X \text{ on } Y))$ is given by the coherator Q_X. For any complex $E^\bullet$ in $T(\mathbf{Perf}(X \text{ on } Y))$, $RQ_X E^\bullet$ is quasi-coherent. Thus it is enough to show that the complex $RQ_X E^\bullet$ is perfect and its cohomological support is in Y. To prove this, we may replace $E^\bullet$ by an object which is quasi-isomorphic to $E^\bullet$ since perfection and cohomological support are invariant under quasi-isomorphism. As X is quasi-compact, $E^\bullet$ is cohomologically bounded. By replacing a truncation of $E^\bullet$ we may assume that $E^\bullet$ is strictly bounded. Then by Proposition 2.6 (iii), $RQ_X E^\bullet$ is quasi-isomorphic to $E^\bullet$. Hence, $RQ_X E^\bullet$ is perfect and its cohomological support is in Y.   □

2.4. Divisorial schemes. On a divisorial scheme (defined below), a pseudo-coherent $\mathcal{O}_X$-Module is just a quasi-coherent $\mathcal{O}_X$-Module which has a resolution by algebraic vector bundles and that such a Module has Tor-dimension $\leq r$ if it has a resolution of length $\leq r$ by algebraic vector bundles (see Cor. 2.16).

DEFINITION 2.12 ([14], Exp. II, 2.2.5; [25], Def. 2.1.1). A quasi-compact scheme X is said to be *divisorial* if it has *an ample family of line bundles*. That is it has a family of line bundles $\{\mathcal{L}_\alpha\}$ which satisfies the following condition (see *op. cit.* for another equivalent conditions): For any $f \in \Gamma(X, \mathcal{L}_\alpha^{\otimes n})$, we put the open set $X_f := \{x \in X \mid f(x) \neq 0\}$. Then $\{X_f\}$ is a basis for the Zariski topology of X where n runs over all positive integer, $\mathcal{L}_\alpha$ runs over the family of line bundles and f runs over all global sections of all of $\mathcal{L}_\alpha^{\otimes n}$.

Immediately, every divisorial scheme is semi-separated and therefore strictly coherent.

EXAMPLE 2.13. (i) A quasi-projective scheme over an affine scheme is divisorial. So classical algebraic varieties are divisorial. Since every scheme is locally affine, every scheme is locally divisorial.

(ii) A separated regular noetherian scheme is divisorial.

(iii) Let k be a field and X an affine space $\mathbb{A}_k^n$ with double origin. Then X is regular noetherian but is not divisorial ([25], Exerc. 8.6).

LEMMA 2.14 ([10], 5.5.8; [14], Exp. II, 2.2.3.1). *For a line bundle $\mathcal{L}$ on a divisorial scheme X and a section $f \in \Gamma(X, \mathcal{L})$, the canonical open immersion $X_f \to X$ is affine.*

THEOREM 2.15 (Global resolution theorem, [14], Exp. II; [25], Prop. 2.3.1). *Let X be a divisorial scheme.*

(i) *Any pseudo-coherent complex of quasi-coherent $\mathcal{O}_X$-Modules is globally quasi-isomorphic to a strictly pseudo-coherent complex.*

(ii) *Any perfect complex is isomorphic to a strictly perfect complex in $\mathcal{D}(\mathbf{Mod}(X))$.*

COROLLARY 2.16. *Let X be a divisorial scheme and $\mathcal{M}$ a pseudo-coherent $\mathcal{O}_X$-Module. Then the Module $\mathcal{M}$ is a quasi-coherent $\mathcal{O}_X$-Module which has a (possibly infinite) resolution*

$$\cdots \to \mathcal{F}_i \to \mathcal{F}_{i-1} \to \cdots \to \mathcal{F}_0 \to \mathcal{M} \to 0$$

by algebraic vector bundles $\mathcal{F}_i$. Furthermore, the Module $\mathcal{M}$ has Tor-dimension $\leq r$ if it has a resolution of length $\leq r$ by algebraic vector bundles.

PROOF. The pseudo-coherent $\mathcal{O}_X$-Module $\mathcal{M}$ is quasi-coherent (Lem. 2.8 (iii)). By Th. 2.15 (i), $\mathcal{M}$ has a resolution $\cdots \to \mathcal{F}_i \to \mathcal{F}_{i-1} \to \cdots \to \mathcal{F}_0 \to \mathcal{M} \to 0$ by algebraic vector bundles $\mathcal{F}_i$. Now we assume that $\mathcal{M}$ has Tor-dimension $\leq r$. By replacing $\mathcal{F}_r$ with the image of $\mathcal{F}_{r+1} \to \mathcal{F}_r$, we have an exact sequence $0 \to \mathcal{N}_r \to \mathcal{F}_{r-1} \to \cdots \to \mathcal{F}_0 \to \mathcal{M} \to 0$ where $\mathcal{N}_r$ is flat (Lem. 2.2 (ii)) and obviously finitely presented and thus $\mathcal{N}_r$ is an algebraic vector bundle (Lem. 2.2 (i)). □

2.5. Regular closed immersion. There are several definitions of regular immersion on a scheme X (see [12] and [14], Exp. VII). Both are equivalent if the scheme X is noetherian. We adopt the definition in [14].

DEFINITION 2.17. Let $u : \mathcal{L} \to \mathcal{O}_X$ be a morphism of $\mathcal{O}_X$-Modules from an algebraic vector bundle $\mathcal{L}$ to $\mathcal{O}_X$. The *Koszul complex* associated with u is the strictly perfect complex $\mathrm{Kos}^\bullet(u)$ defined as follows: For $n > 0$, we put

$$\mathrm{Kos}^{-n}(u)(= \mathrm{Kos}_n(u)) := \bigwedge^n \mathcal{L}, \quad \text{and}$$

$$d_n(x_1 \wedge \cdots \wedge x_n) := \sum_{r=1}^{n} (-1)^{r-1} u(x_r) x_1 \wedge \cdots \wedge \widehat{x}_r \wedge \cdots \wedge x_n.$$

DEFINITION 2.18 ([14], Exp. VII, 1.4). (i) An $\mathcal{O}_X$-Module homomorphism $u : \mathcal{L} \to \mathcal{O}_X$ from an algebraic vector bundle $\mathcal{L}$ to $\mathcal{O}_X$ is said to be *regular* if $\mathrm{Kos}^\bullet(u)$ is a resolution of $\mathcal{O}_X / \mathrm{Im}\, u$.

(ii) An ideal sheaf $\mathcal{I}$ on X is *regular* if locally on X, there is a regular map $u : \mathcal{L} \to \mathcal{O}_X$ such that $\mathrm{Im}\, u = \mathcal{I}$. More precisely, this means that if there is an open covering $\{U_i\}_{i \in I}$ of X and for each $i \in I$, there is a regular map $u_i : \mathcal{L}|_{U_i} \to \mathcal{O}_{U_i}$ such that $\mathrm{Im}\, u_i = \mathcal{I}|_{U_i}$.

(iii) A closed immersion $Y \hookrightarrow X$ is said to be *regular* if the defining ideal of Y is regular. We put $\mathcal{N}_{X/Y} := \mathcal{I}/\mathcal{I}^2$ and call it the *conormal sheaf* of the regular closed immersion.

LEMMA 2.19 ([12]). *Let $Y \hookrightarrow X$ be a regular closed immersion whose defining ideal is $\mathcal{I}$.*

(i) *The ideal sheaf $\mathcal{I}$ satisfies the following conditions:*
   (a) *$\mathcal{I}$ is of finite type.*
(b) *For each n, $\mathcal{I}^n / \mathcal{I}^{n+1}$ is a locally free $\mathcal{O}_X / \mathcal{I}$-Module of finite type.*
(c) *The canonical map*

$$\mathrm{Sym}_{\mathcal{O}_X / \mathcal{I}}(\mathcal{N}_{X/Y}) \to \mathrm{Gr}_{\mathcal{I}}(\mathcal{O}_X)$$

is an isomorphism of $\mathcal{O}_X / \mathcal{I}$-Algebras. Here $\mathrm{Sym}_{\mathcal{O}_X / \mathcal{I}}(\mathcal{N}_{X/Y})$ is the symmetric algebra associated with $\mathcal{N}_{X/Y}$, $\mathrm{Gr}_{\mathcal{I}}(\mathcal{O}_X) = \bigoplus_{n \geq 0} \mathcal{I}^n / \mathcal{I}^{n+1}$ is the graded algebra associated with $\mathcal{I}$-adic filtration in $\mathcal{O}_X$ and the canonical map is defined by the universal property of symmetric algebra.

(ii) *If the scheme X is noetherian, then $\mathcal{I}$ is regular in the sense of op. cit. That is, for any point $x \in X$ there is an open neighborhood U of x, and a regular sequence $f_1, \ldots, f_r \in \Gamma(U, \mathcal{I})$ which generates $\mathcal{I}|_U$.*

3. Weight on pseudo-coherent Modules

DEFINITION 3.1. Let X be a scheme. A pseudo-coherent $\mathcal{O}_X$-Module $\mathcal{F}$ is *of weight* r if it is of Tor-dimension $\leq r$ and there is a regular closed immersion $Y \hookrightarrow X$ of codimension r in X such that $\operatorname{Supp}\mathcal{F} \subset Y$.

We denote by $\mathbf{Wt}^r(X)$ the category of pseudo-coherent $\mathcal{O}_X$-Modules of weight r. For a closed subspace $Y \subset X$ and a non-negative integer r, we denote by $\mathbf{Wt}^r(X \text{ on } Y)$ the category of pseudo-coherent $\mathcal{O}_X$-Modules supported on the subspace Y of weight r. Immediately, a pseudo-coherent $\mathcal{O}_X$-Module of weight 0 is just an algebraic vector bundle.

LEMMA 3.2. *For a regular closed immersion* $Y \hookrightarrow X$ *of codimension* r *in a scheme* X, *the category* $\mathbf{Wt}^r(X \text{ on } Y)$ *is closed under extensions and direct summand in the abelian category* $\mathbf{Mod}(X)$. *In particular,* $\mathbf{Wt}^r(X \text{ on } Y)$ *is an idempotent complete exact category.*

PROOF. The assertion follows from Lemma 2.2 (iii), Lemma 2.10 (i), and Lemma 2.8 (vii). □

A pseudo-coherent $\mathcal{O}_X$-Module $\mathcal{F}$ of weight r has globally finite Tor-amplitude. Thus it is perfect by Lemma 2.8 (v) and we have an inclusion functor $\mathbf{Wt}^r(X \text{ on } Y) \hookrightarrow \mathbf{Perf}(X \text{ on } Y)$. Moreover we have the natural inclusion functor $\mathbf{Ch}^b(\mathbf{Wt}^r(X \text{ on } Y)) \hookrightarrow \mathbf{Perf}(X \text{ on } Y)$ by Lemma 2.8 (viii). Now, we state our main theorem again.

THEOREM 3.3. *Let* X *be a divisorial scheme and* $Y \hookrightarrow X$ *a regular closed immersion of codimension* r. *Then the inclusion* $\mathbf{Ch}^b(\mathbf{Wt}^r(X \text{ on } Y)) \hookrightarrow \mathbf{Perf}(X \text{ on } Y)$ *induces an equivalence of categories between their derived categories.*

Consider the inclusion functor $\mathbf{Wt}^r(X \text{ on } Y) \hookrightarrow \mathbf{Ch}^b(\mathbf{Wt}^r(X \text{ on } Y))$ which sends $\mathcal{F}$ in $\mathbf{Wt}^r(X \text{ on } Y)$ to the complex which is $\mathcal{F}$ in degree 0 and 0 in other degrees. We denote by $K^S(\mathbf{Ch}^b(\mathbf{Wt}^r(X \text{ on } Y)); \mathrm{qis})$ the K-theory spectrum of the Waldhausen category associated with $\mathbf{Ch}^b(\mathbf{Wt}^r(X \text{ on } Y))$ whose weak equivalences are the quasi-isomorphisms (for the definition of the spectrum K^S, see [21]). The inclusion above induces a homotopy equivalence

$$K^S(\mathbf{Wt}^r(X \text{ on } Y)) \xrightarrow{\sim} K^S(\mathbf{Ch}^b(\mathbf{Wt}^r(X \text{ on } Y)); \mathrm{qis})$$

by the Gillet-Waldhausen theorem for non-connected K-theory ([20], Th. 3.4). From the theorem above and Schlichting's approximation theorem ([21], Th. 9), we have

$$K^S(\mathbf{Ch}^b(\mathbf{Wt}^r(X \text{ on } Y)); \mathrm{qis}) \xrightarrow{\sim} K^S(X \text{ on } Y).$$

The comparison theorem (*op. cit.*, Th. 5) says $K^S(X \text{ on } Y) \xrightarrow{\sim} K^B(X \text{ on } Y)$. For the cyclic homology groups, we have similar isomorphisms from our main theorem and the derived invariance by [17]. Hence we get the following corollary.

COROLLARY 3.4. *In the notation above, we have*

$$K^S(\mathbf{Wt}^r(X \text{ on } Y)) \xrightarrow{\sim} K^S(X \text{ on } Y) \xrightarrow{\sim} K^B(X \text{ on } Y),$$

$$HC(\mathbf{Wt}^r(X \text{ on } Y)) \xrightarrow{\sim} HC(X \text{ on } Y).$$

4. Proof of the main theorem

In this section, we prove Theorem 3.3. First we consider the following two categories. Let $\mathcal{B}$ be the category of perfect complexes in $\mathbf{Ch}^-(\mathbf{Wt}^r(X \text{ on } Y))$ and $\mathcal{C}$ the category of perfect complexes of quasi-coherent $\mathcal{O}_X$-Modules supported on Y. By Lemma 2.8, the categories $\mathcal{B}$ and $\mathcal{C}$ are closed under extensions and direct summand in $\mathbf{Ch}(\mathbf{Mod}(X))$. Therefore, they are idempotent complete exact categories. Note that any perfect complex has globally finite Tor-amplitude on X (Rem. 2.4 and Lem. 2.8 (v)). From Lemma 2.8 (iii), we have the following exact inclusion functors:

$$\mathbf{Ch}^b(\mathbf{Wt}^r(X \text{ on } Y)) \xrightarrow{\alpha} \mathcal{B} \xrightarrow{\beta} \mathcal{C} \xrightarrow{\gamma} \mathbf{Perf}(X \text{ on } Y).$$

First we shall prove α induces an equivalence of categories between their derived categories. Recall the following lemmas.

LEMMA 4.1 ([25], 1.9.7 and [23]). *Let* $i : \mathcal{X} \to \mathcal{Y}$ *be a fully faithful complicial exact functor between complicial biWaldhausen categories which closed under the formation of canonical homotopy pullbacks and pushouts and assume their weak equivalence classes are just quasi-isomorphism classes. If* i *satisfies the condition* (**DE**) *or* (**DE**)^{op} *below, then* i *induces category equivalences between their derived categories:*

(**DE**) *For any object* Y *in* $\mathcal{V}$, *there is an object* X *in* $\mathcal{X}$ *and a weak equivalence* $i(X) \to Y$.

(**DE**)^{op} *For any object* Y *in* $\mathcal{Y}$, *there is an object* X *in* $\mathcal{X}$ *and a weak equivalence* $Y \to i(X)$.

LEMMA 4.2 ([2], Lem. 2.6). *Let* $\mathcal{E}$ *be an idempotent complete exact categories and* $f : X^\bullet \to Y^\bullet$ *a quasi-isomorphism between bounded above complexes in* $\mathbf{Ch}(\mathcal{E})$. *Assume* $X^\bullet$ *or* $Y^\bullet$ *is strictly bounded. Say the other one as* $Z^\bullet$. *Then there is a sufficiently small* N *such that* $Z^\bullet \to \tau^{\geq N} Z^\bullet$ *is a quasi-isomorphism and* $\tau^{\geq N} Z^\bullet$ *is in* $\mathbf{Ch}^b(\mathcal{E})$.

LEMMA 4.3. *The inclusion* $\alpha : \mathbf{Ch}^b(\mathbf{Wt}^r(X \text{ on } Y)) \hookrightarrow \mathcal{B}$ *satisfies the condition* (**DE**)^{op} *in Lemma 4.1. In particular, we have an equivalence of categories*

$$T(\mathbf{Ch}^b(\mathbf{Wt}^r(X \text{ on } Y))) \xrightarrow{\sim} T(\mathcal{B}).$$

PROOF. Let $\mathcal{E}$ be the category of pseudo-coherent $\mathcal{O}_X$-Modules of Tor-dimension $\leqq r$. It is closed under extensions (Lem. 2.2 (iii)) and direct summand (Lem. 2.8 (vii)). In particular, it is an idempotent complete exact category. We denote by $\mathcal{D}$ the category of perfect complexes in $\mathbf{Ch}^-(\mathcal{E})$ whose cohomological support is in Y. Fix a complex $P^\bullet$ in $\mathcal{B}$. By the global resolution theorem (Th. 2.15), $P^\bullet$ is quasi-isomorphic to a strict perfect complex. Since we have an inclusion $\mathbf{sPerf}(X \text{ on } Y) \subset \mathcal{D}$, $P^\bullet$ is quasi-isomorphic to a bounded complex in $\mathcal{D}$. Now applying Lemma 4.2 to $\mathcal{E}$, there exists an integer N such that the canonical map $P^\bullet \to \tau^{\geq N} P^\bullet$ is a quasi-isomorphism. Since $\mathrm{Supp}(\mathrm{Im}\, d^{N-1})$ is in Y, $\tau^{\geq N} P^\bullet$ is actually in $\mathbf{Ch}^b(\mathbf{Wt}^r(X \text{ on } Y))$. The assertion follows from it. $\square$

Next, we consider the inclusion $\beta : \mathcal{B} \hookrightarrow \mathcal{C}$.

PROPOSITION 4.4. *The inclusion functor* $\beta : \mathcal{B} \hookrightarrow \mathcal{C}$ *satisfies the condition* (**DE**) *in Lemma 4.1.*

To prove Proposition 4.4, we need the following lemmas.

LEMMA 4.5. (i) *Let $\mathcal{I}$ be the definition ideal of Y. Then $\mathcal{O}_X/\mathcal{I}^p$ is of weight r for any non-negative integer p.*

(ii) *Let $\mathcal{F}$ be a pseudo-coherent $\mathcal{O}_X$-Module of weight r and $\mathcal{L}$ an algebraic vector bundle. Then, $\mathcal{L} \otimes_{\mathcal{O}_X} \mathcal{F}$ is of weight r.*

PROOF. (i) First we notice that $\mathcal{O}_X/\mathcal{I}$ is in $\mathbf{Wt}^r(X \text{ on } Y)$ by Koszul resolution. Next since $\mathcal{I}^n/\mathcal{I}^{n+1}$ is locally isomorphic to direct sum of $\mathcal{O}_X/\mathcal{I}$, we learn that $\mathcal{I}^n/\mathcal{I}^{n+1}$ is also in $\mathbf{Wt}^r(X \text{ on } Y)$ by Lemma 2.2 (iv). Using Lemma 3.2 for

$$0 \to \mathcal{I}^{n+1}/\mathcal{I}^{n+p} \to \mathcal{I}^n/\mathcal{I}^{n+p} \to \mathcal{I}^n/\mathcal{I}^{n+1} \to 0,$$

the dévissage argument shows that $\mathcal{I}^n/\mathcal{I}^{n+p}$ is also in $\mathbf{Wt}^r(X \text{ on } Y)$ for any non-negative integer n and positive integer p.

(ii) Since $\mathcal{L}$ is flat, we have an inequality $\mathrm{Td}(\mathcal{L} \otimes_{\mathcal{O}_X} \mathcal{F}) \leqq r$. We also have a formula

$$\mathrm{Supp}\,\mathcal{L} \otimes_{\mathcal{O}_X} \mathcal{F} = \mathrm{Supp}\,\mathcal{L} \cap \mathrm{Supp}\,\mathcal{F} \subset Y.$$

Therefore $\mathcal{L} \otimes_{\mathcal{O}_X} \mathcal{F}$ is of weight r.   □

LEMMA 4.6 ([25], Lem. 1.9.5). *Let $\mathcal{A}$ be an abelian category and $\mathcal{D}$ a full sub additive category of $\mathcal{A}$. Let $\mathcal{C}$ be a full subcategory of $\mathbf{Ch}(\mathcal{A})$ satisfies the following conditions:*

(a) *$\mathcal{C}$ is closed under quasi-isomorphisms. That is, any complex quasi-isomorphic to an object in $\mathcal{C}$ is also in $\mathcal{C}$.*

(b) *Every complex in $\mathcal{C}$ is cohomologically bounded above.*

(c) *$\mathbf{Ch}^b(\mathcal{D})$ is contained in $\mathcal{C}$.*

(d) *$\mathcal{C}$ contains the mapping cone of any map from an object in $\mathbf{Ch}^b(\mathcal{D})$ to an object in $\mathcal{C}$.*

Finally, Suppose the following condition, so "$\mathcal{D}$ has enough objects to resolve":

(e) *For any integer n, any $C^\bullet$ in $\mathcal{C}$ such that $\mathrm{H}^i(C^\bullet) = 0$ for any $i \geqq n$ and any epimorphism in $\mathcal{A}$, $A \twoheadrightarrow \mathrm{H}^{n-1}(C^\bullet)$, then there exists a D in $\mathcal{D}$ and a morphism $D \to A$ such that the composite $D \twoheadrightarrow \mathrm{H}^{n-1}(C^\bullet)$ is an epimorphism in $\mathcal{A}$.*

Then, for any $D^\bullet$ in $\mathbf{Ch}^-(\mathcal{D}) \cap \mathcal{C}$, any $C^\bullet$ in $\mathcal{C}$, and any morphism $x : D^\bullet \to C^\bullet$, there exist a $E^\bullet$ in $\mathbf{Ch}^-(\mathcal{D}) \cap \mathcal{C}$, a degree-wise split monomorphism $a : D^\bullet \to E^\bullet$ and a quasi-isomorphism $y : E^\bullet \xrightarrow{\sim} C^\bullet$ such that $x = y \circ a$. Moreover if $x : D^\bullet \to C^\bullet$ is an n-quasi-isomorphism for some integer n, then one may choose $E^\bullet$ above so that $a^k : D^k \to E^k$ is an isomorphism for $k \geqq n$.

LEMMA 4.7. *Let X be a divisorial scheme whose ample family of line bundles is $\{\mathcal{L}_\alpha\}$ and $E^\bullet$ a perfect complex on X. Then there are line bundles $\mathcal{L}_{\alpha_k}$ in the ample family, integers m_k and sections $f_k \in \Gamma(X, \mathcal{L}_{\alpha_k}^{\otimes m_k})$ $(1 \leqq k \leqq m)$ such that*

(a) *For each k, X_{f_k} is affine.*

(b) *$\{X_{f_k}\}_{1 \leqq k \leqq m}$ is an open cover of X.*

(c) *For each k, $E^\bullet|_{X_{f_k}}$ is quasi-isomorphic to a strictly perfect complex.*

PROOF. Since $E^\bullet$ is perfect, we can take an affine open covering $\{U_i\}_{i \in I}$ of X such that $E^\bullet|_{U_i}$ is quasi-isomorphic to a strictly perfect complex for each $i \in I$. Since $\{\mathcal{L}_\alpha\}$ is an ample family, for each $x \in X$, there are an $i_x \in I$, a line bundle $\mathcal{L}_{\alpha_x}$ in the ample family, an integer m_x and a section $f_x \in \Gamma(X, \mathcal{L}_{\alpha_x}^{\otimes m_x})$ such that $x \in X_{f_x} \subset U_{i_x}$. Since U_{i_x} is affine, X_{f_x} is affine by Lemma 2.14. Now $\{X_{f_x}\}_{x \in X}$ is an affine open covering of X and has a finite sub covering by quasi-compactness of X. $\qquad\square$

LEMMA 4.8 ([25], Lem. 1.9.4 (b)). *Let $E^\bullet$ be a strictly pseudo-coherent complex on X such that $H^i(E^\bullet) = 0$ for $i \geq m$. Then $\operatorname{Ker} d^{m-1}$ is an algebraic vector bundle. In particular $H^{m-1}(E^\bullet)$ is of finite type.*

PROOF OF PROP. 4.4. Let $\{\mathcal{L}_\alpha\}$ be an ample family of line bundles on X and $\mathcal{I}$ the defining ideal of Y. We denote by $\mathcal{D}$ the additive category generated by all the $\mathcal{L}_\alpha^{\otimes m} \otimes_{\mathcal{O}_X} \mathcal{O}_X / \mathcal{I}^p$ with integer m and positive integer p. By Lemma 4.5, $\mathcal{D} \subset \mathbf{Wt}^r(X \text{ on } Y)$. We intend to apply Lemma 4.6 to $\mathcal{A} = \mathbf{Qcoh}(X \text{ on } Y)$ the category of quasi-coherent $\mathcal{O}_X$-Modules whose support on Y. We have to check the assumptions in Lemma 4.6. Only non-trivial assumption is the condition "having enough objects to resolve". Let $C^\bullet$ be a complex in $\mathcal{C}$ such that $H^i(C^\bullet) = 0$ for $i \geq n$, and $\mathcal{F} \twoheadrightarrow H^{n-1}(C^\bullet)$ an epimorphism in $\mathcal{A}$. By Lemma 4.7, there are line bundles $\mathcal{L}_{\alpha_k}$ integers m_k and their sections $f_k \in \Gamma(X, \mathcal{L}_{\alpha_k}^{\otimes m_k})$ $(1 \leq k \leq m)$ such that they satisfy the following conditions.

(a) For each k, X_{f_k} is affine.

(b) $\{X_{f_k}\}_{1 \leq k \leq m}$ is an open cover of X.

(c) For each k, $C^\bullet|_{X_{f_k}}$ is quasi-isomorphic to a strictly perfect complex.

Fix an integer k. Since $H^{n-1}(C^\bullet)|_{X_{f_k}}$ is of finite type by Lemma 4.8, there is sub $\mathcal{O}_{X_{f_k}}$-Module of finite type $\mathcal{G} \hookrightarrow \mathcal{F}|_{X_{f_k}}$ such that the composition $\mathcal{G} \hookrightarrow \mathcal{F}|_{X_{f_k}} \twoheadrightarrow H^{n-1}(C^\bullet)|_{X_{f_k}}$ is an epimorphism. Now since $\mathcal{G}$ and $\mathcal{I}|_{X_{f_k}}$ are $\mathcal{O}_{X_{f_k}}$-Modules of finite type (Lemma 2.19 (i)), we have $(\mathcal{I}|_{X_{f_k}})^{p_k} \mathcal{G} = 0$ for some p_k. Therefore $\mathcal{G}$ is considered as $\mathcal{O}_X / \mathcal{I}^{p_k}|_{X_{f_k}}$-Module of finite type. Hence we have an epimorphism $(\mathcal{O}_X / \mathcal{I}^{p_k}|_{X_{f_k}})^{\oplus t_k} \twoheadrightarrow \mathcal{G}$. We have an $\mathcal{O}_X$-Modules homomorphism $(\mathcal{O}_X / \mathcal{I}^{p_k})^{\oplus t_k} \rightarrow \mathcal{F} \otimes_{\mathcal{O}_X} \mathcal{L}_{\alpha_k}^{\otimes m_k s_k}$ for some integer s_k ([9], 9.3.1 and [12], 1.7.5). Therefore considering the same argument for every k, we get a morphism

$$\bigoplus_{k=1}^m (\mathcal{O}_X / \mathcal{I}^{p_k} \otimes_{\mathcal{O}_X} \mathcal{L}_{\alpha_k}^{\otimes -m_k s_k})^{\oplus t_k} \rightarrow \mathcal{F}$$

whose composition with $\mathcal{F} \twoheadrightarrow H^{n-1}(C^\bullet)$ is an epimorphism in $\mathbf{Qcoh}(X \text{ on } Y)$. $\qquad\square$

Finally, we shall prove that γ induces an equivalence of categories between their derived categories. Now we consider the following exact inclusion functors:

$$\mathcal{C} \xrightarrow{\gamma_1} \mathbf{Perf}_{qc}(X \text{ on } Y) \xrightarrow{\gamma_2} \mathbf{Perf}(X \text{ on } Y).$$

Lemma 2.11 assert that γ_2 induces a homotopy equivalence on spectra. Thus, it is enough to show that the inclusion functor γ_1 induces an equivalence of categories between their derived categories. More strongly we prove the following proposition:

PROPOSITION 4.9. *The local cohomological functor*

$$R\Gamma_Y = \varinjlim_p \mathcal{EXT}(\mathcal{O}_X / \mathcal{I}^p, ?) : \mathcal{T}(\mathbf{Perf}_{qc}(X \text{ on } Y)) \rightarrow \mathcal{T}(\mathcal{C})$$

gives the inverse functor of the inclusion functor γ_1.

PROOF. Let us consider the functor

$$\Gamma_Y := \varinjlim_p \mathcal{HOM}(\mathcal{O}_X / \mathcal{I}^p, ?) : \mathbf{Qcoh}(X) \rightarrow \mathbf{Qcoh}(X \text{ on } Y).$$

Since $\mathcal{I}$ is of finite type, for any $\mathcal{O}_X$-Module $\mathcal{M}$ in $\mathbf{Qcoh}(X \text{ on } Y)$, we have $\Gamma_Y \mathcal{M} = \mathcal{M}$. This identity and the existence of the canonical natural transformation $\Gamma_Y \to$ id imply that Γ_Y is a right adjoint functor of the inclusion $\mathbf{Qcoh}(X \text{ on } Y) \hookrightarrow \mathbf{Qcoh}(X)$. Therefore $\mathbf{Qcoh}(X \text{ on } Y)$ has enough injective objects and Γ_Y is left exact. Now we will prove that the counit map $R\Gamma_Y E^\bullet \to E^\bullet$ is quasi-isomorphism for any cohomologically bounded complexes of quasi-coherent $\mathcal{O}_X$-Modules $E^\bullet$ with cohomologically supported in Y. For an injective resolution $E^\bullet \hookrightarrow I^\bullet$, we have $R\Gamma_Y E^\bullet = \varinjlim_p \mathcal{HOM}(\mathcal{O}_X/\mathcal{I}^p, I^\bullet)$. To prove the canonical map

$$\varinjlim_p \mathcal{HOM}(\mathcal{O}_X/\mathcal{I}^p, I^\bullet) \to \mathcal{HOM}(\mathcal{O}_X, I^\bullet) = I^\bullet$$

is quasi-isomorphism, for any affine open set $U = \operatorname{Spec} A$ in X, by Lemma 4.10 below, we have an isomorphism

$$R\mathcal{HOM}(\mathcal{O}_X/\mathcal{I}^p, I^\bullet)|_U \xrightarrow{\sim} R\mathcal{HOM}(\mathcal{O}_X/\mathcal{I}^p|_U, I^\bullet|_U).$$

Let $I^\bullet|_U \to \tilde{J}^\bullet$ be an injective resolution in $\mathbf{Qcoh}(U)$, where each J^i is an injective A-module and $\tilde{J}^i$ is the Module associated with J^i. Then we have a short exact sequence of complexes for each non-negative integer p,

$$0 \to \operatorname{Hom}(A/K^p, J^\bullet) \to J^\bullet \to \operatorname{Hom}(K^p, J^\bullet) \to 0,$$

where $\tilde{K} := \mathcal{I}|_U$. Now $\varinjlim_p \mathcal{HOM}(K^p, J^\bullet)$ is acyclic on $Y \cap U$ by Lemma 4.11 below. Next we show that $R\Gamma_Y$ preserves perfection. Let us take a complex $E^\bullet$ in $T(\mathbf{Perf}_{\mathrm{qc}}(X \text{ on } Y))$. Since X is quasi-compact, $E^\bullet$ is cohomologically bounded. From the argument in the previous paragraph, we learn that $R\Gamma_Y E^\bullet$ is quasi-isomorphic to $E^\bullet$. Since perfection is invariant under quasi-isomorphisms, we get the assertion. Combining the obvious fact that γ_1 is fully faithful, we conclude that $R\Gamma_Y$ gives an inverse functor of γ_1. $\square$

LEMMA 4.10 ([**30**], Cor. 2.2). *Let X be a strictly coherent scheme. For any pseudo-coherent $\mathcal{O}_X$-Module $\mathcal{F}$ and any $E^\bullet \in \mathcal{D}^+(\mathbf{Qcoh}(X))$, and any quasi-compact open subscheme $U \subset X$, we have*

$$R\mathcal{HOM}(\mathcal{F}, E^\bullet)|_U \xrightarrow{\sim} R\mathcal{HOM}(\mathcal{F}|_U, E^\bullet|_U).$$

Here, the derived functors are taken in $\mathbf{Qcoh}(X)$.

LEMMA 4.11. *For a commutative ring with unit A, its finitely generated ideal I and a finitely generated A-module M, $\varinjlim_p \operatorname{Hom}_A(I^p, M)$ is supported on $\operatorname{Spec} A \smallsetminus V(I)$.*

PROOF. If I is generated by an element f, we have an isomorphism

$$\varinjlim_p \operatorname{Hom}_A((f^p), M) \xrightarrow{\sim} M_f.$$

defined by $\phi \in \operatorname{Hom}_A((f^p), M)$ maps to $\phi(f^p)/f^p \in M_f$.

If I is generated by elements $f_1, \ldots, f_r$, for any non-negative integer p, we put $I^{(p)} := (f_1^p, \ldots, f_r^p)$. Then we have a surjective map

$$\bigoplus_{k=1}^r A f_k^p \twoheadrightarrow I^{(p)}.$$

Since $\{I^{(p)}\}$ and $\{I^p\}$ are cofinal, therefore we have an injection

$$\varinjlim_p \mathrm{Hom}_A(I^p, M) \hookrightarrow \prod_{k=1}^{r} \mathrm{Hom}_A((f_k^p), M).$$

Hence we get the assertion by the previous paragraph. $\qquad\square$

5. Applications

In this section, we assume that A is a Cohen-Macaulay ring of Krull dimension d and $X = \mathrm{Spec}\,A$. By the very definition, the ring A satisfies the following condition (cf. [3], §2.5, Prop. 7): For any height r ideal J in A, there is an A-regular sequence $x_1, \ldots, x_r$ contained in J. In this case, a coherent A-module of weight d is just a module of finite length and finite projective dimension.

PROPOSITION 5.1. *For any integer $0 \leq r \leq d$ and a closed subset $Y \subset X$, $\mathbf{Wt}^r(X$ on $Y)$ is closed under extensions in $\mathbf{Mod}(X)$. In particular $\mathbf{Wt}^r(X$ on $Y)$ is an idempotent complete exact category.*

PROOF. Let us consider a short exact sequence

$$\mathcal{F} \rightarrowtail \mathcal{G} \twoheadrightarrow \mathcal{H}$$

in $\mathbf{Mod}(X)$ where $\mathcal{F}$ and $\mathcal{H}$ are in $\mathbf{Wt}^r(X$ on $Y)$. Then $\mathcal{G}$ is of Tor-dimension $\leq r$, $\mathrm{Supp}\,\mathcal{G} \subset Y$ and $\mathrm{Codim}_X \mathrm{Supp}\,\mathcal{G} \geq r$. Therefore there is an A-regular sequence $x_1, \ldots, x_r$ with $\mathrm{Supp}\,\mathcal{G} \subset V(x_1, \ldots, x_r)$. Hence we conclude that $\mathcal{G}$ is in $\mathbf{Wt}^r(X$ on $Y)$. $\qquad\square$

THEOREM 5.2. *For any integer $0 \leq r \leq d$, the canonical inclusion functor $\mathbf{Ch}^b(\mathbf{Wt}^r(X)) \hookrightarrow \mathbf{Perf}^r(X)$ induces an equivalence of categories between their derived categories. In particular, we have canonical homotopy equivalences of spectra and mixed complexes*

$$K^S(\mathbf{Wt}^r(X)) \xrightarrow{\sim} K^S(\mathbf{Perf}^r(X)),$$

$$HC(\mathbf{Wt}^r(X)) \xrightarrow{\sim} HC(\mathbf{Perf}^r(X)).$$

PROOF. We can write the categories $\mathbf{Ch}^b(\mathbf{Wt}^r(X))$ and $\mathbf{Perf}^r(X)$ as follows:

$$\mathbf{Ch}^b(\mathbf{Wt}^r(X)) = \varprojlim_{Y \subset X} \mathbf{Ch}^b(\mathbf{Wt}^r(X \text{ on } Y)),$$

$$\mathbf{Perf}^r(X) = \varprojlim_{Y \subset X} \mathbf{Perf}(X \text{ on } Y),$$

where the limits taking over the regular closed immersions of codimension $\geq r$. Hence we get the result by Theorem 3.3 and continuity of the functor $\mathcal{T}$. $\qquad\square$

As another application of our main theorem, we recall Weibel's K-dimensional conjecture.

CONJECTURE 5.3 (K-dimensional conjecture). *For any noetherian scheme Z of finite Krull-dimension n, and integer $q > n$, we have $K_{-q}^B(Z) = 0$.*

This conjecture is recently proved for schemes which are essentially of finite type over a field of characteristic 0 in [4]. According to [1], if for any local ring $\mathcal{O}_{Z,z}$ of Z, we have $K_{-q}^B(\mathrm{Spec}\,\mathcal{O}_{Z,z} \text{ on } \overline{\{z\}}) = 0$ for $q > \dim \mathcal{O}_{Z,z}$, then the above conjecture is true for Z. Now, assume the ring A is local with maximal ideal $\mathfrak{m}$. Since $X = \mathrm{Spec}\,A$ is Cohen-Macaulay, $Y := V(\mathfrak{m}) \hookrightarrow X$ is a regular closed immersion of codimension d. From Corollary 3.4 we have $K^S(\mathbf{Wt}^d(X \text{ on } Y)) \xrightarrow{\sim} K^B(X \text{ on } Y)$, and the conjecture for any Cohen-Macaulay scheme is reduced to vanishing of $K^S_{-q}(\mathbf{Wt}^d(X \text{ on } Y))$ for $q > d$.

References

[1] P. Balmer, *Niveau spectral sequences on singular schemes and failure of generalized Gersten conjecture*, Proc. Amer. Math. Soc. **137**, (2009), 99–106.

[2] P. Balmer and M. Schlichting, *Idempotent completion of triangulated categories*, J. Algebra **236** (2001), 819–834.

[3] N. Bourbaki, *Éléments de mathématique. Algèbre commutative. Chapitre 10*, Springer-Verlag, Berlin, 2007, Reprint of the 1998 original.

[4] G. Cortiñas, C. Haesemeyer, M. Schlichting, and C. Weibel, *Cyclic homology, cdh-cohomology and negative K-theory*, Ann. of Math. (2) **167** (2008), 549–573.

[5] S. M. Gersten, *Some exact sequences in the higher K-theory of rings*, Algebraic K-theory, I: Higher K-theories (Proc. Conf., Battelle Memorial Inst., Seattle, Wash., 1972), Springer, Berlin, 1973, pp. 211–243. Lecture Notes in Math., Vol. 341.

[6] _____, *The localization theorem for projective modules*, Comm. Algebra **2** (1974), 317–350.

[7] D. R. Grayson, *Higher algebraic K-theory. II (after Daniel Quillen)*, Algebraic K-theory (Proc. Conf., Northwestern Univ., Evanston, Ill., 1976), Springer, Berlin, 1976, pp. 217–240. Lecture Notes in Math., Vol. 551.

[8] _____, *Weight filtrations via commuting automorphisms*, K-Theory **9** (1995), 139–172.

[9] A. Grothendieck et al, *Éléments de géométrie algébrique. I. Le langage des schémas*, Inst. Hautes Études Sci. Publ. Math. (1960).

[10] _____, *Éléments de géométrie algébrique. II. Étude globale élémentaire de quelques classes de morphismes*, Inst. Hautes Études Sci. Publ. Math. (1961).

[11] _____, *Éléments de géométrie algébrique. III. Étude cohomologique des faisceaux cohérents*, Inst. Hautes Études Sci. Publ. Math. (1961, 1963).

[12] _____, *Éléments de géométrie algébrique. IV. Étude locale des schémas et des morphismes de schémas*, Inst. Hautes Études Sci. Publ. Math. (1964-67).

[13] _____, *Théorie des topos et cohomologie étale des schémas*, Lecture Notes in Mathematics, Vol. 269, 270, 305, Springer-Verlag, Berlin, 1972, 1972, 1973, Séminaire de Géométrie Algébrique du Bois-Marie (SGA 4) 1963–1964.

[14] _____, *Théorie des intersections et théorème de Riemann-Roch*, Lecture Notes in Mathematics, Vol. 225, Springer-Verlag, Berlin, 1971, Séminaire de Géométrie Algébrique du Bois-Marie (SGA 6) 1966–1967.

[15] H. Gillet and C. Soulé, *Intersection theory using Adams operations*, Invent. Math. **90** (1987), 243–277.

[16] B. Keller, *On the cyclic homology of ringed spaces and schemes*, Doc. Math. **3** (1998), 231–259.

[17] _____, *On the cyclic homology of exact categories*, J. Pure Appl. Algebra **136** (1999), 1–56.

[18] M. Levine, *Localization on singular varieties*, Invent. Math. **91** (1988), 423–464.

[19] P. C. Roberts and V. Srinivas, *Modules of finite length and finite projective dimension*, Invent. Math. **151** (2003), 1–27.

[20] M. Schlichting, *Delooping the K-theory of exact categories*, Topology **43** (2004), 1089–1103.

[21] _____, *Negative K-theory of derived categories*, Math. Z. **253** (2006), 97–134.

[22] _____, *Higher algebraic K-theory (after Quillen, Thomason, and others)*, preprint available at
http://www.math.lsu.edu/~mschlich/research/prelim.html (2007).

[23] R. W. Thomason, *Les K-groupes d'un schéma éclaté et une formule d'intersection excédentaire*, Invent. Math. **112** (1993), 195–215.

[24] _____, *The classification of triangulated subcategories*, Compositio Math. **105** (1997), 1–27.

[25] R. W. Thomason and T. Trobaugh, *Higher algebraic K-theory of schemes and of derived categories*, The Grothendieck Festschrift, Vol. III, Progr. Math., vol. 88, Birkhäuser Boston, Boston, MA, 1990, pp. 247–435.

[26] M. E. Walker, *Motivic complex and the K-theory of automorphisms*, thesis (1996).

[27] _____, *Adams operations for bivariant K-theory and a filtration using projective lines*, K-Theory **21** (2000), 101–140.

[28] C. A. Weibel, *K-theory and analytic isomorphisms*, Invent. Math. **61** (1980), 177–197.

[29] _____, *Cyclic homology for schemes*, Proc. Amer. Math. Soc. **124** (1996), 1655–1662.

[30] D. Yao, *The K-theory of vector bundles with endomorphisms over a scheme*, Jour. of Alg., Vol. **184** (1996), 407–423.

E-mail address: hira@kurims.kyoto-u.ac.jp

E-mail address: mochi81@hotmail.com

Higher localized analytic indices and strict deformation quantization

Paulo Carrillo Rouse

ABSTRACT. This paper is concerned with the localization of higher analytic indices for Lie groupoids. Let $\mathscr{G}$ be a Lie groupoid with Lie algebroid $A\mathscr{G}$. Let τ be a (periodic) cyclic cocycle over the convolution algebra $C_c^\infty(\mathscr{G})$. We say that τ can be localized if there is a morphism

$$K^0(A^*\mathscr{G}) \xrightarrow{Ind_\tau} \mathbb{C}$$

satisfying $Ind_\tau(a) = \langle ind\, D_a, \tau \rangle$ (Connes pairing). In this case, we call Ind_τ the higher localized index associated to τ. In [CR08a] we use the algebra of functions over the tangent groupoid introduced in [CR08b], which is in fact a strict deformation quantization of the Schwartz algebra $\mathscr{S}(A\mathscr{G})$, to prove the following results:
- Every bounded continuous cyclic cocycle can be localized.
- If $\mathscr{G}$ is étale, every cyclic cocycle can be localized.

We will recall this results with the difference that in this paper, a formula for higher localized indices will be given in terms of an asymptotic limit of a pairing at the level of the deformation algebra mentioned above. We will discuss how the higher index formulas of Connes-Moscovici, Gorokhovsky-Lott fit in this unifying setting.

1. Introduction

This paper is concerned with the localization of higher analytic indices for Lie groupoids. In [CM90], Connes and Moscovici defined, for any smooth manifold M and every Alexander-Spanier class $[\bar{\varphi}] \in \bar{H}_c^{ev}(M)$, a localized index morphism

$$(1) \qquad Ind_\varphi : K_c^0(T^*M) \longrightarrow \mathbb{C}.$$

which has as a particular case the analytic index morphism of Atiyah-Singer for $[1] \in \bar{H}^0(M)$.

Indeed, given an Alexander-Spanier cocycle φ on M, Connes-Moscovici construct a cyclic cocycle $\tau(\varphi)$ over the algebra of smoothing operators, $\Psi^{-\infty}(M)$ (lemma 2.1, ref.cit.). Now, if D is an elliptic pseudodifferential operator over M, it defines an index class $ind\, D \in K_0(\Psi^{-\infty}(M)) \approx \mathbb{Z}$. Then they showed (theorem 2.4, ref.cit.) that the pairing

$$(2) \qquad \langle ind\, D, \tau(\varphi) \rangle$$

only depends on the principal symbol class $[\sigma_D] \in K^0(T^*M)$ and on the class of φ, and this defines the localized index morphism (1). Connes-Moscovici go further to prove a localized index formula generalizing the Atiyah-Singer theorem. They used this formula to prove the so called Higher index theorem for coverings which served for proving the Novikov conjecture for Hyperbolic groups.

We discuss now the Lie groupoid case. This concept is central in non commutative geometry. Groupoids generalize the concepts of spaces, groups and equivalence relations. In the late 70's, mainly with the work of Alain Connes, it became clear that groupoids appeared naturally as substitutes of singular spaces [**Con79, Mac87, Ren80, Pat99**]. Many people have contributed to realizing this idea. We can find for instance a groupoid-like treatment in Dixmier's works on transformation groups, [**Dix**], or in Brown-Green-Rieffel's work on orbit classification of relations, [**BGR77**]. In foliation theory, several models for the leaf space of a foliation were realized using groupoids, mainly by people like Haefliger ([**Hae84**]) and Wilkelnkemper ([**Win83**]), to mention some of them. There is also the case of Orbifolds, these can be seen indeed as étale groupoids, (see for example Moerdijk's paper [**Moe02**]). There are also some particular groupoid models for manifolds with corners and conic manifolds worked by people like Monthubert [**Mon03**], Debord-Lescure-Nistor ([**DLN06**]) and Aastrup-Melo-Monthubert-Schrohe ([**AMMS**]) for example. Furthermore, Connes showed that many groupoids and algebras associated to them appeared as 'non commutative analogues' of smooth manifolds to which many tools of geometry such as K-theory and Characteristic classes could be applied [**Con79, Con94**]. Lie groupoids became a very natural place where to perform pseudodifferential calculus and index theory, [**Con79, MP97, NWX99**].

The study of the indices in the groupoid case is, as we will see, more delicate than the classical case. There are new phenomena appearing. If $\mathscr{G}$ is a Lie groupoid, a $\mathscr{G}$-pseudodifferential operator is a differentiable family (see [**MP97, NWX99**]) of operators. Let P be such an operator, the index of P, $ind\, P$, is an element of $K_0(C_c^\infty(\mathscr{G}))$. We have also a Connes-Chern pairing

$$K_0(C_c^\infty(\mathscr{G})) \times HC^{even}(C_c^\infty(\mathscr{G})) \xrightarrow{\langle -,- \rangle} \mathbb{C}.$$

We would like to compute the pairings of the form

(3) $\langle ind\, D, \tau \rangle$

for D a $\mathscr{G}$-pseudodifferential elliptic operator. For instance, the Connes-Mosocovici Higher index theorem gives a formula for the above pairing when the groupoid $\mathscr{G}$ is the groupoid associated to a Γ-covering and for cyclic group cocycles.

Now, the first step in order to give a formula for the pairing (3) above is to localize the pairing, that is, to show that it only depends on the principal symbol class in $K^0(A^*\mathscr{G})$ (this would be the analog of theorem 2.4, [**CM90**]).

Let τ be a (periodic) cyclic cocycle over $C_c^\infty(\mathscr{G})$. We say that τ can be localized if the correspondence

(4) $Ell(\mathscr{G}) \xrightarrow{\ ind\ } K_0(C_c^\infty(\mathscr{G})) \xrightarrow{\langle -, \tau \rangle} \mathbb{C}$

factors through the principal symbol class morphism, where $Ell(\mathscr{G})$ is the set of $\mathscr{G}$-pseudodifferential elliptic operators. In other words, if there is a unique morphism

$K^0(A^*\mathcal{G}) \xrightarrow{Ind_\tau} \mathbb{C}$ which fits in the following commutative diagram

(5)
$$Ell(\mathcal{G}) \xrightarrow{ind} K_0(C_c^\infty(\mathcal{G})) \xrightarrow{\langle -,\tau \rangle} \mathbb{C}$$

$[psymb] \downarrow \qquad \qquad \nearrow Ind_\tau$

$$K^0(A^*\mathcal{G})$$

i.e., satisfying $Ind_\tau(a) = \langle ind\, D_a, \tau \rangle$, and hence completely characterized by this property. In this case, we call Ind_τ the higher localized index associated to τ.

In this paper, we prove a localization result using an appropriate strict deformation quantization algebra. For stating the main theorem we need to introduce some terms.

Let $\mathcal{G}$ be a Lie groupoid. It is known that the topological K-theory group $K^0(A^*\mathcal{G})$, encodes the classes of principal symbols of all $\mathcal{G}$-pseudodifferential elliptic operators, [**AS68**]. On other hand the K-theory of the Schwartz algebra of the Lie algebroid satisfies $K_0(\mathscr{S}(A\mathcal{G})) \approx K^0(A^*\mathcal{G})$ (see for instance [**CR08a**] Proposition 4.5).

In [**CR08b**], we constructed a strict deformation quantization of the algebra $\mathscr{S}(A\mathcal{G})$. This algebra is based on the notion of the tangent groupoid which is a deformation groupoid associated to any Lie groupoid: Indeed, associated to a Lie groupoid $\mathcal{G} \rightrightarrows \mathcal{G}^{(0)}$, there is a Lie groupoid

$$\mathcal{G}^T := A\mathcal{G} \times \{0\} \bigsqcup \mathcal{G} \times (0,1] \rightrightarrows \mathcal{G}^{(0)} \times [0,1],$$

compatible with $A\mathcal{G}$ and $\mathcal{G}$, called the tangent groupoid of $\mathcal{G}$. We can now recall the main theorem in ref.cit.

THEOREM. *There exists an intermediate algebra $\mathscr{S}_c(\mathcal{G}^T)$ consisting of smooth functions over the tangent groupoid*

$$C_c^\infty(\mathcal{G}^T) \subset \mathscr{S}_c(\mathcal{G}^T) \subset C_r^*(\mathcal{G}^T),$$

such that it is a field of algebras over $[0,1]$, whose fibers are

$$\mathscr{S}(A\mathcal{G}) \text{ at } t = 0, \text{ and}$$

$$C_c^\infty(\mathcal{G}) \text{ for } t \neq 0.$$

Let τ be a $(q+1)$-multilinear functional over $C_c^\infty(\mathcal{G})$. For each $t \neq 0$, we let τ_t be the $(q+1)$-multilinear functional over $\mathscr{S}_c(\mathcal{G}^T)$ defined by

(6)
$$\tau_t(f^0, ..., f^q) := \tau(f_t^0, ..., f_t^q).$$

It is immediate that if τ is a (periodic) cyclic cocycle over $C_c^\infty(\mathcal{G})$, then τ_t is a (periodic) cyclic cocycle over $\mathscr{S}_c(\mathcal{G}^T)$ for each $t \neq 0$.

The main result of this paper is the following:

THEOREM 1. *Every bounded cyclic cocycle can be localized. Moreover, in this case, the following formula for the higher localized index holds:*

(7)
$$Ind_\tau(a) = lim_{t \to 0} \langle \tilde{a}, \tau_t \rangle,$$

where $\tilde{a} \in K_0(\mathscr{S}_c(\mathcal{G}^T))$ is such that $e_0(\tilde{a}) = a \in K^0(A^*\mathcal{G})$. In fact the pairing above is constant for $t \neq 0$.

Where, a multilinear map $\tau : \underbrace{C_c^\infty(\mathscr{G}) \times \cdots \times C_c^\infty(\mathscr{G})}_{q+1-times} \to \mathbb{C}$ is bounded if it

extends to a continuous multilinear map $\underbrace{C_c^k(\mathscr{G}) \times \cdots \times C_c^k(\mathscr{G})}_{q+1-times} \xrightarrow{\tau_k} \mathbb{C}$, for some

$k \in \mathbb{N}$. The restriction of taking bounded continuous cyclic cocycles in the last theorem is not at all restrictive. In fact, all the geometrical cocycles are of this kind (Group cocycles, The transverse fundamental class, Godbillon-Vey and all the Gelfand-Fuchs cocycles for instance).

Moreover, for the case of étale groupoids, the explicit calculations of the Periodic cohomologies spaces developed in [**BN94, Cra99**] allow us to conclude that the formula (7) above holds for every cyclic cocycle (Corollary 4.8).

At the end of this work we will discuss how the higher index formulas of Connes-Moscovici, Gorokhovsky-Lott ([**CM90, GL03**]) fit in this unifying setting.

Acknowledgments I would like to thank Georges Skandalis for reading an earlier version of this paper and for the very useful comments and remarks he did to it.

I would also like to thank the referee for his remarks to improve this work.

2. Index theory for Lie groupoids

2.1. Lie groupoids. Let us recall what a groupoid is:

DEFINITION 2.1. A *groupoid* consists of the following data: two sets $\mathscr{G}$ and $\mathscr{G}^{(0)}$, and maps

· $s, r : \mathscr{G} \to \mathscr{G}^{(0)}$ called the source and target map respectively,
· $m : \mathscr{G}^{(2)} \to \mathscr{G}$ called the product map (where $\mathscr{G}^{(2)} = \{(\gamma, \eta) \in \mathscr{G} \times \mathscr{G} : s(\gamma) = r(\eta)\}$),

such that there exist two maps, $u : \mathscr{G}^{(0)} \to \mathscr{G}$ (the unit map) and $i : \mathscr{G} \to \mathscr{G}$ (the inverse map), such that, if we denote $m(\gamma, \eta) = \gamma \cdot \eta$, $u(x) = x$ and $i(\gamma) = \gamma^{-1}$, we have

1. $r(\gamma \cdot \eta) = r(\gamma)$ and $s(\gamma \cdot \eta) = s(\eta)$.
2. $\gamma \cdot (\eta \cdot \delta) = (\gamma \cdot \eta) \cdot \delta$, $\forall \gamma, \eta, \delta \in \mathscr{G}$ when this is possible.
3. $\gamma \cdot x = \gamma$ and $x \cdot \eta = \eta$, $\forall \gamma, \eta \in \mathscr{G}$ with $s(\gamma) = x$ and $r(\eta) = x$.
4. $\gamma \cdot \gamma^{-1} = u(r(\gamma))$ and $\gamma^{-1} \cdot \gamma = u(s(\gamma))$, $\forall \gamma \in \mathscr{G}$.

Generally, we denote a groupoid by $\mathscr{G} \rightrightarrows \mathscr{G}^{(0)}$.

Along this paper we will only deal with Lie groupoids, that is, a groupoid in which $\mathscr{G}$ and $\mathscr{G}^{(0)}$ are smooth manifolds (possibly with boundary), and s, r, m, u are smooth maps (with s and r submersions, see [**Mac87, Pat99**]). For A, B subsets of $\mathscr{G}^{(0)}$ we use the notation $\mathscr{G}_A^B$ for the subset $\{\gamma \in \mathscr{G} : s(\gamma) \in A, r(\gamma) \in B\}$.

Our first example of Lie groupoids will be the Lie groups, we will give other examples below.

EXAMPLE 2.2 (Lie Groups). Let G be a Lie group. Then

$$G \rightrightarrows \{e\}$$

is a Lie groupoid with product given by the group product, the unit is the unit element of the group and the inverse is the group inverse

Lie groupoids generalize Lie groups. Now, for Lie groupoids there is also a notion playing the role of the Lie algebra:

DEFINITION 2.3 (The Lie algebroid of a Lie groupoid). Let $\mathscr{G} \to \mathscr{G}^{(0)}$ be a Lie groupoid. The Lie algebroid of $\mathscr{G}$ is the vector bundle

$$A\mathscr{G} \to \mathscr{G}^{(0)}$$

given by definition as the normal vector bundle associated to the inclusion $\mathscr{G}^{(0)} \subset \mathscr{G}$ (we identify $\mathscr{G}^{(0)}$ with its image by u).

For the case when a Lie groupoid is given by a Lie group as above $G \rightrightarrows \{e\}$, we recover $AG = T_e G$. Now, in the Lie theory is very important that this vector space, $T_e G$, has a Lie algebra structure. In the setting of Lie groupoids the Lie algebroid $A\mathscr{G}$ has a structure of Lie algebroid. We will not need this in this paper.

Let us put some classical examples of Lie groupoids.

EXAMPLE 2.4 (Manifolds). Let M be a C^∞-manifold. We can consider the groupoid

$$M \rightrightarrows M$$

where every morphism is the identity over M.

EXAMPLE 2.5 (Groupoid associated to a manifold). Let M be a C^∞-manifold. We can consider the groupoid

$$M \times M \rightrightarrows M$$

with $s(x, y) = y$, $r(x, y) = x$ and the product given by $(x, y) \circ (y, z) = (x, z)$. We denote this groupoid by $\mathscr{G}_M$.

EXAMPLE 2.6. [Fiber product groupoid associated to a submersion] This is a generalization of the example above. Let $N \xrightarrow{p} M$ be a submersion. We consider the fiber product $N \times_M N := \{(n, n') \in N \times N : p(n) = p(n')\}$, which is a manifold because p is a submersion. We can then take the groupoid

$$N \times_M N \rightrightarrows N$$

which is only a subgroupoid of $N \times N$.

EXAMPLE 2.7 (G-spaces). Let G be a Lie group acting by diffeomorphisms in a manifold M. The transformation groupoid associated to this action is

$$M \rtimes G \rightrightarrows M.$$

As a set $M \rtimes G = M \times G$, and the maps are given by $s(x, g) = x \cdot g$, $r(x, g) = x$, the product given by $(x, g) \circ (x \cdot g, h) = (x, gh)$, the unit is $u(x) = (x, e)$ and with inverse $(x, g)^{-1} = (x \cdot g, g^{-1})$.

EXAMPLE 2.8 (Vector bundles). Let $E \xrightarrow{p} X$ be a smooth vector bundle over a manifold X. We consider the groupoid

$$E \rightrightarrows X$$

with $s(\xi) = p(\xi)$, $r(\xi) = p(\xi)$, the product uses the vector space structure and it is given by $\xi \circ \eta = \xi + \eta$, the unit is zero section and the inverse is the additive inverse at each fiber.

EXAMPLE 2.9 (Haefliger's groupoid). Let q be a positive integer. The Haefliger's groupoid Γ^q has as space of objects $\mathbb{R}^q$. A morphism (or arrow) $x \mapsto y$ in Γ^q is the germ of a (local) diffeomorphism $(\mathbb{R}^q, x) \to (\mathbb{R}^q, y)$. This Lie groupoid and its classifying space play a vey important role in the theory of foliations, [Hae84].

EXAMPLE 2.10 (Orbifolds). An Orbifold is an étale groupoid for which (s, r) :
$\mathscr{G} \to \mathscr{G}^{(0)} \times \mathscr{G}^{(0)}$ is a proper map. See [**Moe02**] for further details.

EXAMPLE 2.11 (Groupoid associated to a covering). Let Γ be a discret group
acting freely and properly in $\widetilde{M}$ with compact quotient $\widetilde{M}/\Gamma := M$. We denote
by $\mathscr{G}$ the quotient $\widetilde{M} \times \widetilde{M}$ by the diagonal action of Γ. We have a Lie groupoid
$\mathscr{G} \rightrightarrows \mathscr{G}^{(0)} = M$ with $s(\widetilde{x}, \widetilde{y}) = y$, $r(\widetilde{x}, \widetilde{y}) = x$ and product $(\widetilde{x}, \widetilde{y}) \circ (\widetilde{y}, \widetilde{z}) = (\widetilde{x}, \widetilde{z})$.

A particular of this situation is when $\Gamma = \pi_1(M)$ and $\widetilde{M}$ is the universal
covering. This groupoid played a main role in the Novikov's conjecture proof for
hyperbolic groups given by Connes and Moscovici, [**CM90**].

EXAMPLE 2.12 (Holonomy groupoid of a Foliation). Let M be a compact manifold of dimension n. Let F be a subvector bundle of the tangent bundle TM. We
say that F is integrable if $C^\infty(F) := \{X \in C^\infty(M, TM) : \forall x \in M, X_x \in F_x\}$ is a
Lie subalgebra of $C^\infty(M, TM)$. This induces a partition in embedded submanifolds
(the leaves of the foliation), given by the solution of integrating F.

The holonomy groupoid of (M, F) is a Lie groupoid

$$\mathscr{G}(M, F) \rightrightarrows M$$

with Lie algebroid $A\mathscr{G} = F$ and minimal in the following sense: any Lie groupoid
integrating the foliation [1] contains an open subgroupoid which maps onto the holonomy groupoid by a smooth morphism of Lie groupoids.

The holonomy groupoid was constructed by Ehresmann [**Ehr65**] and Winkelnkemper [**Win83**] (see also [**CC00**], [**God91**], [**Pat99**]).

2.1.1. *The convolution algebra of a Lie groupoid.* We recall how to define an
algebra structure in $C_c^\infty(\mathscr{G})$ using smooth Haar systems.

DEFINITION 2.13. A *smooth Haar system* over a Lie groupoid is a family of
measures μ_x in $\mathscr{G}_x$ for each $x \in \mathscr{G}^{(0)}$ such that,

- for $\eta \in \mathscr{G}_x^y$ we have the following compatibility condition:

$$\int_{\mathscr{G}_x} f(\gamma) d\mu_x(\gamma) = \int_{\mathscr{G}_y} f(\gamma \circ \eta) d\mu_y(\gamma)$$

- for each $f \in C_c^\infty(\mathscr{G})$ the map

$$x \mapsto \int_{\mathscr{G}_x} f(\gamma) d\mu_x(\gamma)$$

belongs to $C_c^\infty(\mathscr{G}^{(0)})$

A Lie groupoid always posses a smooth Haar system. In fact, if we fix a smooth
(positive) section of the 1-density bundle associated to the Lie algebroid we obtain
a smooth Haar system in a canonical way. We suppose for the rest of the paper a
given smooth Haar system given by 1-densities (for complete details see [**Pat99**]).
We can now define a convolution product on $C_c^\infty(\mathscr{G})$: Let $f, g \in C_c^\infty(\mathscr{G})$, we set

$$(f * g)(\gamma) = \int_{\mathscr{G}_{s(\gamma)}} f(\gamma \cdot \eta^{-1}) g(\eta) d\mu_{s(\gamma)}(\eta)$$

This gives a well defined associative product.

---

[1]having F as Lie algebroid

REMARK 2.14. There is a way to define the convolution algebra using half densities (see Connes book [**Con94**]).

2.2. Analytic indices for Lie groupoids. As we mentioned in the introduction, we are going to consider some elements in the K-theory group $K_0(C_c^\infty(\mathcal{G}))$. We recall how these elements are usually defined (See [**NWX99**] for complete details): First we recall a few facts about $\mathcal{G}$-Pseudodifferential calculus:

A $\mathcal{G}$-*Pseudodifferential operator* is a family of pseudodifferential operators $\{P_x\}_{x \in \mathcal{G}^{(0)}}$ acting in $C_c^\infty(\mathcal{G}_x)$ such that if $\gamma \in \mathcal{G}$ and

$$U_\gamma : C_c^\infty(\mathcal{G}_{s(\gamma)}) \to C_c^\infty(\mathcal{G}_{r(\gamma)})$$

the induced operator, then we have the following compatibility condition

$$P_{r(\gamma)} \circ U_\gamma = U_\gamma \circ P_{s(\gamma)}.$$

We also admit, as usual, operators acting in sections of vector bundles $E \to \mathcal{G}^{(0)}$. There is also a differentiability condition with respect to x that can be found in [**NWX99**].

In this work we are going to work exclusively with uniformly supported operators, let us recall this notion. Let $P = (P_x, x \in \mathcal{G}^{(0)})$ be a $\mathcal{G}$-operator, we denote by k_x the Schwartz kernel pf P_x. Let

$$supp\, P := \overline{\cup_x supp\, k_x}, \text{ and}$$

$$supp_\mu P := \mu_1(supp\, P),$$

where $\mu_1(\rho', \rho) = \rho'\rho^{-1}$. We say that P is uniformly supported if $supp_\mu P$ is compact.

We denote by $\Psi^m(\mathcal{G}, E)$ the space of uniformly supported $\mathcal{G}$-operators, acting on sections of a vector bundle E. We denote also

$$\Psi^\infty(\mathcal{G}, E) = \bigcup_m \Psi^m(\mathcal{G}, E) \text{ et } \Psi^{-\infty}(\mathcal{G}, E) = \bigcap_m \Psi^m(\mathcal{G}, E).$$

The composition of two such operators is again of this kind (lemma 3, [**NWX99**]). In fact, $\Psi^\infty(\mathcal{G}, E)$ is a filtered algebra (theorem 1, rf.cit.), *i.e.*,

$$\Psi^m(\mathcal{G}, E)\Psi^{m'}(\mathcal{G}, E) \subset \Psi^{m+m'}(\mathcal{G}, E).$$

In particular, $\Psi^{-\infty}(\mathcal{G}, E)$ is a bilateral ideal.

REMARK 2.15. The choice on the support justifies on the fact that $\Psi^{-\infty}(\mathcal{G}, E)$ is identified with $C_c^\infty(\mathcal{G}, End(E))$, thanks the Schwartz kernel theorem (theorem 6 [**NWX99**]).

The notion of principal symbol extends also to this setting. Let us denote by $\pi : A^*\mathcal{G} \to \mathcal{G}^{(0)}$ the projection. For $P = (P_x, x \in \mathcal{G}^{(0)}) \in \Psi^m(\mathcal{G}, E, F)$, the principal symbol of P_x, $\sigma_m(P_x)$, is a C^∞ section of the vector bundle $End(\pi_x^* r^* E, \pi_x^* r^* F)$ over $T^*\mathcal{G}_x$ (where $\pi_x : T^*\mathcal{G}_x \to \mathcal{G}_x$), such that at each fiber the morphism is homogeneous of degree m (see [**AS68**] for more details). There is a section $\sigma_m(P)$ of $End(\pi^* E, \pi^* F)$ over $A^*\mathcal{G}$ such that

$$(8) \qquad \sigma_m(P)(\xi) = \sigma_m(P_x)(\xi) \in End(E_x, F_x) \text{ si } \xi \in A_x^*\mathcal{G}$$

Hence (8) above, induces a unique surjective linear map

$$(9) \qquad \sigma_m : \Psi^m(\mathcal{G}, E) \to \mathscr{S}^m(A^*\mathcal{G}, End(E, F)),$$

with kernel $\Psi^{m-1}(\mathcal{G}, E)$ (see for instance proposition 2 [**NWX99**]) and where $\mathscr{S}^m(A^*\mathcal{G}, End(E, F))$ denotes the sections of the fiber $End(\pi^* E, \pi^* F)$ over $A^*\mathcal{G}$ homogeneous of degree m at each fiber.

DEFINITION 2.16 ($\mathscr{G}$-Elliptic operators). Let $P = (P_x, x \in \mathscr{G}^{(0)})$ be a $\mathscr{G}$-pseudodifferential operator. We will say that P is elliptic if each P_x is elliptic.

We denote by $Ell(\mathscr{G})$ the set of $\mathscr{G}$-pseudodifferential elliptic operators.

The linear map (9) defines a principal symbol class $[\sigma(P)] \in K^0(A^*\mathscr{G})$:

$$(10) \qquad\qquad Ell(\mathscr{G}) \xrightarrow{\sigma} K^0(A^*\mathscr{G}).$$

Connes, [**Con79**], proved that if $P = (P_x, x \in \mathscr{G}^{(0)}) \in Ell(\mathscr{G})$, then it exists $Q \in \Psi^{-m}(\mathscr{G}, E)$ such that

$$I_E - PQ \in \Psi^{-\infty}(\mathscr{G}, E) \text{ et } I_E - QP \in \Psi^{-\infty}(\mathscr{G}, E),$$

where I_E denotes the identity operator over E. In other words, P defines a quasi-isomorphism in $(\Psi^{+\infty}, C_c^\infty(\mathscr{G}))$ and thus an element in $K_0(C_c^\infty(\mathscr{G}))$ explicitly (when E is trivial) given by

$$(11) \qquad\qquad \left[T \begin{pmatrix} 1 & 0 \\ 0 & 0 \end{pmatrix} T^{-1} \right] - \left[\begin{pmatrix} 1 & 0 \\ 0 & 0 \end{pmatrix} \right] \in K_0(\widetilde{C_c^\infty(\mathscr{G})}),$$

where 1 is the unit in $\widetilde{C_c^\infty(\mathscr{G})}$ (unitarisation of $C_c^\infty(\mathscr{G})$), and where T is given by

$$T = \begin{pmatrix} (1 - PQ)P + P & PQ - 1 \\ 1 - QP & Q \end{pmatrix}$$

with inverse

$$T^{-1} = \begin{pmatrix} Q & 1 - QP \\ PQ - 1 & (1 - PQ)P + P \end{pmatrix}.$$

If E is not trivial we obtain in the same way an element of $K_0(C_c^\infty(\mathscr{G}, End(E, F)))$ $\approx K_0(C_c^\infty(\mathscr{G}))$ since $C_c^\infty(\mathscr{G}, End(E, F)))$ is Morita equivalent to $C_c^\infty(\mathscr{G})$.

DEFINITION 2.17 ($\mathscr{G}$-Index). Let P be a $\mathscr{G}$-pseudodifferential elliptic operator. We denote by $ind\, P \in K_0(C_c^\infty(\mathscr{G}))$ the element defined by P as above. It is called the index of P. It defines a correspondence

$$(12) \qquad\qquad Ell(\mathscr{G}) \xrightarrow{ind} K_0(C_c^\infty(\mathscr{G})).$$

EXAMPLE 2.18 (The principal symbol class as a Groupoid index). Let $\mathscr{G}$ be a Lie groupoid. We can consider the Lie algebroid as Lie groupoid with its vector bundle structure $A\mathscr{G} \rightrightarrows \mathscr{G}^{(0)}$. Let P be a $\mathscr{G}$-pseudodifferential elliptic operator, then the principal symbol $\sigma(P)$ is a $A\mathscr{G}$-pseudodifferential elliptic operator. Its index, $ind(\sigma(P)) \in K_0(C_c^\infty(A\mathscr{G}))$ can be pushforward to $K_0(C_0(A^*\mathscr{G}))$ using the inclusion of algebras $C_c^\infty(A\mathscr{G}) \xrightarrow{j} C_0(A^*\mathscr{G})$ (modulo Fourier), the resulting image gives precisely the map (10) above, i.e., $j_*(ind\,\sigma(P)) = [\sigma(P)] \in K_0(C_0(A^*\mathscr{G})) \approx K^0(A^*\mathscr{G})$.

We have a diagram

$$\begin{array}{ccc} Ell(\mathscr{G}) & \xrightarrow{ind} & K_0(C_c^\infty(\mathscr{G})) \\ \sigma \downarrow & \nearrow & \\ K^0(A^*\mathscr{G}) & & \end{array},$$

where the pointed arrow does not always exist. It does in the classical cases, but not for general Lie groupoids as shown by the next example ([**Con94**] pp. 142).

EXAMPLE 2.19. Let $\mathbb{R} \to \{0\}$ be the groupoid given by the group structure in $\mathbb{R}$. In [**Con94**] (proposition 12, II.10.γ), Connes shows that the map

$$D \mapsto ind\, D \in K_0(C_c^\infty(\mathbb{R}))$$

defines an injection of the projective space of no zero polynomials $D = P(\frac{\partial}{\partial x})$ into $K_0(C_c^\infty(\mathbb{R}))$.

We could consider the morphism

(13) $$K_0(C_c^\infty(\mathscr{G})) \xrightarrow{j} K_0(C_r^*(\mathscr{G}))$$

induced by the inclusion $C_c^\infty(\mathscr{G}) \subset C_r^*(\mathscr{G})$, then the composition

$$Ell(\mathscr{G}) \xrightarrow{ind} K_0(C_c^\infty(\mathscr{G})) \xrightarrow{j} K_0(C_r^*(\mathscr{G}))$$

does factors through the principal symbol class. In other words, we have the following commutative diagram

$$
\begin{array}{ccc}
Ell(\mathscr{G}) & \xrightarrow{ind} & K_0(C_c^\infty(\mathscr{G})) \\
{\scriptstyle\sigma}\downarrow & & \downarrow{\scriptstyle j} \\
K^0(A^*\mathscr{G}) & \xrightarrow[ind_a]{} & K_0(C_r^*(\mathscr{G})).
\end{array}
$$

Indeed, ind_a is the index morphism associated to the exact sequence of C^*-algebras ([**Con79**], [**CS84**], [**MP97**], [**NWX99**])

(14) $$0 \to C_r^*(\mathscr{G}) \longrightarrow \overline{\Psi^0(\mathscr{G})} \xrightarrow{\sigma} C_0(S^*\mathscr{G}) \to 0$$

where $\overline{\Psi^0(\mathscr{G})}$ is a certain C^*-completion of $\Psi^0(\mathscr{G})$, $S^*\mathscr{G}$ is the sphere vector bundle of $A^*\mathscr{G}$ and σ is the extension of the principal symbol.

DEFINITION 2.20. [$\mathscr{G}$-Analytic index] Let $\mathscr{G} \rightrightarrows \mathscr{G}^{(0)}$ be a Lie groupoid. The morphism

(15) $$K^0(A^*\mathscr{G}) \xrightarrow{ind_a} K_0(C_r^*(\mathscr{G}))$$

is called the analytic index of $\mathscr{G}$.

The K-theory of C^*-algebras has very good cohomological properties, however as we are going to discuss in the next subsection, it is sometimes preferable to work with the indices at the level of C_c^∞-algebras.

2.2.1. *Pairing with Cyclic cohomology: Index formulas.* The interest to keep track on the C_c^∞-indices is because at this level we can make explicit calculations via the Chern-Weil-Connes theory. In fact there is a pairing [**Con85, Con94, Kar87**]

(16) $$\langle \_ , \_ \rangle : K_0(C_c^\infty(\mathscr{G})) \times HP^*(C_c^\infty(\mathscr{G})) \to \mathbb{C}$$

There are several known cocycles over $C_c^\infty(\mathscr{G})$. An important problem in Noncommutative Geometry is to compute the above pairing in order to obtain numerical invariants from the indices in $K_0(C_c^\infty(\mathscr{G}))$, [**Con94, CM90, GL06**]. Let us illustrate this affirmation with the following example.

EXAMPLE 2.21. [**CM90, Con94**] Let Γ be a discrete group acting properly and freely on a smooth manifold $\tilde{M}$ with compact quotient $\tilde{M}/\Gamma := M$. Let $\mathscr{G} \rightrightarrows \mathscr{G}^{(0)} = M$ be the Lie groupoid quotient of $\tilde{M} \times \tilde{M}$ by the diagonal action of Γ.

Let $c \in H^*(\Gamma) := H^*(B\Gamma)$. Connes-Moscovici showed in [**CM90**] that the higher Novikov signature, $Sign_c(M)$, can be obtained with the pairing of the signature operator D_{sign} and a cyclic cocycle τ_c associated to c:

(17) $$\langle \tau_c, ind\, D_{sgn} \rangle = Sign_c(M, \psi).$$

The Novikov conjecture states that these higher signatures are oriented homotopy invariants of M. Hence, if $ind\, D_{sign} \in K_0(C_c^\infty(\mathscr{G}))$ is a homotopy invariant of (M, ψ) then the Novikov conjecture would follow. We only know that $j(ind\, D_{sign}) \in K_0(C_r^*(\mathscr{G}))$ is a homotopy invariant. But then we have to extend the action of τ_c to $K_0(C_r^*(\mathscr{G}))$. Connes-Moscovici show that this action extends for Hyperbolic groups.

The pairing (16) above is not interesting for C^*-algebras. Indeed, the Cyclic cohomology for C^*-algebras is trivial (see [**CST04**] 5.2 for an explanation). In fact, as shown by the example above, a very interesting problem is to compute the pairing at the C_c^∞-level and then extend the action of the cyclic cocycles to the K-theory of the C^*-algebra. This problem is known as the extension problem and it was solved by Connes for some cyclic cocycles associated to foliations, [**Con86**], and by Connes-Moscovici, [**CM90**], for group cocycles when the group is hyperbolic.

The most general formula for the pairing (16), known until these days (as far the author is aware), is the one of Gorokhovsky-Lott for Foliation groupoids ([**GL06**], theorem 5.) which generalized a previous Connes formula for étale groupoids ([**Con94**], theorem 12, III.7.γ, see also [**GL03**] for a superconnection proof). It basically says the following: Let $\mathscr{G} \rightrightarrows M$ be a foliation groupoid (Morita equivalent to an étale groupoid). It carries a foliation $\mathscr{F}$. Let ρ be a closed holonomy-invariant transverse current on M. Suppose $\mathscr{G}$ acts freely, properly and cocompactly on a manifold P. Let D be a $\mathscr{G}$-elliptic differential operator on P. Then the following formula holds:

(18) $$\langle Ind\, D, \rho \rangle = \int_{P/\mathscr{G}} \hat{A}(TF)ch([\sigma_D])\nu^*(\omega_\rho),$$

where $\omega_\rho \in H^*(B\mathscr{G}, o)$ is a universal class associated to ρ and $\nu : P/\mathscr{G} \to B\mathscr{G}$ is a classifying map.

Now, we can expect an easy (topological) calculation only if the map $D \mapsto \langle D, \tau \rangle$ ($\tau \in HP^*(C_c^\infty(\mathscr{G}))$ fix) factors through the symbol class of D, $[\sigma(D)] \in K^0(A^*\mathscr{G})$: we want to have a diagram of the following kind:

$$
\begin{array}{ccc}
Ell(\mathscr{G}) & \xrightarrow{ind} & K_0(C_c^\infty(\mathscr{G})) \xrightarrow{\langle \cdot, \tau \rangle} \mathbb{C} \\
\sigma \downarrow & \nearrow & \\
K^0(A^*\mathscr{G}) & \tau &
\end{array}
$$

This paper is concerned with the solution of the factorization problem just described. Our approach will use a geometrical deformation associated to any Lie groupoid, known as the tangent groupoid. We will discuss this in the next section.

3. Index theory and strict deformation quantization

3.1. Deformation to the normal cone. The tangent groupoid is a particular case of a geometric construction that we describe here.

Let M be a C^∞ manifold and $X \subset M$ be a C^∞ submanifold. We denote by $\mathscr{N}_X^M$ the normal bundle to X in M, i.e., $\mathscr{N}_X^M := T_X M / TX$.

We define the following set

(19) $$\mathscr{D}_X^M := \mathscr{N}_X^M \times 0 \bigsqcup M \times \mathbb{R}^*$$

The purpose of this section is to recall how to define a C^∞-structure in $\mathscr{D}_X^M$. This is more or less classical, for example it was extensively used in [**HS87**].

Let us first consider the case where $M = \mathbb{R}^p \times \mathbb{R}^q$ and $X = \mathbb{R}^p \times \{0\}$ (where we identify canonically $X = \mathbb{R}^p$). We denote by $q = n - p$ and by $\mathscr{D}_p^n$ for $\mathscr{D}_{\mathbb{R}^p}^{\mathbb{R}^n}$ as above. In this case we clearly have that $\mathscr{D}_p^n = \mathbb{R}^p \times \mathbb{R}^q \times \mathbb{R}$ (as a set). Consider the bijection $\psi : \mathbb{R}^p \times \mathbb{R}^q \times \mathbb{R} \to \mathscr{D}_p^n$ given by

(20) $$\psi(x, \xi, t) = \begin{cases} (x, \xi, 0) & \text{if } t = 0 \\ (x, t\xi, t) & \text{if } t \neq 0 \end{cases}$$

which inverse is given explicitly by

$$\psi^{-1}(x, \xi, t) = \begin{cases} (x, \xi, 0) & \text{if } t = 0 \\ (x, \frac{1}{t}\xi, t) & \text{if } t \neq 0 \end{cases}$$

We can consider the C^∞-structure on $\mathscr{D}_p^n$ induced by this bijection.

We pass now to the general case. A local chart $(\mathscr{U}, \phi)$ in M is said to be a X-slice if

1) $\phi : \mathscr{U} \xrightarrow{\cong} U \subset \mathbb{R}^p \times \mathbb{R}^q$
2) If $\mathscr{U} \cap X = \mathscr{V}$, $\mathscr{V} = \phi^{-1}(U \cap \mathbb{R}^p \times \{0\})$ (we denote $V = U \cap \mathbb{R}^p \times \{0\}$)

With this notation, $\mathscr{D}_V^U \subset \mathscr{D}_p^n$ as an open subset. We may define a function

(21) $$\tilde{\phi} : \mathscr{D}_{\mathscr{V}}^{\mathscr{U}} \to \mathscr{D}_V^U$$

in the following way: For $x \in \mathscr{V}$ we have $\phi(x) \in \mathbb{R}^p \times \{0\}$. If we write $\phi(x) = (\phi_1(x), 0)$, then

$$\phi_1 : \mathscr{V} \to V \subset \mathbb{R}^p$$

is a diffeomorphism. We set $\tilde{\phi}(v, \xi, 0) = (\phi_1(v), d_N \phi_v(\xi), 0)$ and $\tilde{\phi}(u, t) = (\phi(u), t)$ for $t \neq 0$. Here $d_N \phi_v : N_v \to \mathbb{R}^q$ is the normal component of the derivative $d\phi_v$ for $v \in \mathscr{V}$. It is clear that $\tilde{\phi}$ is also a bijection (in particular it induces a C^∞ structure on $\mathscr{D}_{\mathscr{V}}^{\mathscr{U}}$). Now, let us consider an atlas $\{(\mathscr{U}_\alpha, \phi_\alpha)\}_{\alpha \in \Delta}$ of M consisting of X-slices. Then the collection $\{(\mathscr{D}_{\mathscr{V}_\alpha}^{\mathscr{U}_\alpha}, \tilde{\phi}_\alpha)\}_{\alpha \in \Delta}$ is a C^∞-atlas of $\mathscr{D}_X^M$ (proposition 3.1 in [**CR08b**]).

DEFINITION 3.1 (Deformation to the normal cone). Let $X \subset M$ be as above. The set $\mathscr{D}_X^M$ equipped with the C^∞ structure induced by the atlas described in the last proposition is called "*The deformation to normal cone associated to* $X \subset M$".

REMARK 3.2. Following the same steps, we can define a deformation to the normal cone associated to an injective immersion $X \hookrightarrow M$.

One important feature about this construction is that it is in some sense functorial. More explicitly, let (M, X) and (M', X') be C^∞-pairs as above and let $F : (M, X) \to (M', X')$ be a pair morphism, i.e., a C^∞ map $F : M \to M'$, with $F(X) \subset X'$. We define $\mathscr{D}(F) : \mathscr{D}_X^M \to \mathscr{D}_{X'}^{M'}$ by the following formulas:

$$\mathscr{D}(F)(x, \xi, 0) = (F(x), d_N F_x(\xi), 0) \text{ and}$$

$\mathscr{D}(F)(m,t) = (F(m),t)$ for $t \neq 0$, where $d_N F_x$ is by definition the map

$$(\mathscr{N}_X^M)_x \xrightarrow{d_N F_x} (\mathscr{N}_{X'}^{M'})_{F(x)}$$

induced by $T_x M \xrightarrow{dF_x} T_{F(x)} M'$.

Then $\mathscr{D}(F) : \mathscr{D}_X^M \rightarrow \mathscr{D}_{X'}^{M'}$ is a C^∞-map (proposition 3.4 in [**CR08b**]).

3.2. The tangent groupoid.

DEFINITION 3.3 (Tangent groupoid). *Let $\mathscr{G} \rightrightarrows \mathscr{G}^{(0)}$ be a Lie groupoid. The tangent groupoid associated to $\mathscr{G}$ is the groupoid that has $\mathscr{D}_{\mathscr{G}^{(0)}}^{\mathscr{G}}$ as the set of arrows and $\mathscr{G}^{(0)} \times \mathbb{R}$ as the units, with:*

- $s^T(x, \eta, 0) = (x, 0)$ *and* $r^T(x, \eta, 0) = (x, 0)$ *at* $t = 0$.
- $s^T(\gamma, t) = (s(\gamma), t)$ *and* $r^T(\gamma, t) = (r(\gamma), t)$ *at* $t \neq 0$.
- *The product is given by* $m^T((x, \eta, 0), (x, \xi, 0)) = (x, \eta + \xi, 0)$ *and* $m^T((\gamma, t), (\beta, t)) = (m(\gamma, \beta), t)$ *if* $t \neq 0$ *and if* $r(\beta) = s(\gamma)$.
- *The unit map* $u^T : \mathscr{G}^{(0)} \rightarrow \mathscr{G}^T$ *is given by* $u^T(x, 0) = (x, 0)$ *and* $u^T(x, t) = (u(x), t)$ *for* $t \neq 0$.

We denote $\mathscr{G}^T := \mathscr{D}_{\mathscr{G}^{(0)}}^{\mathscr{G}}$ *and* $A\mathscr{G} := \mathscr{N}_{\mathscr{G}^{(0)}}^{\mathscr{G}}$.

As we have seen above $\mathscr{G}^T$ can be considered as a C^∞ manifold. As a consequence of the functoriality of the Deformation to the normal cone, one can show that the tangent groupoid is in fact a Lie groupoid. Indeed, it is easy to check that if we identify in a canonical way $\mathscr{D}_{\mathscr{G}^{(0)}}^{\mathscr{G}^{(2)}}$ with $(\mathscr{G}^T)^{(2)}$, then

$$m^T = \mathscr{D}(m), \ s^T = \mathscr{D}(s), \ r^T = \mathscr{D}(r), \ u^T = \mathscr{D}(u)$$

where we are considering the following pair morphisms:

$$m : ((\mathscr{G})^{(2)}, \mathscr{G}^{(0)}) \rightarrow (\mathscr{G}, \mathscr{G}^{(0)}),$$

$$s, r : (\mathscr{G}, \mathscr{G}^{(0)}) \rightarrow (\mathscr{G}^{(0)}, \mathscr{G}^{(0)}),$$

$$u : (\mathscr{G}^{(0)}, \mathscr{G}^{(0)}) \rightarrow (\mathscr{G}, \mathscr{G}^{(0)}).$$

REMARK 3.4. Finally, let $\{\mu_x\}$ be a smooth Haar system on $\mathscr{G}$, i.e., a choice of $\mathscr{G}$-invariant Lebesgue measures. In particular we have an associated smooth Haar system on $A\mathscr{G}$ (groupoid given by the vector bundle structure), which we denote again by $\{\mu_x\}$. Then the following family $\{\mu_{(x,t)}\}$ is a smooth Haar system for the tangent groupoid of $\mathscr{G}$ (details may be found in [**Pat99**]):

- $\mu_{(x,0)} := \mu_x$ at $(\mathscr{G}^T)_{(x,0)} = A_x \mathscr{G}$ and
- $\mu_{(x,t)} := t^{-q} \cdot \mu_x$ at $(\mathscr{G}^T)_{(x,t)} = \mathscr{G}_x$ for $t \neq 0$, where $q = dim \, \mathscr{G}_x$.

In this article, we are only going to consider these Haar systems for the tangent groupoids.

3.2.1. *Analytic indices for Lie groupoids as deformations.* Let $\mathscr{G} \rightrightarrows \mathscr{G}^{(0)}$ be a Lie groupoid and

$$K^0(A^*\mathscr{G}) \xrightarrow{ind_a} K_0(C_r^*(\mathscr{G})),$$

its analytic index. This morphism can also be constructed using the tangent groupoid and its C^*-algebra.

It is easy to check that the evaluation morphisms extend to the C^*-algebras:

$$C_r^*(\mathscr{G}^T) \xrightarrow{ev_0} C_r^*(A\mathscr{G}) \text{ and}$$

$$C_r^*(\mathscr{G}^T) \xrightarrow{ev_t} C_r^*(\mathscr{G}) \text{ for } t \neq 0.$$

Moreover, since $\mathscr{G} \times (0,1]$ is an open saturated subset of $\mathscr{G}^T$ and $A\mathscr{G}$ an open saturated closed subset, we have the following exact sequence ([**HS87**])

$$(22) \qquad 0 \to C_r^*(\mathscr{G} \times (0,1]) \longrightarrow C_r^*(\mathscr{G}^T) \xrightarrow{ev_0} C_r^*(A\mathscr{G}) \to 0.$$

Now, the C^*-algebra $C_r^*(\mathscr{G} \times (0,1]) \cong C_0((0,1], C_r^*(\mathscr{G}))$ is contractible. This implies that the groups $K_i(C_r^*(\mathscr{G} \times (0,1]))$ vanish, for $i = 0, 1$. Then, applying the K-theory functor to the exact sequence above, we obtain that

$$K_i(C_r^*(\mathscr{G}^T)) \xrightarrow{(ev_0)_*} K_i(C_r^*(A\mathscr{G}))$$

is an isomorphism, for $i = 0, 1$. In [**MP97**], Monthubert-Pierrot show that

$$(23) \qquad ind_a = (ev_1)_* \circ (ev_0)_*^{-1},$$

modulo the Fourier isomorphism identifying $C_r^*(A\mathscr{G}) \cong C_0(A^*\mathscr{G})$ (see also [**HS87**] and [**NWX99**]). Putting this in a commutative diagram, we have

$$(24)$$

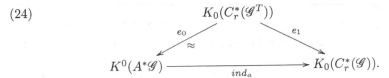

Compare the last diagram with (31) above.

The algebra $C_r^*(\mathscr{G}^T)$ is a strict deformation quantization of $C_0(A^*\mathscr{G})$, and the analytic index morphism of $\mathscr{G}$ can be constructed by means of this deformation. In the next section we are going to discuss the existence of a strict deformation quantization algebra associated the tangent groupoid but in more primitive level, that is, not a C^*-algebra but a Schwartz type algebra. We will use afterwards to define other index morphisms as deformations.

3.3. A Schwartz algebra for the tangent groupoid. In this section we will recall how to construct the deformation algebra mentioned at the introduction. For complete details, we refer the reader to [**CR08b**].

The Schwartz algebra for the Tangent groupoid will be a particular case of a construction associated to any deformation to the normal cone.

DEFINITION 3.5. Let $p, q \in \mathbb{N}$ and $U \subset \mathbb{R}^p \times \mathbb{R}^q$ an open subset, and let $V = U \cap (\mathbb{R}^p \times \{0\})$.

(1) Let $K \subset U \times \mathbb{R}$ be a compact subset. We say that K is a conic compact subset of $U \times \mathbb{R}$ relative to V if

$$K_0 = K \cap (U \times \{0\}) \subset V$$

(2) Let $\Omega_V^U = \{(x, \xi, t) \in \mathbb{R}^p \times \mathbb{R}^q \times \mathbb{R} : (x, t \cdot \xi) \in U\}$, which is an open subset of $\mathbb{R}^p \times \mathbb{R}^q \times \mathbb{R}$ and thus a C^∞ manifold. Let $g \in C^\infty(\Omega_V^U)$. We say that g has compact conic support, if there exists a conic compact K of $U \times \mathbb{R}$ relative to V such that if $(x, t\xi, t) \notin K$ then $g(x, \xi, t) = 0$.

(3) We denote by $\mathscr{S}_c(\Omega_V^U)$ the set of functions $g \in C^\infty(\Omega_V^U)$ that have compact conic support and that satisfy the following condition:

(s_1) $\forall\, k, m \in \mathbb{N}$, $l \in \mathbb{N}^p$ and $\alpha \in \mathbb{N}^q$ there exist $C_{(k,m,l,\alpha)} > 0$ such that

$$(1 + \|\xi\|^2)^k \|\partial_x^l \partial_\xi^\alpha \partial_t^m g(x, \xi, t)\| \leq C_{(k,m,l,\alpha)}$$

Now, the spaces $\mathscr{S}_c(\Omega_V^U)$ are invariant under diffeomorphisms. More precisely: Let $F : U \to U'$ be a C^∞-diffeomorphism such that $F(V) = V'$; let $\tilde{F} : \Omega_V^U \to \Omega_{V'}^{U'}$ be the induced map. Then, for every $g \in \mathscr{S}_c(\Omega_{V'}^{U'})$, we have that $\tilde{g} := g \circ \tilde{F} \in \mathscr{S}_c(\Omega_V^U)$ (proposition 4.2 in [**CR08b**]).

This compatibility result allows to give the following definition.

DEFINITION 3.6. Let $g \in C^\infty(\mathscr{D}_X^M)$.

(a) We say that g has conic compact support K, if there exists a compact subset $K \subset M \times \mathbb{R}$ with $K_0 := K \cap (M \times \{0\}) \subset X$ (conic compact relative to X) such that if $t \neq 0$ and $(m, t) \notin K$ then $g(m, t) = 0$.

(b) We say that g is rapidly decaying at zero if for every $(\mathscr{U}, \phi)$ X-slice chart and for every $\chi \in C_c^\infty(\mathscr{U} \times \mathbb{R})$, the map $g_\chi \in C^\infty(\Omega_V^U)$ (Ω_V^U as in definition 3.5.) given by

$$g_\chi(x, \xi, t) = (g \circ \varphi^{-1})(x, \xi, t) \cdot (\chi \circ p \circ \varphi^{-1})(x, \xi, t)$$

is in $\mathscr{S}_c(\Omega_V^U)$, where

· p is the deformation of the pair map $(M, X) \xrightarrow{Id} (M, M)$, i.e., $p : \mathscr{D}_X^M \to M \times \mathbb{R}$ is given by $(x, \xi, 0) \mapsto (x, 0)$, and $(m, t) \mapsto (m, t)$ for $t \neq 0$, and

· $\varphi := \tilde{\phi}^{-1} \circ \psi : \Omega_V^U \to \mathscr{D}_V^{\mathscr{U}}$, where ψ and $\tilde{\phi}$ are defined at (20) and (21) above.

Finally, we denote by $\mathscr{S}_c(\mathscr{D}_X^M)$ the set of functions $g \in C^\infty(\mathscr{D}_X^M)$ that are rapidly decaying at zero with conic compact support.

REMARK 3.7.

(a) Obviously $C_c^\infty(\mathscr{D}_X^M)$ is a subspace of $\mathscr{S}_c(\mathscr{D}_X^M)$.

(b) Let $\{(\mathscr{U}_\alpha, \phi_\alpha)\}_{\alpha \in \Delta}$ be a family of X-slices covering X. We have a decomposition of $\mathscr{S}_c(\mathscr{D}_X^M)$ as follows (see remark 4.5 in [**CR08b**] and discussion below it):

$$(25) \qquad \mathscr{S}_c(\mathscr{D}_X^M) = \sum_{\alpha \in \Lambda} \mathscr{S}_c(\mathscr{D}_{V_\alpha}^{\mathscr{U}_\alpha}) + C_c^\infty(M \times \mathbb{R}^*).$$

The main theorem in [**CR08b**] (Theorem 4.10) is the following

THEOREM 3.8. *The space $\mathscr{S}_c(\mathscr{G}^T)$ is stable under convolution, and we have the following inclusions of algebras*

$$C_c^\infty(\mathscr{G}^T) \subset \mathscr{S}_c(\mathscr{G}^T) \subset C_r^*(\mathscr{G}^T)$$

Moreover, $\mathscr{S}_c(\mathscr{G}^T)$ is a field of algebras over $\mathbb{R}$, whose fibers are

$$\mathscr{S}(A\mathscr{G}) \text{ at } t = 0, \text{ and}$$

$$C_c^\infty(\mathscr{G}) \text{ for } t \neq 0.$$

In the statement of this theorem, $\mathscr{S}(A\mathscr{G})$ denotes the Schwartz algebra over the Lie algebroid. Let us briefly recall the notion of Schwartz space associated to a vector bundle: For a trivial bundle $X \times \mathbb{R}^q \to X$, $\mathscr{S}(X \times \mathbb{R}^q) := C_c^\infty(X, \mathscr{S}(\mathbb{R}^q))$ (see [**Trè06**]). In general, $\mathscr{S}(E)$ is defined using local charts. More precisely, a partition of the unity argument, allows to see that if we take a covering of X,

$\{(\mathcal{V}_\alpha, \tau_\alpha)\}_{\alpha \in \Delta}$, consisting on trivializing charts, then we have a decomposition of the following kind:

$$(26) \qquad \mathscr{S}(E) = \sum_\alpha \mathscr{S}(\mathcal{V}_\alpha \times \mathbb{R}^q).$$

The "Schwartz algebras" have in general the good K–theory groups. As we said in the introduction, we are interested in the group $K^0(A^*\mathcal{G}) = K_0(C_0(A^*\mathcal{G}))$. It is not enough to take the K–theory of $C_c^\infty(A\mathcal{G})$ (see example 2.19). As we showed in [**CR08a**] (proposition 4.5), $\mathscr{S}(A^*\mathcal{G})$ has the wanted K-theory, i.e., $K^0(A^*\mathcal{G}) \cong K_0(\mathscr{S}(A\mathcal{G}))$. In particular, our deformation algebra restricts at zero to the right algebra.

From now on it will be important to restrict our functions on the tangent groupoid to the closed interval $[0, 1]$. We keep the notation $\mathscr{S}_c(\mathscr{D}_X^M)$ for the restricted space. All the results above remain true. So for instance $\mathscr{S}_c(\mathcal{G}^T)$ is an algebra which is a field of algebras over the closed interval $[0, 1]$ with 0-fiber $\mathscr{S}(A\mathcal{G})$ and $C_c^\infty(\mathcal{G})$ otherwise.

We have the following short exact sequence of algebras ([**CR08a**], proposition 4.6):

$$(27) \qquad 0 \longrightarrow J \longrightarrow \mathscr{S}_c(\mathcal{G}^T) \xrightarrow{e_0} \mathscr{S}(A\mathcal{G}) \longrightarrow 0,$$

where $J = Ker(e_0)$ by definition.

4. Higher localized indices

DEFINITION 4.1. Let τ be a (periodic) cyclic cocycle over $\mathcal{U}_c^\infty(\mathcal{G})$. We say that τ can be localized if the correspondence

$$(28) \qquad Ell(\mathcal{G}) \xrightarrow{ind} K_0(C_c^\infty(\mathcal{G})) \xrightarrow{\langle -, \tau \rangle} \mathbb{C}$$

factors through the principal symbol class morphism. In other words, if there is a unique morphism $K^0(A^*\mathcal{G}) \xrightarrow{Ind_\tau} \mathbb{C}$ which fits in the following commutative diagram

$$(29) \qquad Ell(\mathcal{G}) \xrightarrow{ind} K_0(C_c^\infty(\mathcal{G})) \xrightarrow{\langle -, \tau \rangle} \mathbb{C}$$

$$[psymb] \downarrow \qquad \nearrow Ind_\tau$$

$$K^0(A^*\mathcal{G})$$

i.e., satisfying $Ind_\tau(a) = \langle ind\, D_a, \tau \rangle$, and hence completely characterized by this property. In this case, we call Ind_τ the higher localized index associated to τ.

REMARK 4.2. If a cyclic cocycle can be localized then the higher localized index Ind_τ is completely characterized by the property: $Ind_\tau([\sigma_D]) = \langle ind\, D, \tau \rangle$, $\forall D \in Ell(\mathcal{G})$.

We are going to prove first a localization result for Bounded cyclic cocycles, we recall its definition.

DEFINITION 4.3. A multilinear map $\tau : \underbrace{C_c^\infty(\mathcal{G}) \times \cdots \times C_c^\infty(\mathcal{G})}_{q+1-times} \to \mathbb{C}$ is bounded if it extends to a continuous multilinear map $\underbrace{C_c^k(\mathcal{G}) \times \cdots \times C_c^k(\mathcal{G})}_{q+1-times} \xrightarrow{\tau_k} \mathbb{C}$, for some $k \in \mathbb{N}$.

We can re-state theorem 6.9 in [**CR08a**] in the following way:

THEOREM 4.4. *Let $\mathscr{G} \rightrightarrows \mathscr{G}^{(0)}$ be a Lie groupoid, then*

(i) *Every bounded cyclic cocycle over $C_c^\infty(\mathscr{G})$ can be localized.*

(ii) *Moreover, if the groupoid is étale, then every cyclic cocycle can be localized.*

We will recall the main steps for proving this result. For this purpose we need to define the intermediate group

$$K_0(C_c^\infty(\mathscr{G})) \rightarrow K_0^B(\mathscr{G}) \rightarrow K_0(C_r^*(\mathscr{G})).$$

Let us denote, for each $k \in \mathbb{N}$, $K_0^{h,k}(\mathscr{G})$ the quotient group of $K_0(C_c^k(\mathscr{G}))$ by the equivalence relation induced by $K_0(C_c^k(\mathscr{G} \times [0,1])) \overset{e_0}{\underset{e_1}{\rightrightarrows}} K_0(C_c^k(\mathscr{G}))$. Let $K_0^F(\mathscr{G}) = \varprojlim_k K_0^{h,k}(\mathscr{G})$ be the projective limit relative to the inclusions $C_c^k(\mathscr{G}) \subset C_c^{k-1}(\mathscr{G})$. We can take the inductive limit

$$\varinjlim_m K_0^F(\mathscr{G} \times \mathbb{R}^{2m})$$

induced by $K_0^F(\mathscr{G} \times \mathbb{R}^{2m}) \overset{Bott}{\longrightarrow} K_0^F(\mathscr{G} \times \mathbb{R}^{2(m+1)})$ (the Bott morphism). We denote this group by

$$(30) \qquad\qquad K_0^B(\mathscr{G}) := \varinjlim_m K_0^F(\mathscr{G} \times \mathbb{R}^{2m}),$$

Now, theorem 5.4 in [**CR08a**] establish the following two assertions:

(1) There is a unique group morphism

$$ind_a^B : K^0(A^*\mathscr{G}) \rightarrow K_0^B(\mathscr{G})$$

that fits in the following commutative diagram

(31)
$$
\begin{array}{ccc}
 & K_0(\mathscr{S}_c(\mathscr{G}^T)) & \\
 \overset{e_0}{\swarrow} & & \overset{e_1^B}{\searrow} \\
K^0(A^*\mathscr{G}) \xrightarrow{\quad ind_a^B \quad} & & K_0^B(\mathscr{G}),
\end{array}
$$

where e_1^B is the evaluation at one $K_0(\mathscr{S}_c(\mathscr{G}^T)) \overset{e_1}{\longrightarrow} K_0(C_c^\infty(\mathscr{G}))$ followed by the canonical map $K_0(C_c^\infty(\mathscr{G})) \rightarrow K_0^B(\mathscr{G})$.

(2) This morphism also fits in the following commutative diagram

(32)
$$
\begin{array}{ccc}
Ell(\mathscr{G}) & \xrightarrow{\quad ind \quad} & K_0(C_c^\infty(\mathscr{G})) \\
\downarrow{\scriptstyle \sigma} & & \downarrow \\
K^0(A^*\mathscr{G}) & \xrightarrow{\quad ind_a^B \quad} & K_0^B(\mathscr{G}) \\
\downarrow{\scriptstyle id} & & \downarrow \\
K^0(A^*\mathscr{G}) & \xrightarrow{\quad ind_a \quad} & K_0(C_r^*(\mathscr{G}))
\end{array}
$$

Next, it is very easy to check (see [CR08a] Proposition 6.7) that if τ is a bounded cyclic cocycle, then the pairing morphism $K_0(C_c^\infty(\mathscr{G})) \xrightarrow{\langle,\tau\rangle}$ extends to $K_0^B(\mathscr{G})$, i.e., we have a commutative diagram of the following type:

(33)
$$
\begin{array}{ccc}
K_0(C_c^\infty(\mathscr{G})) & \xrightarrow{\ <,\tau>\ } & \mathbf{C} \\
{\scriptstyle\iota}\big\downarrow & \nearrow{\scriptstyle\tau_B} & \\
K_0^B(\mathscr{G}) & &
\end{array}
$$

Now, theorem 4.4 follows immediately because we can put together diagrams (32) and (33) to get the following commutative diagram

(34)
$$
\begin{array}{ccccc}
Ell(\mathscr{G}) & \xrightarrow{\ ind\ } & K_0(C_c^\infty(\mathscr{G})) & \xrightarrow{\ \langle\_,\tau\rangle\ } & \mathbb{C} \\
{\scriptstyle\sigma}\big\downarrow & & \big\downarrow & \nearrow{\scriptstyle\tau_B} & \\
K^0(A^*\mathscr{G}) & \xrightarrow[\ ind_a^B\ ]{} & K_0^B(\mathscr{G}) & &
\end{array}\ .
$$

4.1. Higher localized index formula. In this section we will give a formula for the Higher localized indices in terms of a pairing in the strict deformation quantization algebra $\mathscr{S}_c(\mathscr{G}^T)$. We have first to introduce some notation :

Let τ be a $(q+1)$-multilinear functional over $C_c^\infty(\mathscr{G})$. For each $t \neq 0$, we let τ_t be the $(q+1)$ multilinear functional over $\mathscr{S}_c(\mathscr{G}^T)$ defined by

(35)
$$\tau_t(f^0, ..., f^q) := \tau(f_t^0, ..., f_t^q).$$

In fact, if we consider the evaluation morphisms

$$e_t : \mathscr{S}_c(\mathscr{G}^T) \to C_c^\infty(\mathscr{G}),$$

for $t \neq 0$, then it is obvious that τ_t is a (b, B)-cocycle (periodic cyclic cocycle) over $\mathscr{S}_c(\mathscr{G}^T)$ if τ is a (b, B)-cocycle over $C_c^\infty(\mathscr{G})$. Indeed, $\tau_t = e_t^*(\tau)$ by definition.

We can now state the main theorem of this article.

THEOREM 4.5. *Let τ be a bounded cyclic cocycle then the higher localized index of τ, $K^0(A^*\mathscr{G}) \xrightarrow{Ind_\tau} \mathbb{C}$, is given by*

(36)
$$Ind_\tau(a) = lim_{t \to 0}\langle \tilde{a}, \tau_t \rangle,$$

where $\tilde{a} \in K_0(\mathscr{S}_c(\mathscr{G}^T))$ is such that $e_0(\tilde{a}) = a \in K^0(A^*\mathscr{G})$. In fact the pairing above is constant for $t \neq 0$.

REMARK 4.6. Hence, if τ is a bounded cyclic cocycle and D is a $\mathscr{G}$-pseudodifferential elliptic operator, then we have the following formula for the pairing:

(37)
$$\langle ind\, D, \tau \rangle = \langle \widetilde{\sigma_D}, \tau_t \rangle,$$

for each $t \neq 0$, and where $\widetilde{\sigma_D} \in K_0(\mathscr{S}_c(\mathscr{G}^T))$ is such that $e_0(\tilde{\sigma}) = \sigma_D$. In particular,

(38)
$$\langle ind\, D, \tau \rangle = lim_{t \to 0}\langle \widetilde{\sigma_D}, \tau_t \rangle.$$

For the proof of the theorem above we will need the following lemma.

LEMMA 4.7. *For $s, t \in (0, 1]$, τ_s and τ_t define the same pairing map*

$$K_0(\mathscr{S}_c(\mathscr{G}^T)) \longrightarrow \mathbb{C}.$$

PROOF. Let p be an idempotent in $\widetilde{\mathscr{S}_c(\mathscr{G}^T)} = \mathscr{S}_c(\mathscr{G}^T) \oplus \mathbb{C}$. It defines a smooth family of idempotents p_t in $\mathscr{S}_c(\mathscr{G}^T)$. We set $a_t := \frac{dp_t}{dt}(2p_t - 1)$. Hence, a simple calculation shows

$$\frac{d}{dt}\langle \tau, p_t \rangle = \sum_{i=0}^{2n} \tau(p_t, ..., [a_t, p_t], ..., p_t) =: L_{a_t}\tau(p_t, ..., p_t).$$

Now, *the Lie derivatives* L_{x_t} act trivially on $HP^0(\mathscr{S}_c(\mathscr{G}^T))$ (see [**Con85, Goo85**]), then $\langle \tau, p_t \rangle$ is constant in t. Finally, by definition, $\langle \tau_t, p \rangle = \langle \tau, p_t \rangle$. Hence $t \mapsto \langle \tau_t, p \rangle$ is a constant function for $t \in (0, 1]$.   □

PROOF OF THEOREM 4.5. Putting together diagrams (31) and (34), we get the following commutative diagram

(39)

In other words, for $a \in K^0(A^*\mathscr{G})$, $Ind_\tau(a) = \langle \tilde{a}, \tau_1 \rangle$. Now, by lemma 4.7 we can conclude that

$$Ind_\tau(a) = \langle \tilde{a}, \tau_t \rangle,$$

for each $t \neq 0$. In particular the limit when t tends to zero is given by

$$Ind_\tau(a) = lim_{t \to 0}\langle \tilde{a}, \tau_t \rangle.$$

                                                                 □

For étale groupoids, we can state the following corollary.

COROLLARY 4.8. *If* $\mathscr{G} \rightrightarrows \mathscr{G}^{(0)}$ *is an étale groupoid, then formula (36) holds for every cyclic cocycle.*

PROOF. Thanks to the works of Burghelea, Brylinski-Nistor and Crainic ([**Bur85, BN94, Cra99**]), we know a very explicit description of the Periodic cyclic cohomology for étale groupoids. For instance, we have a decomposition of the following kind (see for example [**Cra99**] theorems 4.1.2. and 4.2.5)

(40) $HP^*(C_c^\infty(\mathscr{G})) = \Pi_\mathscr{O} H_\tau^{*+r}(B\mathscr{N}_\mathscr{O}),$

where $\mathscr{N}_\mathscr{O}$ is an étale groupoid associated to $\mathscr{O}$ (the normalizer of $\mathscr{O}$, see 3.4.8 in ref.cit.). For instance, when $\mathscr{O} = \mathscr{G}^{(0)}$, $\mathscr{N}_\mathscr{O} = \mathscr{G}$.

Now, all the cyclic cocycles coming from the cohomology of the classifying space are bounded. Indeed, we know that each factor of $HP^*(C_c^\infty(\mathscr{G}))$ in the decomposition (40) consists of bounded cyclic cocycles (see last section of [**CR08a**]). Now, the pairing

$$HP^*(C_c^\infty(\mathscr{G})) \times K_0(C_c^\infty(\mathscr{G})) \longrightarrow \mathbb{C}$$

is well defined. In particular, the restriction of the pairing to $HP^*(C_c^\infty(\mathscr{G}))|_\mathscr{O}$ vanishes for almost every $\mathscr{O}$. The conclusion is now immediate from the theorem above.   □

Once we have the formula (36) above, it is well worth it to recall why the evaluation morphism

$$(41) \qquad K_0(\mathscr{S}_c(\mathscr{G}^T)) \xrightarrow{e_0} K^0(A^*\mathscr{G})$$

is surjective. Let $[\sigma] \in K_0(\mathscr{S}(A\mathscr{G})) = K^0(A^*\mathscr{G})$. We know from the $\mathscr{G}$-pseudo-differential calculus that $[\sigma]$ can be represented by a smooth homogeneous elliptic symbol (see [**AS68, CH90, MP97, NWX99**]). We can consider the symbol over $A^*\mathscr{G} \times [0,1]$ that coincides with σ for all t, we denote it by $\tilde{\sigma}$. Now, since $A\mathscr{G}^T = A\mathscr{G} \times [0,1]$, we can take $\tilde{P} = (P_t)_{t \in [0,1]}$ a $\mathscr{G}^T$-elliptic pseudodifferential operator associated to σ, that is, $\sigma_{\tilde{P}} = \tilde{\sigma}$. Let $i : C_c^\infty(\mathscr{G}^T) \to \mathscr{S}_c(\mathscr{G}^T)$ be the inclusion (which is an algebra morphism), then $i_*(ind\,\tilde{P}) \in K_0(\mathscr{S}_c(\mathscr{G}^T))$ is such that $e_{0,*}(i_*(ind\,\tilde{P})) - [\sigma]$. Hence, the lifting of a principal symbol class is given by the index of $\tilde{P} = (P_t)_{t \in [0,1]}$ and theorem 4.5 says that the pairing with a bounded cyclic cocycle does not depend on the choice of the operator P. Now, for compute this index, as in formula (11), one should find first a parametrix for the family $\tilde{P} = (P_t)_{t \in [0,1]}$.

For instance, in [**CM90**] (section 2), Connes-Moscovici consider elliptic differential operators over compact manifolds, let us say an operator $D \in DO^r(M; E, F)^{-1}$. Then they consider the family of operators tD (multiplication by t in the normal direction) for $t > 0$ and they construct a family of parametrix $\tilde{Q}(t)$. The corresponding idempotent is then homotopic to $W(tD)$, where

$$(42) \qquad W(D) = \begin{pmatrix} e^{-D^*D} & e^{-\frac{1}{2}D^*D}(\frac{I-e^{-D^*D}}{D^*D})^{\frac{1}{2}}()^* \\ e^{-\frac{1}{2}DD^*}(\frac{I-e^{-DD^*}}{DD^*})^{\frac{1}{2}}D & I - e^{-D^*D} \end{pmatrix}$$

is the Wasserman idempotent. In the language of the tangent groupoid, the family $\tilde{D} = \{D_t\}_{t \in [0,1]}$ where $D_0 = \sigma_D$ and $D_t = tD$ for $t > 0$, defines a $\mathscr{G}^T$-differential elliptic operator. What Connes and Moscovici compute is precisely the limit on right hand side of formula (36).

Also, in [**MW94**] (section 2), Moscovici-Wu proceed in a similar way by using the finite propagation speed property to construct a parametrix for operators $\tilde{D} = \{D_t\}_{t \in [0,1]}$ over the tangent groupoid. Then they obtain as associated idempotent the so called graph projector. What they compute after is again a particular case of the right hand side of (36).

Finally, in [**GL03**] (section 5.1), Gorokhovsky-Lott use the same technics as the two previous examples in order to obtain their index formula.

REMARK 4.9. As a final remark is interesting to mention that in the formula (36) both sides make always sense. In fact the pairing

$$\langle \tilde{a}, \tau_t \rangle$$

is constant for $t \neq 0$.

We could then consider the differences

$$\eta_\tau(D) := \langle ind\,D, \tau \rangle - lim_{t \to 0}\langle \widetilde{\sigma_D}, \tau_t \rangle$$

for any D pseudodifferential elliptic $\mathscr{G}$-operator.

For Lie groupoids (the base is a smooth manifold) these differences do not seem to be very interesting, however it would be interesting to adapt the methods and results of this paper to other kind of groupoids or higher structures, for example to Continuous families groupoids, groupoids associated to manifolds with boundary or with conical singularities ([**Mon03, DLN06**]). Then probably these kind of differences could give interesting data. See [**MW94, LP05**] for related discussions.

References

[AMMS] J. Aastrup, S.T. Melo, B. Monthubert, and E. Schrohe, *Boutet de monvel's calculus and groupoids i.*

[AS68] M. F. Atiyah and I. M. Singer, *The index of elliptic operators. I*, Ann. of Math. (2) **87** (1968), 484–530.

[BGR77] Lawrence G. Brown, Philip Green, and Marc A. Rieffel, *Stable isomorphism and strong Morita equivalence of C^*-algebras*, Pacific J. Math. **71** (1977), no. 2, 349–363.

[BN94] J.-L. Brylinski and Victor Nistor, *Cyclic cohomology of étale groupoids*, K-Theory **8** (1994), no. 4, 341–365.

[Bur85] Dan Burghelea, *The cyclic homology of the group rings*, Comment. Math. Helv. **60** (1985), no. 3, 354–365.

[CC00] Alberto Candel and Lawrence Conlon, *Foliations. I*, Graduate Studies in Mathematics, vol. 23, American Mathematical Society, Providence, RI, 2000.

[CH90] Alain Connes and Nigel Higson, *Déformations, morphismes asymptotiques et K-théorie bivariante*, C. R. Acad. Sci. Paris Sér. I Math. **311** (1990), no. 2, 101–106.

[CM90] Alain Connes and Henri Moscovici, *Cyclic cohomology, the Novikov conjecture and hyperbolic groups*, Topology **29** (1990), no. 3, 345–388.

[Con79] Alain Connes, *Sur la théorie non commutative de l'intégration*, Algèbres d'opérateurs (Sém., Les Plans-sur-Bex, 1978), Lecture Notes in Math., vol. 725, Springer, Berlin, 1979, pp. 19–143.

[Con85] ———, *Noncommutative differential geometry*, Inst. Hautes Études Sci. Publ. Math. (1985), no. 62, 257–360.

[Con86] ———, *Cyclic cohomology and the transverse fundamental class of a foliation*, Geometric methods in operator algebras (Kyoto, 1983), Pitman Res. Notes Math. Ser., vol. 123, Longman Sci. Tech., Harlow, 1986, pp. 52–144.

[Con94] ———, *Noncommutative geometry*, Academic Press Inc., San Diego, CA, 1994.

[CR08a] Paulo Carrillo-Rouse, *Compactly supported analytic indices for lie groupoids*, Accepted in Journal of K-theory. Arxiv:math.KT/0803.2060 (2008).

[CR08b] ———, *A Schwartz type algebra for the tangent groupoid*, "K-theory and Noncommutative Geometry" Book EMS, edited by G. Cortinas, J. Cuntz, M. Karoubi, R. Nest and C. Weibel (2008).

[Cra99] Marius Crainic, *Cyclic cohomology of étale groupoids: the general case*, K-Theory **17** (1999), no. 4, 319–362.

[CS84] Alain Connes and Georges Skandalis, *The longitudinal index theorem for foliations*, Publ. Res. Inst. Math. Sci. **20** (1984), no. 6, 1139–1183.

[CST04] Joachim Cuntz, Georges Skandalis, and Boris Tsygan, *Cyclic homology in non-commutative geometry*, Encyclopaedia of Mathematical Sciences, vol. 121, Springer-Verlag, Berlin, 2004, , Operator Algebras and Non-commutative Geometry, II.

[Dix] Jacques Dixmier, *Les C^*-algbres et leurs reprsentations*, Gauthier-Villars.

[DLN06] Claire Debord, Jean-Marie Lescure, and Victor Nistor, *Groupoids and an index theorem for conical pseudomanifolds*, arxiv:math.OA/0609438 (2006).

[Ehr65] Charles Ehresmann, *Catégories et structures*, Dunod, Paris, 1965.

[GL03] Alexander Gorokhovsky and John Lott, *Local index theory over étale groupoids*, J. Reine Angew. Math. **560** (2003), 151–198.

[GL06] ———, *Local index theory over foliation groupoids*, Adv. Math. **204** (2006), no. 2, 413–447.

[God91] Claude Godbillon, *Feuilletages*, Progress in Mathematics, vol. 98, Birkhäuser Verlag, Basel, 1991, Études géométriques. [Geometric studies], With a preface by G. Reeb.

[Goo85] Thomas G. Goodwillie, *Cyclic homology, derivations, and the free loopspace*, Topology **24** (1985), no. 2, 187–215.

[Hae84] André Haefliger, *Groupoïdes d'holonomie et classifiants*, Astérisque (1984), no. 116, 70–97, Transversal structure of foliations (Toulouse, 1982).

[HS87]   Michel Hilsum and Georges Skandalis, *Morphismes K-orientés d'espaces de feuilles et fonctorialité en théorie de Kasparov (d'après une conjecture d'A. Connes)*, Ann. Sci. École Norm. Sup. (4) **20** (1987), no. 3, 325–390.

[Kar87]   Max Karoubi, *Homologie cyclique et K-théorie*, Astérisque (1987), no. 149, 147.

[LP05]   Eric Leichtnam and Paolo Piazza, *Étale groupoids, eta invariants and index theory*, J. Reine Angew. Math. **587** (2005), 169–233.

[Mac87]   K. Mackenzie, *Lie groupoids and Lie algebroids in differential geometry*, London Mathematical Society Lecture Note Series, vol. 124, Cambridge University Press, Cambridge, 1987.

[Moe02]   Ieke Moerdijk, *Orbifolds as groupoids: an introduction*, Orbifolds in mathematics and physics (Madison, WI, 2001), Contemp. Math., vol. 310, Amer. Math. Soc., Providence, RI, 2002, pp. 205–222.

[Mon03]   Bertrand Monthubert, *Groupoids and pseudodifferential calculus on manifolds with corners*, J. Funct. Anal. **199** (2003), no. 1, 243–286.

[MP97]   Bertrand Monthubert and François Pierrot, *Indice analytique et groupoïdes de Lie*, C. R. Acad. Sci. Paris Sér. I Math. **325** (1997), no. 2, 193–198.

[MW94]   H. Moscovici and F.-B. Wu, *Localization of topological Pontryagin classes via finite propagation speed*, Geom. Funct. Anal. **4** (1994), no. 1, 52–92.

[NWX99]   Victor Nistor, Alan Weinstein, and Ping Xu, *Pseudodifferential operators on differential groupoids*, Pacific J. Math. **189** (1999), no. 1, 117–152.

[Pat99]   Alan L. T. Paterson, *Groupoids, inverse semigroups, and their operator algebras*, Progress in Mathematics, vol. 170, Birkhäuser Boston Inc., Boston, MA, 1999.

[Ren80]   Jean Renault, *A groupoid approach to C*-algebras*, Lecture Notes in Mathematics, vol. 793, Springer, Berlin, 1980.

[Trè06]   François Trèves, *Topological vector spaces, distributions and kernels*, Dover Publications Inc., Mineola, NY, 2006, Unabridged republication of the 1967 original.

[Win83]   H. E. Winkelnkemper, *The graph of a foliation*, Ann. Global Anal. Geom. **1** (1983), no. 3, 51–75.

An algebraic proof of Bogomolov-Tian-Todorov theorem

Donatella Iacono and Marco Manetti

ABSTRACT. We give a completely algebraic proof of the Bogomolov-Tian-Todorov theorem. More precisely, we shall prove that if X is a smooth projective variety with trivial canonical bundle defined over an algebraically closed field of characteristic 0, then the L_∞-algebra governing infinitesimal deformations of X is quasi-isomorphic to an abelian differential graded Lie algebra.

CONTENTS

Introduction

Let X be a smooth projective variety over an algebraically closed field $\mathbb{K}$ of characteristic 0, with tangent sheaf Θ_X. Given an affine open cover $\mathcal{U} = \{U_i\}$ of X, we can consider the Čech complex $\check{C}(\mathcal{U}, \Theta_X)$. By classical deformation theory [**Kod86, Se06**], the group $H^1(\mathcal{U}, \Theta_X)$ classifies first order deformations of X, while $H^2(\mathcal{U}, \Theta_X)$ is an obstruction space for X. Moreover, as a consequence of the results contained in [**Hin97, HiS97a, HiS97b, FMM08**], there exists a canonical sequence of higher brackets on $\check{C}(\mathcal{U}, \Theta_X)$, defining an L_∞ structure and governing deformations of X over local Artinian $\mathbb{K}$-algebras, via Maurer-Cartan equation. When $\mathbb{K} = \mathbb{C}$, such L_∞ structure is canonically quasi-isomorphic to the Kodaira-Spencer differential graded Lie algebra of X, whose Maurer-Cartan equation corresponds to the integrability condition of almost complex structures [**GM90**].

Assume now that X has trivial canonical bundle, the well known Bogomolov-Tian-Todorov (BTT) theorem states that X has unobstructed deformations. This was first proved by Bogolomov in [**Bo78**] in the particular case of complex hamiltonian manifolds; then, Tian [**Ti87**] and Todorov [**To89**] proved independently the theorem for compact Kähler manifolds with trivial canonical bundle. Their proofs are transcendental and make a deep use of the underlying differentiable structure as well of the $\partial\bar{\partial}$-Lemma.

More algebraic proofs of BTT theorem, based on T^1-lifting theorem and degeneration of the Hodge spectral sequence, were given in [**Ra92**] for $\mathbb{K} = \mathbb{C}$ and in [**Kaw92, FM99**] for any $\mathbb{K}$ as above.

For $\mathbb{K} = \mathbb{C}$, the BTT theorem is also a consequence of the stronger result [**GM90, Ma04b**] that the Kodaira-Spencer differential graded Lie algebra of X is quasi-abelian, i.e., quasi-isomorphic to an abelian differential graded Lie algebra. This result, also proved with transcendental methods, is important for applications to Mirror symmetry [**BK98**].

The main result of this paper is to give a purely algebraic proof of the quasi-abelianity of the L_∞-algebra $\check{C}(\mathcal{U}, \Theta_X)$ when X is projective with trivial canonical bundle. This is achieved using the degeneration of the Hodge-de Rham spectral sequence, proved algebraically by Faltings, Deligne and Illusie [**Fa88, DeIl87**], and the L_∞ description of the period map [**FiMa08**].

More precisely, we prove the following theorems.

THEOREM (A). *Let X be a smooth projective variety of dimension n defined over an algebraically closed field of characteristic 0. If the contraction map*

$$H^*(\Theta_X) \xrightarrow{i} \mathrm{Hom}^*(H^*(\Omega_X^n), H^*(\Omega_X^{n-1}))$$

is injective, then for every affine open cover $\mathcal{U}$ of X, the L_∞-algebra $\check{C}(\mathcal{U}, \Theta_X)$ is quasi-abelian.

THEOREM (B). *Let X be a smooth projective variety defined over an algebraically closed field of characteristic 0. If the canonical bundle of X is trivial or torsion, then the L_∞-algebra $\check{C}(\mathcal{U}, \Theta_X)$ is quasi-abelian.*

The paper goes as follows: the first section is intended for the non expert reader and is devoted to recall the basic notions of differential graded Lie algebras, L_∞-algebras and their role in deformation theory.

In Section 2, we review the construction of the Thom-Whitney complex associated with a semicosimplicial complex.

In Section 3, following [**FMM08**], we introduce semicosimplicial differential graded Lie algebras, the associated Thom-Whitney DGLAs and the L_∞ structure on the associated total complexes. We also investigate some properties of mapping cones associated with a morphism of DGLAs.

In Sections 4, we collect some technical results about Cartan homotopies and contractions.

In Section 5, we give the definition of the deformation functor $H^1_{\mathrm{sc}}(\exp \mathfrak{g}^\Delta)$, associated with a semicosimplicial Lie algebra $\mathfrak{g}^\Delta$, introduced essentially in [**Hin97, Pr03**] and described in more detailed way in [**FMM08**]. Moreover, following [**FMM08**], we prove, in a complete algebraic way, that the infinitesimal deformations of a smooth variety X, defined over a field of characteristic 0, are controlled by the L_∞-algebra $\check{C}(\mathcal{U}, \Theta_X)$, where $\mathcal{U}$ is an open affine cover of X.

Section 6 is devoted to the algebraic proof of the previous Theorems A and B together with some applications to deformation theory.

Acknowledgement. The first draft of this paper was written during the workshop on "Algebraic and Geometric Deformation Spaces", Bonn August 11-15, 2008. Both authors thank Max Planck Institute and Hausdorff Center for Mathematics for financial support and warm hospitality. The first author wish to thank the Mathematical Department "G. Castelnuovo" of Sapienza Università di Roma for the hospitality. Thanks also to D. Fiorenza and E. Martinengo for useful discussions on the subject of this paper. We are grateful to the referee for improvements in the exposition of the paper.

1. Review of DGLAs and L_∞-algebras

Let $\mathbb{K}$ be a fixed algebraically closed field of characteristic zero. A *differential graded vector space* is a pair (V, d), where $V = \oplus_i V^i$ is a $\mathbb{Z}$-graded vector space and $d\colon V^i \to V^{i+1}$ is a differential of degree $+1$. For every integer n, we define a new differential graded vector space $V[n]$ by setting

$$V[n]^i = V^{n+i}, \qquad d_{V[n]} = (-1)^n d_V.$$

A *differential graded Lie algebra* (DGLA for short) is the data of a differential graded vector space (L, d) together with a bilinear map $[-, -]\colon L \times L \to L$ (called bracket) of degree 0 such that:

(1) (graded skewsymmetry) $[a, b] = -(-1)^{\deg(a)\deg(b)}[b, a]$.
(2) (graded Jacobi identity) $[a, [b, c]] = [[a, b], c] + (-1)^{\deg(a)\deg(b)}[b, [a, c]]$.
(3) (graded Leibniz rule) $d[a, b] = [da, b] + (-1)^{\deg(a)}[a, db]$.

In particular, the Loibniz rule implies that the bracket of a DGLA L induces a structure of graded Lie algebra on its cohomology $H^*(L) = \oplus_i H^i(L)$.

EXAMPLE 1.1. Let (V, d_V) be a differential graded vector space and $\operatorname{Hom}^i(V, V)$ the space of morphisms $V \to V$ of degree i. Then, $\operatorname{Hom}^*(V, V) = \bigoplus_i \operatorname{Hom}^i(V, V)$ is a DGLA with bracket

$$[f, g] = fg - (-1)^{\deg(f)\deg(g)}gf,$$

and differential d given by

$$d(f) = [d_V, f] = d_V f - (-1)^{\deg(f)}f d_V.$$

For later use, we point out that there exists a natural isomorphism

$$H^*(\operatorname{Hom}^*(V, V)) \xrightarrow{\simeq} \operatorname{Hom}^*(H^*(V), H^*(V)).$$

A morphism of differential graded Lie algebras $\varphi\colon L \to M$ is a linear map that preserves degrees and commutes with brackets and differentials. A *quasi-isomorphism* of DGLAs is a morphism that induces an isomorphism in cohomology. Two DGLAs L and M are said to be *quasi-isomorphic* if they are equivalent under the equivalence relation generated by: $L \sim M$ if there exists a quasi-isomorphism $\phi\colon L \to M$.

Next, denote by **Set** the category of sets (in a fixed universe) and by **Art** = **Art**$_\mathbb{K}$ the category of local Artinian $\mathbb{K}$-algebras with residue field $\mathbb{K}$. Unless otherwise specified, for every objects $A \in$ **Art**, we denote by $\mathfrak{m}_A$ its maximal ideal. Given a DGLA L, we define the Maurer-Cartan functor $\operatorname{MC}_L\colon$ **Art** $\to$ **Set** by setting [**Ma99**]:

$$\operatorname{MC}_L(A) = \left\{ x \in L^1 \otimes \mathfrak{m}_A \mid dx + \frac{1}{2}[x, x] = 0 \right\},$$

where the DGLA structure on $L \otimes \mathfrak{m}_A$ is the natural extension of the DGLA structure on L. The gauge action $* : \exp(L^0 \otimes \mathfrak{m}_A) \times \mathrm{MC}_L(A) \longrightarrow \mathrm{MC}_L(A)$ may be defined by the explicit formula

$$e^a * x := x + \sum_{n \geq 0} \frac{[a, -]^n}{(n+1)!} ([a, x] - da).$$

The deformation functor $\mathrm{Def}_L : \mathbf{Art} \longrightarrow \mathbf{Set}$ associated to a DGLA L is:

$$\mathrm{Def}_L(A) = \frac{\mathrm{MC}_L(A)}{\text{gauge}} = \frac{\{x \in L^1 \otimes \mathfrak{m}_A \mid dx + \frac{1}{2}[x, x] = 0\}}{\exp(L^0 \otimes \mathfrak{m}_A)}.$$

REMARK 1.2. Every morphism of DGLAs induces a natural transformation of the associated deformation functors. A basic result asserts that if L and M are quasi-isomorphic DGLAs, then the associated functor Def_L and Def_M are isomorphic [**SS79, GM88, GM90**], [**Ma99**, Corollary 3.2], [**Ma04b**, Corollary 5.52].

Next, we briefly recall the definition of an L_∞ structures on a graded vector space V. For a more detailed description of such structures we refer to [**SS79, LS93, LM95, Ma02, Fu03, Kon03, Get04, FiMa07**] and [**Ma04b**, Chapter IX].

Let V be a graded vector space: we denote by $\odot^n V$ its graded symmetric n-th power. Given $v_1, \ldots, v_n$ homogeneous elements of V, for every permutation σ we have

$$v_1 \odot \cdots \odot v_n = \epsilon(\sigma; v_1, \ldots, v_n)\, v_{\sigma(1)} \odot \cdots \odot v_{\sigma(n)},$$

where $\epsilon(\sigma; v_1, \ldots, v_n)$ is the Koszul sign. When the sequence $v_1, \ldots, v_n$ is clear from the context, we simply write $\epsilon(\sigma)$ instead of $\epsilon(\sigma; v_1, \ldots, v_n)$.

DEFINITION 1.3. Denote by Σ_n the group of permutations of the set $\{1, 2, \ldots, n\}$. The set of *unshuffles* of type $(p, n-p)$ is the subset $S(p, n-p) \subset \Sigma_n$ of permutations σ such that $\sigma(1) < \sigma(2) < \cdots < \sigma(p)$ and $\sigma(p+1) < \sigma(p+2) < \cdots < \sigma(n)$.

DEFINITION 1.4. An L_∞ structure on a graded vector space V is a sequence $\{q_k\}_{k \geq 1}$ of linear maps $q_k \in \mathrm{Hom}^1(\odot^k(V[1]), V[1])$ such that the map

$$Q : \bigoplus_{n \geq 1} \odot^n V[1] \to \bigoplus_{n \geq 1} \odot^n V[1],$$

defined as

$$Q(v_1 \odot \cdots \odot v_n) = \sum_{k=1}^n \sum_{\sigma \in S(k, n-k)} \epsilon(\sigma) q_k(v_{\sigma(1)} \odot \cdots \odot v_{\sigma(k)}) \odot v_{\sigma(k+1)} \odot \cdots \odot v_{\sigma(n)},$$

is a codifferential on the reduced symmetric graded coalgebra $\bigoplus_{n \geq 1} \odot^n V[1]$: in other words, the datum $(V, q_1, q_2, q_3, \ldots)$ is called an L_∞-algebra if $QQ = 0$.

If $(V, q_1, q_2, q_3, \ldots)$ is an L_∞-algebra, then $q_1 q_1 = 0$ and therefore $(V[1], q_1)$ is a differential graded vector space. The relation between DGLAs and L_∞-algebras is given by the following example.

EXAMPLE 1.5 ([**Qui69**]). Let $(L, d, [\,,\,])$ be a differential graded Lie algebra and define:

$$q_1 = -d : L[1] \to L[1],$$

$$q_2 \in \mathrm{Hom}^1(\odot^2(L[1]), L[1]), \qquad q_2(v \odot w) = (-1)^{\deg(v)}[v, w],$$

and $q_k = 0$ for every $k \geq 3$. Then $(L, q_1, q_2, 0, \ldots)$ is an L_∞-algebra.

An L_∞-*morphism* $(V, q_1, q_2, \ldots) \to (W, p_1, p_2, \ldots)$ of L_∞-algebras is a sequence $f_\infty = \{f_n\}$ of degree zero linear maps

$$f_n : \overset{n}{\bigodot} V[1] \to W[1], \qquad n \geq 1,$$

such that the (unique) morphism of graded coalgebras

$$F : \bigoplus_{n \geq 1} \overset{n}{\bigodot} V[1] \to \bigoplus_{n \geq 1} \overset{n}{\bigodot} W[1],$$

lifting $\sum_n f_n : \bigoplus_{n \geq 1} \bigodot^n V[1] \to W[1]$, commutes with the codifferentials. This condition implies that the *linear part* $f_1 : V[1] \to W[1]$ of an L_∞-morphism $f_\infty :$ $(V, q_1, q_2, \ldots) \to (W, p_1, p_2, \ldots)$ satisfies the condition $f_1 \circ q_1 = p_1 \circ f_1$, and therefore f_1 is a map of differential complexes $(V[1], q_1) \to (W[1], p_1)$.

An L_∞-morphism $f_\infty = \{f_n\}$ is said to be linear if $f_n = 0$, for every $n \geq 2$. Notice that an L_∞-morphism between two DGLAs is linear if and only if it is a morphism of differential graded Lie algebras.

A *quasi-isomorphism* of L_∞-algebra is an L_∞-morphism, whose linear part is a quasi-isomorphism of complexes. Two L_∞-algebras are *quasi-isomorphic* if they are equivalent under the equivalence relation generated by the relation: $L \sim M$ if there exists a quasi-isomorphism $\phi : L \to M$.

DEFINITION 1.6. An L_∞-algebra $(V, q_1, q_2, \ldots)$ is called *abelian* if $q_i = 0$ for every $i \geq 2$. An L_∞-algebra is called *quasi-abelian* if it is quasi-isomorphic to an abelian L_∞-algebra.

Given an L_∞-algebra $(V, q_1, q_2, \ldots)$ and a differential graded commutative algebra A, there exists a natural L_∞ structure on the tensor product $V \otimes A$. The *Maurer-Cartan functor* MC_V *associated with the L_∞-algebra V* is the functor [**SS79**, **Fu03**, **Kon03**]:

$$\mathrm{MC}_V : \mathbf{Art} \to \mathbf{Set}$$

$$\mathrm{MC}_V(A) = \left\{ \gamma \in V[1]^0 \otimes \mathfrak{m}_A \;\middle|\; \sum_{j \geq 1} \frac{q_j(\gamma^{\odot j})}{j!} = 0 \right\}.$$

Two elements x and $y \in \mathrm{MC}_V(A)$ are *homotopy equivalent* if there exists $g(s) \in$ $\mathrm{MC}_{V \otimes \mathbb{K}[s, ds]}(A)$ such that $g(0) = x$ and $g(1) = y$. Then, the *deformation functor* Def_V associated with the L_∞-algebra V is

$$\mathrm{Def}_V : \mathbf{Art} \to \mathbf{Set}, \qquad \mathrm{Def}_V(A) = \frac{\mathrm{MC}_V(A)}{\text{homotopy}}.$$

EXAMPLE 1.7. Given an abelian L_∞-algebra $(V, q_1, 0, 0, \ldots)$, for every $A \in \mathbf{Art}$ there exists a canonical isomorphism

$$\mathrm{Def}_V(A) = H^0(V[1], q_1) \otimes \mathfrak{m}_A.$$

REMARK 1.8. If L is a DGLA, then the deformation functor associated with L, viewed as an L_∞-algebra, is isomorphic to the previous one (Maurer-Cartan modulo gauge equivalence) [**SS79**, **Ma02**, **Fu03**, **Kon03**].

REMARK 1.9. As for the DGLAs, every morphism of L_∞-algebras induces a natural transformation of the associated deformation functors. If two L_∞-algebras are quasi-isomorphic, then there exists an isomorphism between the associated deformation functors [**Fu03, Kon03**], [**Ma04b**, Corollary IX.22]. In particular, the deformation functor of a quasi-abelian L_∞-algebra is unobstructed.

LEMMA 1.10. *Let* $(V, q_1, q_2, \ldots)$, $(W, r_1, r_2, \ldots)$ *be* L_∞*-algebras with* W *quasi-abelian and* $f_\infty \colon V \to W$ *an* L_∞*-morphism. If* f_1 *is injective in cohomology, then* V *is quasi-abelian.*

PROOF. According to homotopy classification of L_∞-algebras (see e.g. [**Kon03**]), if H is the cohomology of the complex (W, r_1), there exists an L_∞ structure on H and a surjective quasi-isomorphism of L_∞-algebras $p_\infty \colon W \to H$. The L_∞ structure on H depends, up to isomorphism, by the quasi-isomorphism class of W and therefore every bracket on H is trivial.

Replacing W with H and f_∞ with $p_\infty f_\infty$ it is not restrictive to assume $r_i = 0$ for every i. There exist a graded vector space K and a morphism of graded vector spaces $\beta \colon W[1] \to K[1]$ such that the composition

$$(V[1], q_1) \xrightarrow{f_1} (W[1], 0) \xrightarrow{\beta} (K[1], 0)$$

is a quasi-isomorphism of complexes. Then, β is a linear L_∞-morphism and the composition

$$(V, q_1, q_2, \ldots) \xrightarrow{f_\infty} (W, 0, 0, \ldots) \xrightarrow{\beta} (K, 0, 0, \ldots)$$

is a quasi-isomorphism of L_∞-algebras. $\qquad\square$

2. The Thom-Whitney complex

Let $\mathbf{\Delta}_{\mathrm{mon}}$ be the category whose objects are the finite ordinal sets $[n] = \{0, 1, \ldots, n\}$, $n = 0, 1, \ldots$, and whose morphisms are order-preserving injective maps among them. Every morphism in $\mathbf{\Delta}_{\mathrm{mon}}$, different from the identity, is a finite composition of *coface* morphisms:

$$\partial_k \colon [i-1] \to [i], \qquad \partial_k(p) = \begin{cases} p & \text{if } p < k \\ p+1 & \text{if } k \le p \end{cases}, \qquad k = 0, \ldots, i.$$

The relations about compositions of them are generated by

$$\partial_l \partial_k = \partial_{k+1} \partial_l, \qquad \text{for every } l \le k.$$

According to [**EZ50, We94**], a *semicosimplicial* object in a category $\mathbf{C}$ is a covariant functor $A^\Delta \colon \mathbf{\Delta}_{\mathrm{mon}} \to \mathbf{C}$. Equivalently, a semicosimplicial object A^Δ is a diagram in $\mathbf{C}$:

$$A_0 \rightrightarrows A_1 \substack{\longrightarrow \\ \longrightarrow \\ \longrightarrow} A_2 \substack{\longrightarrow \\ \longrightarrow \\ \longrightarrow \\ \longrightarrow} \cdots,$$

where each A_i is in $\mathbf{C}$, and, for each $i > 0$, there are $i + 1$ morphisms

$$\partial_k \colon A_{i-1} \to A_i, \qquad k = 0, \ldots, i,$$

such that $\partial_l \partial_k = \partial_{k+1} \partial_l$, for any $l \le k$.

Given a semicosimplicial differential graded vector space

$$V^\Delta \colon \quad V_0 \rightrightarrows V_1 \substack{\longrightarrow \\ \longrightarrow \\ \longrightarrow} V_2 \substack{\longrightarrow \\ \longrightarrow \\ \longrightarrow \\ \longrightarrow} \cdots,$$

the graded vector space $\bigoplus_{n\geq 0} V_n[-n]$ has two differentials

$$d = \sum_n (-1)^n d_n, \qquad \text{where} \quad d_n \text{ is the differential of } V_n,$$

and

$$\partial = \sum_i (-1)^i \partial_i, \qquad \text{where} \quad \partial_i \text{ are the coface maps.}$$

More explicitly, if $v \in V_n^i$, then the degree of v is $i + n$ and

$$d(v) = (-1)^n d_n(v) \in V_n^{i+1}, \qquad \partial(v) = \partial_0(v) - \partial_1(v) + \cdots + (-1)^{n+1}\partial_{n+1}(v) \in V_{n+1}^i.$$

Since $d\partial + \partial d = 0$, we define $\text{Tot}(V^\Delta)$ as the graded vector space $\bigoplus_{n\geq 0} V_n[-n]$, endowed with the differential $d + \partial$.

EXAMPLE 2.1. Let $\mathcal{U} = \{U_i\}$ be an affine open cover of a smooth variety X, defined over an algebraically closed field of characteristic 0; denote by Θ_X the tangent sheaf of X. Then, we can define the Čech semicosimplicial Lie algebra $\Theta_X(\mathcal{U})$ as the semicosimplicial Lie algebra

$$\Theta_X(\mathcal{U}): \quad \prod_i \Theta_X(U_i) \Longrightarrow \prod_{i<j} \Theta_X(U_{ij}) \Longrightarrow \prod_{i<j<k} \Theta_X(U_{ijk}) \Longrightarrow \cdots,$$

where the coface maps $\partial_h: \prod_{i_0 < \cdots < i_{k-1}} \Theta_X(U_{i_0 \cdots i_{k-1}}) \to \prod_{i_0 < \cdots < i_k} \Theta_X(U_{i_0 \cdots i_k})$ are given by

$$\partial_h(x)_{i_0 \ldots i_k} = x_{i_0 \ldots \hat{i_h} \ldots i_k \,|U_{i_0 \ldots i_k}}, \qquad \text{for } h = 0, \ldots, k.$$

Since every Lie algebra is, in particular, a differential graded vector space (concentrated in degree 0), it makes sense to consider the total complex $\text{Tot}(\Theta_X(\mathcal{U}))$, which coincides with the Čech complex $\check{C}(\mathcal{U}, \Theta_X)$.

EXAMPLE 2.2. Let Ω_X^* be the algebraic de Rham complex of a smooth variety X of dimension n:

$$\Omega_X^*: \quad 0 \to \mathcal{O}_X = \Omega_X^0 \xrightarrow{d} \Omega_X^1 \xrightarrow{d} \cdots \xrightarrow{d} \Omega_X^n \to 0.$$

Given an affine open cover $\mathcal{U} = \{U_i\}$ of X, we can define a semicosimplicial differential graded vector space

$$\Omega_X^*(\mathcal{U}): \quad \prod_i \Omega_X^*(U_i) \Longrightarrow \prod_{i<j} \Omega_X^*(U_{ij}) \Longrightarrow \prod_{i<j<k} \Omega_X^*(U_{ijk}) \Longrightarrow \cdots,$$

where $\prod_{i_0 < \cdots < i_k} \Omega_X^*(U_{i_0 \cdots i_k})$ is the complex

$$\prod_{i_0 < i_1 < \cdots < i_k} \mathcal{O}_X(U_{i_0 \cdots i_k}) \xrightarrow{d} \prod_{i_0 < i_1 < \cdots < i_k} \Omega_X^1(U_{i_0 \cdots i_k}) \xrightarrow{d} \cdots \prod_{i_0 < \cdots < i_n} \Omega_X^n(U_{i_0 \cdots i_n}).$$

Here the total complex $\text{Tot}(\Omega_X^*(\mathcal{U}))$ is the Čech complex $\check{C}(\mathcal{U}, \Omega_X^*)$ of Ω_X^*, with respect to the affine cover $\mathcal{U}$.

Let V^Δ be a semicosimplicial differential graded vector space and $(A_{PL})_n$ the differential graded commutative algebra of polynomial differential forms on the standard n-simplex $\{(t_0, \ldots, t_n) \in \mathbb{K}^{n+1} \mid \sum t_i = 1\}$ [**FHT01**]:

$$(A_{PL})_n = \frac{\mathbb{K}[t_0, \ldots, t_n, dt_0, \ldots, dt_n]}{(1 - \sum t_i, \sum dt_i)}.$$

For every n, m the tensor product $V_n \otimes (A_{PL})_m$ is a differential graded vector space and then also $\prod_n V_n \otimes (A_{PL})_n$ is a differential graded vector space.

Denoting by

$$\delta^k \colon (A_{PL})_n \to (A_{PL})_{n-1}, \quad \delta^k(t_i) = \begin{cases} t_i & \text{if } 0 \le i < k \\ 0 & \text{if } i = k \\ t_{i-1} & \text{if } k < i \end{cases}, \quad k = 0, \ldots, n,$$

the face maps, for every $0 \le k \le n$, there are well-defined morphisms of differential graded vector spaces

$$Id \otimes \delta^k \colon V_n \otimes (A_{PL})_n \to V_n \otimes (A_{PL})_{n-1},$$

$$\partial_k \otimes Id \colon V_{n-1} \otimes (A_{PL})_{n-1} \to V_n \otimes (A_{PL})_{n-1}.$$

The Thom-Whitney differential graded vector space $\mathrm{Tot}_{TW}(V^\Delta)$ of V^Δ is the differential graded subvector space of $\prod_n V_n \otimes (A_{PL})_n$, whose elements are the sequences $(x_n)_{n \in \mathbb{N}}$ satisfying the equations

$$(Id \otimes \delta^k)x_n = (\partial_k \otimes Id)x_{n-1}, \quad \text{for every } 0 \le k \le n.$$

In [**Whi57**], Whitney noted that the integration maps

$$\int_{\Delta^n} \otimes \, Id \colon (A_{PL})_n \otimes V_n \to \mathbb{C}[n] \otimes V_n = V_n[n]$$

give a quasi-isomorphism of differential graded vector spaces

$$I \colon (\mathrm{Tot}_{TW}(V^\Delta), d_{TW}) \to (\mathrm{Tot}(V^\Delta), d_{\mathrm{Tot}}).$$

Moreover, there exist an explicit injective quasi-isomorphism of differential graded vector spaces

$$E \colon \mathrm{Tot}(V^\Delta) \to \mathrm{Tot}_{TW}(V^\Delta)$$

and an explicit homotopy

$$h \colon \mathrm{Tot}_{TW}(V^\Delta) \to \mathrm{Tot}_{TW}(V^\Delta)[-1],$$

such that

$$IE = \mathrm{Id}_{\mathrm{Tot}(V^\Delta)}; \qquad EI - \mathrm{Id}_{\mathrm{Tot}_{TW}(V^\Delta)} = h d_{TW} + d_{TW} h.$$

Moreover, the morphisms I, E, h are functorial and commute with morphisms of semicosimplicial differential graded vector spaces. For more details and explicit description of E and h, we refer to [**Dup76, Dup78, NaA87, Get04, FMM08, CG08**].

Let V^Δ and W^Δ be two semicosimplicial graded vector spaces, with degeneracy maps $\partial_{k,n}$ and $\partial'_{k,n}$, respectively. Then, we define the tensor product $(V \otimes W)^\Delta$ as the semicosimplicial graded vector space, such that $(V \otimes W)_n^\Delta = V_n \otimes W_n$ and the degeneracy maps are defined on each factor, i.e., $\partial_{k,n} = \partial'_{k,n} \otimes \partial''_{k,n} \colon (V \otimes W)_{n-1}^\Delta \to (V \otimes W)_n^\Delta$.

REMARK 2.3. The Thom-Withney construction is compatible with tensor product, i.e., there exists a natural transformation

$$\Phi \colon \mathrm{Tot}_{TW}(V^\Delta) \otimes \mathrm{Tot}_{TW}(W^\Delta) \to \mathrm{Tot}_{TW}((V \otimes W)^\Delta).$$

Indeed, we have to prove that $x \otimes y \in \mathrm{Tot}_{TW}((V \otimes W)^\Delta)$, for every pair of sequences $x = (x_n)_{n \in \mathbb{N}} \in \mathrm{Tot}_{TW}(V^\Delta)$ and $y = (y_n)_{n \in \mathbb{N}} \in \mathrm{Tot}_{TW}(W^\Delta)$.

We have

$$\partial_k \otimes Id(x_n \otimes y_n) = (\partial_k \otimes Id(x_n)) \otimes (\partial_k \otimes Id(y_n)),$$

and,

$$Id \otimes \delta^k(x_{n+1} \otimes y_{n+1}) = (Id \otimes \delta^k(x_{n+1})) \otimes (Id \otimes \delta^k(y_{n+1})).$$

It is sufficient to observe that the right parts of the above two equation are the same for every $k \le n$.

In particular, every bilinear map of semicosimplicial graded vector spaces $V^\Delta \times W^\Delta \to Z^\Delta$ induces a bilinear map $\mathrm{Tot}_{TW}(V^\Delta) \times \mathrm{Tot}_{TW}(W^\Delta) \to \mathrm{Tot}_{TW}(Z^\Delta)$.

3. Semicosimplicial differential graded Lie algebras and mapping cones

Let

$$\mathfrak{g}^\Delta : \quad \mathfrak{g}_0 \rightrightarrows \mathfrak{g}_1 \Rrightarrow \mathfrak{g}_2 \Rrightarrow \cdots ,$$

be a *semicosimplicial differential graded Lie algebra*. Every $\mathfrak{g}_i$ is a DGLA and so, in particular, a differential graded vector space, thus we can consider the total complex $\mathrm{Tot}(\mathfrak{g}^\Delta)$.

EXAMPLE 3.1. Every morphism $\chi \colon L \to M$ of differential graded Lie algebras can be interpreted as the semicosimplicial DGLA

$$\chi^\Delta : \quad L \rightrightarrows M \Rrightarrow 0 , \qquad \partial_0 = \chi, \partial_1 = 0,$$

and the total complex $\mathrm{Tot}(\chi^\Delta)$ coincides with the mapping cone of χ, i.e.,

$$\mathrm{Tot}(\chi^\Delta)^i = L^i \oplus M^{i-1}, \qquad d(l, m) = (dl, \chi(l) - dm).$$

Even in the case of L and M Lie algebras, it is not possible to define a canonical bracket on the mapping cone, making $\mathrm{Tot}(\chi^\Delta)$ a DGLA and the projection $\mathrm{Tot}(\chi^\Delta) \to L$ a morphism of DGLAs. To see this it is sufficient to consider $L = M$ the Lie subalgebra of $sl(2, \mathbb{K})$ generated by the matrices

$$A = \begin{pmatrix} 0 & 1 \\ 0 & 0 \end{pmatrix}, \qquad B = \begin{pmatrix} 1 & 0 \\ 0 & -1 \end{pmatrix},$$

and χ equal to the identity. If $\mathrm{Tot}(\chi^\Delta)$ is a DGLA then, by functoriality, for every $x \in L$ the subspace generated by $x, \chi(x)$ is a subalgebra and, for every $\lambda \in \mathbb{K}$, the linear map $B \mapsto B$, $A \mapsto \lambda A$ is an automorphism of DGLA. It is an easy exercise to prove that these properties imply the failure of Jacobi identity.

Even if the complex $\mathrm{Tot}(\mathfrak{g}^\Delta)$ has no natural DGLA structure, it can be endowed with a canonical L_∞ structure by homological perturbation theory [**FiMa07, FMM08**].

Indeed, as in the previous section, we can consider the Thom-Whitney construction also for semicosimplicial differential graded Lie algebras: it is evident that in such case we have $\mathrm{Tot}_{TW}(V^\Delta)$ a differential graded lie algebra. In this case, the morphisms I, E, h are functorial and commute with morphisms of semicosimplicial DGLAs. Moreover, these morphisms I, E, h can be used to apply the following basic result about L_∞-algebras, dating back to Kadeishvili's work on the cohomology of A_∞ algebras [**Kad82**]; see also [**HK91**].

THEOREM 3.2 (Homotopy transfer). *Let $(V, q_1, q_2, q_3, \ldots)$ be an L_∞-algebra and (C, δ) a differential graded vector space. Assume we have two morphisms of complexes*

$$\pi : (V[1], q_1) \to (C[1], \delta_{[1]}), \qquad \imath_1 : (C[1], \delta_{[1]}) \to (V[1], q_1)$$

and a linear map $h \in \operatorname{Hom}^{-1}(V[1], V[1])$ such that $hq_1 + q_1 h = \imath_1 \pi - Id$.

Then, there exist a canonical L_∞-algebra structure $(C, \langle \rangle_1, \langle \rangle_2, \ldots)$ on C extending its differential complex structure, and a canonical L_∞-morphism $\imath_\infty : (C, \langle \rangle_1, \langle \rangle_2, \ldots) \to (V, q_1, q_2, q_3, \ldots)$ extending $\imath_1$. In particular, if $\imath_1$ is an injective quasi isomorphism of complexes, then also $\imath_\infty$ is an injective quasi-isomorphism of L_∞-algebras.

PROOF. See [**FiMa07**] and references therein; explicit formulas for the quasi-isomorphism $\imath_\infty$ and the brackets $\langle \rangle_n$ have been described by Merkulov in [**Me99**]; then, it has been remarked by Kontsevich and Soibelman in [**KS00, KS01**] (see also [**Fu03**]) that Merkulov's formulas can be nicely written as summations over rooted trees.   □

COROLLARY 3.3 ([**FiMa07, FMM08**]). *There exists a canonical L_∞-algebra structure $\widetilde{\operatorname{Tot}}(\mathfrak{g}^\Delta)$ on the differential graded vector space $\operatorname{Tot}(\mathfrak{g}^\Delta)$, together with an injective quasi-isomorphism $E_\infty : \widetilde{\operatorname{Tot}}(\mathfrak{g}^\Delta) \to \operatorname{Tot}_{TW}(\mathfrak{g}^\Delta)$.*

PROOF. It is sufficient to apply Theorem 3.2 to the morphisms I, E, h, in order to define a canonical L_∞-algebra structure $\widetilde{\operatorname{Tot}}(\mathfrak{g}^\Delta)$ on $\operatorname{Tot}(\mathfrak{g}^\Delta)$, and an injective quasi-isomorphism $E_\infty : \widetilde{\operatorname{Tot}}(\mathfrak{g}^\Delta) \to \operatorname{Tot}_{TW}(\mathfrak{g}^\Delta)$ extending E.

Notice that E_∞ induces an isomorphism of functors $\operatorname{Def}_{\widetilde{\operatorname{Tot}}(\mathfrak{g}^\Delta)} \xrightarrow{\cong} \operatorname{Def}_{\operatorname{Tot}_{TW}(\mathfrak{g}^\Delta)}$.   □

Let $\chi : L \to M$ be a morphism of differential graded Lie algebras over a field $\mathbb{K}$ of characteristic 0. We have already seen, in Example 3.1, that χ can be interpreted as the semicosimplicial DGLA

$$\chi^\Delta : \qquad L \underset{\chi}{\overset{0}{\rightrightarrows}} M \rightrightarrows 0 \cdots,$$

and therefore we have a canonical L_∞ structure on the total complex

$$\operatorname{Tot}(\chi^\Delta) = \bigoplus_i \operatorname{Tot}(\chi^\Delta)^i, \qquad \operatorname{Tot}(\chi^\Delta)^i = L^i \oplus M^{i-1}.$$

The brackets

$$\mu_n : \bigwedge^n \operatorname{Tot}(\chi^\Delta) \to \operatorname{Tot}(\chi^\Delta)[2-n], \qquad n \geq 1,$$

have been explicitly described in [**FiMa07**]. Namely, one has

$$\mu_1(l, m) = (dl, \chi(l) - dm), \qquad l \in L, m \in M,$$

$$\mu_2((l_1, m_1) \wedge (l_2, m_2)) = \left([l_1, l_2], \frac{1}{2}[m_1, \chi(l_2)] + \frac{(-1)^{\deg(l_1)}}{2}[\chi(l_1), m_2] \right)$$

and for $n \geq 3$

$$\mu_n((l_1, m_1) \wedge \cdots \wedge (l_n, m_n)) = \pm \frac{B_{n-1}}{(n-1)!} \sum_{\sigma \in S_n} \varepsilon(\sigma)[m_{\sigma(1)}, [\cdots, [m_{\sigma(n-1)}, \chi(l_{\sigma(n)})] \cdots]].$$

Here the B_n's are the Bernoulli numbers, c is the Koszul sign and we refer to [FiMa07] for the exact determination of the overall $\pm$ sign in the formulas (it will not be needed in the present paper).

PROPOSITION 3.4. *In the notation above, assume that:*

(1) $\chi \colon L \to M$ *is injective,*
(2) $\chi \colon H^*(L) \to H^*(M)$ *is injective.*

Then, the L_∞-algebra $\widetilde{\mathrm{Tot}}(\chi^\Delta)$ is quasi-abelian.

PROOF. Since $H^0(\mathrm{Hom}^*(L, L)) = \mathrm{Hom}^0(H^*(L), H^*(L))$ and $H^0(\mathrm{Hom}^*(M, L)) = \mathrm{Hom}^0(H^*(M), H^*(L))$, the surjective morphism

$$\mathrm{Hom}^*(M, L) \to \mathrm{Hom}^*(L, L), \qquad \phi \mapsto \phi\chi,$$

induces surjective maps

$$H^0(\mathrm{Hom}^*(M, L)) \to H^0(\mathrm{Hom}^*(L, L)), \qquad Z^0(\mathrm{Hom}^*(M, L)) \to Z^0(\mathrm{Hom}^*(L, L))$$

and then the identity on L can be lifted to a morphism $\pi \colon M \to L$ of differential graded vector spaces. Denoting $V = \ker(\pi)$, we have a direct sum decomposition of differential graded vector spaces $M = \chi(L) \oplus V$.

Denote by d the differential on M, by assumption $d(V) \subset V$ and the inclusion $V[-1] \hookrightarrow \widetilde{\mathrm{Tot}}(\chi^\Delta)$ is an injective quasi-isomorphism. Let H be a graded vector space, endowed with trivial differential, and $g \colon H \to V[-1]$ an injective morphism of differential graded vector spaces, inducing an isomorphism $H \xrightarrow{\;\;} H^*(V[-1])$. The linear map

$$f \colon H \to \widetilde{\mathrm{Tot}}(\chi^\Delta)), \qquad f(h) = (0, g(h)),$$

is annihilated by every bracket of $\widetilde{\mathrm{Tot}}(\chi^\Delta)$ and then it is an injective quasi isomorphism of L_∞-algebras. $\qquad\square$

EXAMPLE 3.5. Let W be a differential graded vector space and let $U \subset W$ be a differential graded subspace. Assume that the induced morphism $H^*(U) \to H^*(W)$ is injective, then the injective morphism of DGLAs

$$\chi \colon \{f \in \mathrm{Hom}^*(W, W) \mid f(U) \subset U\} \to \mathrm{Hom}^*(W, W)$$

satisfies the hypothesis of Proposition 3.4. In fact, the same argument used above shows that there exists a direct sum decomposition of differential graded vector spaces $W = U \oplus V$. Next, consider the subspace

$$K = \{f \in \mathrm{Hom}^*(W, W) \mid f(W) \subset V, \; f(V) = 0\}.$$

It is straightforward to check that K is a complementary subcomplex of the image of χ inside $\mathrm{Hom}^*(W, W)$.

REMARK 3.6. Let W be a differential graded vector space and let $U \subset W$ be a differential graded subspace. It is showed in [FiMa06], that the deformation functor associated with the morphism of DGLAs

$$\chi \colon \{f \in \mathrm{Hom}^*(W, W) \mid f(U) \subset U\} \to \mathrm{Hom}^*(W, W),$$

i.e., the deformation functor associated to the L_∞-algebra $\widetilde{\mathrm{Tot}}(\chi^\Delta)$, has a natural interpretation as the local structure of the derived Grassmannian of W at the point U. Moreover, Proposition 3.4 implies that the derived Grassmannian of W is smooth at the points corresponding to subspaces U such that $H^*(U) \to H^*(W)$ is injective.

4. Semicosimplicial Cartan homotopies

The abstract notion of Cartan homotopy has been introduced in [**FiMa06, FiMa08**] as a powerful tool for the construction of L_∞ morphisms.

DEFINITION 4.1. Let L and M be two differential graded Lie algebras. A linear map of degree -1

$$i \colon L \to M$$

is called a *Cartan homotopy* if, for every $a, b \in L$, we have:

$$i_{[a,b]} = [i_a, d_M i_b + i_{d_L b}] \qquad \text{and} \qquad [i_a, i_b] = 0.$$

For every Cartan homotopy i, it is convenient to consider the map

$$l \colon L \to M, \qquad l_a = d_M i_a + i_{d_L a}.$$

It is straightforward to check that l is a morphism of DGLAs and the conditions of Definition 4.1 become

$$i_{[a,b]} = [i_a, l_b] \qquad \text{and} \qquad [i_a, i_b] = 0.$$

As a morphism of complexes, l is homotopic to 0 (with homotopy i).

EXAMPLE 4.2. Let X be a smooth algebraic variety, Θ_X the tangent sheaf and (Ω_X^*, d) the algebraic de Rham complex. Then, for every open subset $U \subset X$, the contraction

$$\Theta_X(U) \otimes \Omega_X^k(U) \xrightarrow{\;\lrcorner\;} \Omega_X^{k-1}(U)$$

induces a linear map of degree -1

$$i \colon \Theta_X(U) \to \operatorname{Hom}^*(\Omega_X^*(U), \Omega_X^*(U)), \qquad i_\xi(\omega) = \xi \lrcorner \omega$$

that is a Cartan homotopy. In fact, the differential of $\Theta_X(U)$ is trivial and the differential on the DGLA $\operatorname{Hom}^*(\Omega_X^*(U), \Omega_X^*(U))$ is $\phi \mapsto [d, \phi] = d\phi - (-1)^{\deg(\phi)}\phi d$. Therefore, since i_a has degree -1, we have that $l_a = di_a + i_a d$ is the Lie derivative and the above conditions reduce to the classical Cartan's homotopy formulas:

(1) $[i_a, i_b] = 0$;
(2) $i_{[a,b]} = l_a i_b - i_b l_a = [l_a, i_b] = [i_a, l_b]$,

The relation between Cartan's homotopy and L_∞ algebras is given by the following theorem.

THEOREM 4.3 ([**FiMa08**, Corollary 3.7]). *Let* $\chi \colon N \to M$ *be a morphism of DGLAs and* $i \colon L \to M$ *be a Cartan homotopy. Assume that* $\phi \colon L \to N$ *is a morphism of DGLAs, such that* $\chi\phi = l = d_M i + i d_L$. *Then, the linear map*

$$\Phi \colon L \longrightarrow \widetilde{\operatorname{Tot}}(\chi^\Delta) \qquad \Phi(a) = (\phi(a), i_a)$$

is a linear L_∞-*morphism. In particular, if* N *is a subalgebra of* M *containing* $l(L)$ *and* χ *is the inclusion, then the map*

$$L \longrightarrow \widetilde{\operatorname{Tot}}(\chi^\Delta) \qquad a \mapsto (l_a, i_a)$$

is a linear L_∞-*morphism.*

PROOF. Straightforward consequence of the explicit description of the L_∞ structure of $\widetilde{\operatorname{Tot}}(\chi^\Delta)$. $\qquad \square$

REMARK 4.4. It is plain from definition that Cartan homotopies are stable under composition with morphisms of DGLAs. More precisely, if $f\colon L' \to L$ and $g\colon M \to M'$ are morphisms of differential graded Lie algebras and $i\colon L \to M$ is a Cartan homotopy, then also $gif\colon L' \to M'$ is a Cartan homotopy.

LEMMA 4.5. *Let $i\colon L \to M$ be a Cartan homotopy and A be a differential graded commutative algebra. Then the map*

$$i \otimes \mathrm{Id}\colon L \otimes A \to M \otimes A$$

$$(x \otimes a) \mapsto i_x \otimes a$$

is a Cartan homotopy.

PROOF. By definition, for any $x \otimes a$ and $y \otimes b \in L \otimes A$, we have

$$[(i \otimes \mathrm{Id})_{x \otimes a}, (i \otimes \mathrm{Id})_{y \otimes b}] = [i_x \otimes a, i_y \otimes b] = (-1)^{\deg(a)(\deg(y)-1)}[i_x, i_y] \otimes ab = 0.$$

Moreover, denoting $l = d_M i + i d_L$ and $\tilde{l} = d_{M \otimes A}(i \otimes \mathrm{Id}) + (i \otimes \mathrm{Id})d_{L \otimes A}$ we have

$$\tilde{l}_{x \otimes a} = d_{M \otimes A}(i_x \otimes a) + (i \otimes \mathrm{Id})(d_L x \otimes a + (-1)^{\deg(x)}x \otimes d_A a) =$$

$$= d_M i_x \otimes a - (-1)^{\deg(x)}x \otimes d_A a + i_{d_L x} \otimes a + (-1)^{\deg(x)}x \otimes d_A a = l_x \otimes a.$$

Thus, for any $x \otimes a$ and $y \otimes b \in L \otimes A$, we get

$$[(i \otimes \mathrm{Id})_{x \otimes a}, (i \otimes \mathrm{Id})_{y \otimes b}] = (-1)^{\deg(a)\deg(y)}i_{[x,y]} \otimes ab =$$

$$(-1)^{\deg(a)\deg(y)}[l_x, l_y] \otimes ab = [i_x \otimes a, l_y \otimes b] = [(i \otimes \mathrm{Id})_{x \otimes a}, \tilde{l}_{y \otimes b}].$$

□

DEFINITION 4.6. Let L be a differential graded Lie algebra and V a differential graded vector space. A bilinear map

$$L \times V \xrightarrow{\lrcorner} V$$

of degree -1 is called a *contraction* if the induced map

$$i\colon L \to \mathrm{Hom}^*(V,V), \qquad i_l(v) = l \lrcorner v,$$

is a Cartan Homotopy.

The notion of contraction is stable under scalar extensions, more precisely:

LEMMA 4.7. *Let V be a differential graded vector space and*

$$L \times V \xrightarrow{\lrcorner} V$$

a contraction. Then, for every differential graded commutative algebra A, the natural extension

$$(L \otimes A) \times (V \otimes A) \xrightarrow{\lrcorner} (V \otimes A) \qquad (l \otimes a) \lrcorner (v \otimes b) = (-1)^{\deg(a)\deg(v)}l \lrcorner v \otimes ab,$$

is a contraction.

PROOF. According to Remark 4.4, Lemma 4.5 and Definition 4.6, it is sufficient to prove that the natural map $\alpha\colon \mathrm{Hom}^*(V,V) \otimes A \to \mathrm{Hom}^*(V \otimes A, V \otimes A)$,

$$\alpha(\phi \otimes a)(v \otimes b) = (-1)^{\deg(a)\deg(v)}\phi(v) \otimes ab,$$

is a morphism of DGLAs. This is completely straightforward and it is left to the reader.

□

The notions of Cartan homotopy and contraction extend naturally to the semi-cosimplicial setting. Here, we consider only the case of contractions.

DEFINITION 4.8. Let $\mathfrak{g}^\Delta$ be a semicosimplicial DGLA and V^Δ a semicosimplicial differential graded vector space. A semicosimplicial contraction

$$\mathfrak{g}^\Delta \times V^\Delta \overset{\lrcorner}{\longrightarrow} V^\Delta,$$

is a sequence of contractions $\mathfrak{g}_n \times V_n \overset{\lrcorner}{\longrightarrow} V_n$, $n \geq 0$, commuting with coface maps, i.e., $\partial_k(l \lrcorner v) = \partial_k(l) \lrcorner \partial_k(v)$, for every k.

PROPOSITION 4.9. *Every semicosimplicial contraction*

$$\mathfrak{g}^\Delta \times V^\Delta \overset{\lrcorner}{\longrightarrow} V^\Delta$$

extends naturally to a contraction

$$\mathrm{Tot}_{TW}(\mathfrak{g}^\Delta) \times \mathrm{Tot}_{TW}(V^\Delta) \overset{\lrcorner}{\longrightarrow} \mathrm{Tot}_{TW}(V^\Delta).$$

PROOF. By definition, for every n, we have a Cartan homotopy

$$i : \mathfrak{g}_n \to \mathrm{Hom}^*(V_n, V_n),$$

and so, by Lemma 4.7, a Cartan homotopy

$$i : \mathfrak{g}_n \otimes (A_{PL})_n \to \mathrm{Hom}^*(V_n \otimes (A_{PL})_n, V_n \otimes (A_{PL})_n).$$

Therefore, it is enough to prove that $i_x(y) \in \mathrm{Tot}_{TW}(V^\Delta)$, for every pair of sequences $x = (x_n)_{n \in \mathbb{N}} \in \mathrm{Tot}_{TW}(\mathfrak{g}^\Delta)$ and $y = (y_n)_{n \in \mathbb{N}} \in \mathrm{Tot}_{TW}(V^\Delta)$; it follows from Remark 2.3.

$\square$

5. Semicosimplicial Lie algebras and deformations of smooth varieties

Let $\mathfrak{g}^\Delta$ be a semicosimplicial Lie algebra, then we can apply the construction of Section 3 in order to construct the Thom-Whitney DGLA $\mathrm{Tot}_{TW}(\mathfrak{g}^\Delta)$, the L_∞-algebra $\widetilde{\mathrm{Tot}}(\mathfrak{g}^\Delta)$ and their associated (and isomorphic) deformation functors $\mathrm{Def}_{\widetilde{\mathrm{Tot}}(\mathfrak{g}^\Delta)} \simeq \mathrm{Def}_{\mathrm{Tot}_{TW}(\mathfrak{g}^\Delta)}$.

Beyond this way, there is another natural, and more geometric, way to define a deformation functor, see [Pr03, Definitions 1.4 and 1.6] and [**FMM08**, Section 3]. More precisely, if $\mathfrak{g}^\Delta$ is a semicosimplicial Lie algebra, then we denote

$$Z^1_{sc}(\exp \mathfrak{g}^\Delta) : \mathbf{Art} \to \mathbf{Set}$$

as

$$Z^1_{sc}(\exp \mathfrak{g}^\Delta)(A) = \{x \in \mathfrak{g}_1 \otimes \mathfrak{m}_A \mid e^{\partial_0(x)} e^{-\partial_1(x)} e^{\partial_2(x)} = 1\},$$

and

$$H^1_{sc}(\exp \mathfrak{g}^\Delta) : \mathbf{Art} \to \mathbf{Set}$$

such that

$$H^1_{sc}(\exp \mathfrak{g}^\Delta)(A) = Z^1_{sc}(\exp \mathfrak{g}^\Delta)(A)/ \sim,$$

where $x \sim y$ if and only if there exists $a \in \mathfrak{g}_0 \otimes \mathfrak{m}_A$, such that $e^{-\partial_1(a)} e^x e^{\partial_0(a)} = e^y$.

EXAMPLE 5.1. Let $\mathcal{L}$ be a sheaf of Lie algebras on a paracompact topological space X, and $\mathcal{U}$ an open covering of X; it is naturally defined the Čech semicosimplicial Lie algebra $\mathcal{L}(\mathcal{U})$

$$\mathcal{L}(\mathcal{U}): \qquad \prod_i \mathcal{L}(U_i) \Longrightarrow \prod_{i<j} \mathcal{L}(U_{ij}) \Longrightarrow \prod_{i<j<k} \mathcal{L}(U_{ijk}) \Longrightarrow \cdots,$$

and, for every $A \in \mathbf{Art}$, the set $H^1_{sc}(\exp \mathcal{L}(\mathcal{U}))(A)$ is exactly the cohomology set $H^1(\mathcal{U}, \exp(\mathcal{L} \otimes \mathfrak{m}_A))$ [**Hir78**].

The relation between the above functors is given by the following theorem.

THEOREM 5.2. *Let $\mathfrak{g}^\Delta$ be a semicosimplicial Lie algebra. Then, for every $A \in$* **Art**,

$$\mathrm{MC}_{\widetilde{\mathrm{Tot}}(\mathfrak{g}^\Delta)}(A) = Z^1_{sc}(\exp \mathfrak{g}^\Delta)(A),$$

as subsets of $\mathfrak{g}_1 \otimes \mathfrak{m}_A$. Moreover, we have natural isomorphisms of deformation functors

$$\mathrm{Def}_{\mathrm{Tot}_{TW}(\mathfrak{g}^\Delta)} \simeq \mathrm{Def}_{\widetilde{\mathrm{Tot}}(\mathfrak{g}^\Delta)} \xrightarrow{\sim} H^1_{sc}(\exp \mathfrak{g}^\Delta),$$

PROOF. For the proof, we refer to [**FMM08**]. $\qquad \square$

Next, assume that X is a smooth algebraic variety over a field $\mathbb{K}$ of characteristic 0, with tangent sheaf Θ_X, and let $\mathcal{U} = \{U_i\}_{i \in I}$ be an affine open covering of X.

Since every infinitesimal deformation of a smooth affine scheme is trivial [**So06**, Lemma II.1.3], every infinitesimal deformation X_A of X over $\mathrm{Spec}(A)$ is obtained by gluing the trivial deformations $U_i \times \mathrm{Spec}(A)$ along the double intersections U_{ij}, and therefore it is determined by the sequence $\{\theta_{ij}\}_{i<j}$ of automorphisms of sheaves of A-algebras

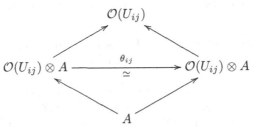

satisfying the cocycle condition

(1) $$\qquad \theta_{jk}\theta_{ik}^{-1}\theta_{ij} = \mathrm{Id}_{\mathcal{O}(U_{ijk}) \otimes A}, \qquad \forall\, i < j < k \in I.$$

Since we are in characteristic zero, we can take the logarithms and write $\theta_{ij} = e^{d_{ij}}$, where $d_{ij} \in \Theta_X(U_{ij}) \otimes \mathfrak{m}_A$. Therefore, the Equation (1) is equivalent to

$$e^{d_{jk}}e^{-d_{ik}}e^{d_{ij}} = 1 \in \exp(\Theta_X(U_{ijk}) \otimes \mathfrak{m}_A), \qquad \forall\, i < j < k \in I.$$

Next, let X'_A be another deformation of X over $\mathrm{Spec}(A)$, defined by the cocycle θ'_{ij}. To give an isomorphism of deformations $X'_A \simeq X_A$ is the same to give, for every i, an automorphism α_i of $\mathcal{O}(U_i) \otimes A$ such that $\theta_{ij} = \alpha_i^{-1}\theta'_{ij}\alpha_j$, for every $i < j$. Taking again logarithms, we can write $\alpha_i = e^{a_i}$, with $a_i \in \Theta_X(U_i) \otimes \mathfrak{m}_A$, and so $e^{-a_i}e^{d'_{ij}}e^{a_j} = e^{d_{ij}}$.

THEOREM 5.3. *Let $\mathcal{U}$ be an affine open cover of a smooth algebraic variety X defined over an algebraically closed field of characteristic 0. Denoting by Def_X the functor of infinitesimal deformations of X, there exist isomorphisms of functors*

$$\mathrm{Def}_X \cong H^1_{sc}(\exp \Theta_X(\mathcal{U})) \cong \mathrm{Def}_{\mathrm{Tot}_{TW}(\Theta_X(\mathcal{U}))} \cong \mathrm{Def}_{\widetilde{\mathrm{Tot}}(\Theta_X(\mathcal{U}))},$$

where $\Theta_X(\mathcal{U})$ is the semicosimplicial Lie algebra defined in Example 2.1.

PROOF. By Theorem 5.2, it is sufficient to prove $\mathrm{Def}_X \cong H^1_{sc}(\exp \Theta_X(\mathcal{U}))$. By definition,

$$Z^1_{sc}(\Theta_X(\mathcal{U}))(A) = \{\{x_{ij}\} \in \prod_{i<j} \Theta_X(U_{ij}) \otimes \mathfrak{m}_A \mid e^{x_{jk}} e^{-x_{ik}} e^{x_{ij}} = 1 \ \forall \ i < j < k\},$$

for each $A \in \mathbf{Art}$. Moreover, given $x = \{x_{ij}\}$ and $y = \{y_{ij}\} \in \prod_{i<j} \Theta_X(U_{ij}) \otimes \mathfrak{m}_A$, we have $x \sim y$ if and only if there exists $a = \{a_i\} \in \prod_i \Theta_X(U_i) \otimes \mathfrak{m}_A$ such that $e^{-a_j} e^{x_{ij}} e^{a_i} = e^{y_{ij}}$ for all $i < j$. $\square$

REMARK 5.4. Note that, when $\mathbb{K} = \mathbb{C}$, the L_∞-algebra $\widetilde{\mathrm{Tot}}(\Theta_X(\mathcal{U}))$ is quasi-isomorphic to the Kodaira-Spencer differential graded Lie algebra of X.

6. Proof of the main theorem

In this section, we use the results developed before to give a complete algebraic proof of the following theorem.

THEOREM 6.1. *Let X be a smooth projective variety of dimension n, defined over an algebraically closed field of characteristic 0. If the contraction map*

$$H^*(\Theta_X) \xrightarrow{i} \mathrm{Hom}^*(H^*(\Omega_X^n), H^*(\Omega_X^{n-1}))$$

is injective, then, for every affine open cover $\mathcal{U}$ of X, the DGLA $\mathrm{Tot}_{TW}(\Theta_X(\mathcal{U}))$ is quasi-abelian.

PROOF. According to Lemma 1.10, it is sufficient to prove that there exist a quasi abelian L_∞-algebra H and a morphism $\mathrm{Tot}_{TW}(\Theta_X(\mathcal{U})) \to H$ that is injective in cohomology.

Let n be the dimension of X and denote by Ω_X^* the algebraic de Rham complex. For every $i \leq n$, let $\check{C}(\mathcal{U}, \Omega_X^i)$ be the Čech complex of the coherent sheaf Ω_X^i, with respect to the affine cover $\mathcal{U}$, and $\check{C}(\mathcal{U}, \Omega_X^*)$ the total complex of the semicosimplicial differential graded vector space $\Omega_X^*(\mathcal{U})$ (Example 2.2). Notice that

$$\check{C}(\mathcal{U}, \Omega_X^*)^i = \bigoplus_{a+b=i} \check{C}(\mathcal{U}, \Omega_X^a)^b.$$

and $\check{C}(\mathcal{U}, \Omega_X^n)$ is a subcomplex of $\check{C}(\mathcal{U}, \Omega_X^*)$.

Then, we have a commutative diagram of complexes with horizontal quasi-isomorphisms:

$$\check{C}(\mathcal{U}, \Omega_X^n) = \mathrm{Tot}(\Omega_X^n(\mathcal{U})) \xrightarrow{E} \mathrm{Tot}_{TW}(\Omega_X^n(\mathcal{U}))$$

$$\check{C}(\mathcal{U}, \Omega_X^*) = \mathrm{Tot}(\Omega_X^*(\mathcal{U})) \xrightarrow{E} \mathrm{Tot}_{TW}(\Omega_X^*(\mathcal{U})).$$

Since $\mathbb{K}$ has characteristic 0 and X is smooth and proper, the Hodge spectral sequence degenerates at E_1 (we refer to [**Fa88, DeIl87**] for a purely algebraic proof of this fact). Therefore, we have injective maps

$$H^*(X, \Omega_X^n) = H^*(\check{C}(\mathcal{U}, \Omega_X^n)) \hookrightarrow H^*(\check{C}(\mathcal{U}, \Omega_X^*)) = H_{DR}^*(X/\mathbb{K}).$$

$$H^*(X, \Omega_X^{n-1}) = H^*(\check{C}(\mathcal{U}, \Omega_X^{n-1})) \hookrightarrow H^* \left(\frac{\check{C}(\mathcal{U}, \Omega_X^*)}{\check{C}(\mathcal{U}, \Omega_X^n)} \right).$$

Thus, the natural inclusions of complexes

$$\mathrm{Tot}_{TW}(\Omega_X^n(\mathcal{U})) \to \mathrm{Tot}_{TW}(\Omega_X^*(\mathcal{U})),$$

$$\mathrm{Tot}_{TW}(\Omega_X^{n-1}(\mathcal{U})) \to \frac{\mathrm{Tot}_{TW}(\Omega_X^*(\mathcal{U}))}{\mathrm{Tot}_{TW}(\Omega_X^n(\mathcal{U}))},$$

are injective in cohomology.

According to Example 4.2, for every open subset $U \subset X$, the contraction of vector fields with differential forms defines a Cartan homotopy:

$$i \colon \Theta_X(U) \to \mathrm{Hom}^*(\Omega_X^*(U), \Omega_X^*(U)), \qquad i_\xi(\omega) = \xi \lrcorner \omega.$$

Since the contraction $\lrcorner$ commutes with restrictions to open subsets, we have a semicosimplicial contraction

$$\Theta_X(\mathcal{U}) \times \Omega_X^*(\mathcal{U}) \xrightarrow{\lrcorner} \Omega_X^*(\mathcal{U}),$$

and, by Proposition 4.9, this induces naturally a Cartan homotopy

$$i \colon \mathrm{Tot}_{TW}(\Theta_X(\mathcal{U})) \longrightarrow \mathrm{Hom}^*(\mathrm{Tot}_{TW}(\Omega^*(\mathcal{U})), \mathrm{Tot}_{TW}(\Omega^*(\mathcal{U}))).$$

Notice that, for every $\xi \in \mathrm{Tot}_{TW}(\Theta_X(\mathcal{U}))$ and every i, we have

$$i_\xi(\mathrm{Tot}_{TW}(\Omega^i(\mathcal{U}))) \subset \mathrm{Tot}_{TW}(\Omega^{i-1}(\mathcal{U})),$$

$$l_\xi(\mathrm{Tot}_{TW}(\Omega^i(\mathcal{U}))) \subset \mathrm{Tot}_{TW}(\Omega^i(\mathcal{U})), \qquad l_\xi = di_\xi + i_{d\xi}.$$

Moreover, the assumption of the theorem, together with [**NaA87**, 3.1], implies that the map

$$\mathrm{Tot}_{TW}(\Theta_X(\mathcal{U})) \xrightarrow{i} \mathrm{Hom}^* \left(\mathrm{Tot}_{TW}(\Omega_X^n(\mathcal{U})), \mathrm{Tot}_{TW}(\Omega_X^{n-1}(\mathcal{U})) \right)$$

is injective in cohomology.

Next, consider the differential graded Lie algebras

$$M = \mathrm{Hom}^*(\mathrm{Tot}_{TW}(\Omega_X^*(\mathcal{U})), \mathrm{Tot}_{TW}\Omega_X^*(\mathcal{U})),$$

$$L = \{f \in M \mid f(\mathrm{Tot}_{TW}(\Omega_X^n(\mathcal{U}))) \subset \mathrm{Tot}_{TW}(\Omega_X^n(\mathcal{U}))\},$$

and let $\chi \colon L \to M$ be the inclusion. According to Example 3.5, the L_∞-algebras $\widetilde{\mathrm{Tot}}(\chi^\Delta)$ is quasi-abelian.

Moreover, we have $l(\mathrm{Tot}_{TW}(\Theta_X(\mathcal{U}))) \subset L$ and so, by Theorem 4.3, there exists a linear L_∞-morphism

$$\mathrm{Tot}_{TW}(\Theta_X(\mathcal{U})) \xrightarrow{(l,i)} \widetilde{\mathrm{Tot}}(\chi^\Delta), \qquad x \mapsto (l_x, i_x).$$

Since the map χ in injective, its mapping cone $\mathrm{Tot}(\chi^\Delta)$ is quasi-isomorphic to its cokernel

$$\mathrm{Coker}\, \chi = \mathrm{Hom}^* \left(\mathrm{Tot}_{TW}(\Omega_X^n(\mathcal{U})), \frac{\mathrm{Tot}_{TW}(\Omega_X^*(\mathcal{U}))}{\mathrm{Tot}_{TW}(\Omega_X^n(\mathcal{U}))} \right)$$

and we have a commutative diagram of complexes

$$\begin{array}{ccc}
\mathrm{Tot}_{TW}(\Theta_X(\mathcal{U})) & \xrightarrow{\;(l,i)\;} & \mathrm{Tot}(\chi^\Delta) \\
\downarrow{\scriptstyle i} & & \downarrow{\scriptstyle q-iso} \\
\mathrm{Hom}^*\big(\mathrm{Tot}_{TW}(\Omega_X^n(\mathcal{U})),\mathrm{Tot}_{TW}(\Omega_X^{n-1}(\mathcal{U}))\big) & \xrightarrow{\;\;\alpha\;\;} & \mathrm{Coker}\,\chi.
\end{array}$$

Since both i and α are injective in cohomology, also the L_∞-morphism (l,i) is injective in cohomology.

$\square$

THEOREM 6.2. *Let* $\mathcal{U} = \{U_i\}$ *be an affine open cover of a smooth projective variety* X *defined over an algebraically closed field of characteristic 0. If the canonical bundle of* X *is trivial or torsion, then the DGLA* $\mathrm{Tot}_{TW}(\Theta_X(\mathcal{U}))$ *is quasi-abelian.*

PROOF. Assume first that X has trivial canonical bundle. If n is the dimension of X, the cup product with a nontrivial section of the canonical bundle gives the isomorphisms $H^i(\Theta_X) \simeq H^i(\Omega_X^{n-1})$ and the conclusion follows immediately from Theorem 6.1. If X has torsion canonical bundle we may consider the canonical cyclic cover $\pi\colon Y \to X$ and the affine open cover $\mathcal{V} = \{\pi^{-1}(U_i)\}$. Now the variety Y has trivial canonical bundle and then the L_∞-algebra $\widetilde{\mathrm{Tot}}(\Theta_Y(\mathcal{V}))$ is quasi-abelian. Since π is an unramified cover the natural injective map $\widetilde{\mathrm{Tot}}(\Theta_X(\mathcal{U})) \to \widetilde{\mathrm{Tot}}(\Theta_Y(\mathcal{V}))$ is also injective in cohomology and we conclude the proof by using the same argument of Theorem 6.1.

$\square$

REMARK 6.3. When $\mathbb{K} = \mathbb{C}$, the previous theorems together with Remark 5.4, implies that the Kodaira-Spencer DGLA is quasi abelian, for a projective manifold with trivial or torsion canonical bundle.

THEOREM 6.4. *Let* X *be a smooth projective variety of dimension* n *defined over an algebraically closed field of characteristic 0. Then, the obstructions to deformations of* X *are contained in the kernel of the contraction map*

$$H^2(\Theta_X) \xrightarrow{\;i\;} \prod_p \mathrm{Hom}(H^p(\Omega_X^n), H^{p+2}(\Omega_X^{n-1})).$$

PROOF. We have seen in the proof of Theorem 6.1 that, for every affine open cover $\mathcal{U}$ of X, there exists an L_∞-morphism $\widetilde{\mathrm{Tot}}(\Theta_X(\mathcal{U})) \to \widetilde{\mathrm{Tot}}(\chi^\Delta)$ and that $\widetilde{\mathrm{Tot}}(\chi^\Delta)$ is quasi-abelian. Therefore, we are in the condition to apply the general strategy used in [**Ma04a, Ma09, Ia07**]: the deformation functor associated to $\widetilde{\mathrm{Tot}}(\chi^\Delta)$ is unobstructed and the obstructions of $\mathrm{Def}_X \simeq \mathrm{Def}_{\widetilde{\mathrm{Tot}}(\Theta_X(\mathcal{U}))}$ are contained in the kernel of the obstruction map $H^2(\mathrm{Tot}(\Theta_X(\mathcal{U}))) \to H^2(\mathrm{Tot}(\chi^\Delta))$. $\square$

COROLLARY 6.5. *Let* X *be a smooth projective variety defined over an algebraically closed field of characteristic 0. If the canonical bundle of* X *is trivial, then* X *has unobstructed deformations.*

PROOF. The previous Corollary 6.2 implies that $\widetilde{\mathrm{Tot}}(\Theta_X(\mathcal{U}))$ is quasi-abelian and so $\mathrm{Def}_{\widetilde{\mathrm{Tot}}(\Theta_X(\mathcal{U}))}$ is smooth. By Theorem 5.3, $\mathrm{Def}_X \cong \mathrm{Def}_{\widetilde{\mathrm{Tot}}(\Theta_X(\mathcal{U}))}$. $\square$

REMARK 6.6. Transcendental proofs of the analogue of Theorem 6.4 for compact Kähler manifolds can be found in [**Ma04a, Cle05, Ma09**], while we refer to

[Ia07, Ma09] for the proof that the T^1-lifting is definitely insufficient for proving Theorem 6.4.

References

[BK98]   S. Barannikov, M. Kontsevich: *Frobenius manifolds and formality of Lie algebras of polyvector fields.* Internat. Math. Res. Notices **4** (1998) 201-215.

[Bo78]   F. Bogomolov: *Hamiltonian Kählerian manifolds.* Dokl. Akad. Nauk SSSR **243** (1978) 1101-1104. Soviet Math. Dokl. **19** (1979) 1462-1465 .

[CG08]   X.Z. Cheng, E. Getzler: *Homotopy commutative algebraic structures.* J. Pure Appl. Algebra **212** (2008) 2535-2542; arXiv:math.AT/0610912.

[Cle05]   H. Clemens: *Geometry of formal Kuranishi theory.* Adv. Math. **198** (2005) 311-365.

[DeIl87]   P. Deligne, L. Illusie: *Relévements modulo p^2 et décomposition du complexe de de Rham.* Invent. Math. **89** (1987) 247-270.

[Dup76]   J.L. Dupont: *Simplicial de Rham cohomology and characteristic classes of flat bundles.* Topology **15** (1976) 233-245.

[Dup78]   J.L. Dupont: *Curvature and characteristic classes.* Lecture Notes in Mathematics **640**, Springer-Verlag, New York Berlin, (1978).

[EZ50]   S. Eilenberg, J.A. Zilber: *Semi-simplicial complexes and singular homology.* Ann. of Math. (2) **51** (1950) 499-513.

[Fa88]   G. Faltings: *p-adic Hodge theory.* J. Amer. Math. Soc. **1** (1988) 255-299.

[FM99]   B. Fantechi, M. Manetti: *On the T^1-lifting theorem.* J. Algebraic Geom. **8** (1999) 31-99.

[FHT01]   Y. Félix, S. Halperin, J. Thomas: *Rational homotopy theory.* Graduate texts in mathematics **205**, Springer-Verlag, New York Berlin, (2001).

[FiMa06]   D. Fiorenza, M. Manetti: L_∞ *algebras, Cartan homotopies and period maps.* Preprint arXiv:math/0605297.

[FiMa07]   D. Fiorenza, M. Manetti: L_∞ *structures on mapping cones.* Algebra Number Theory, **1**, (2007), 301-330; arXiv:math.QA/0601312.

[FiMa08]   D. Fiorenza, M. Manetti: *A period map for generalized deformations.* J. Noncommut. Geom. (to appear); arXiv:0808.0140v1.

[FMM08]   D. Fiorenza, M. Manetti, E. Martinengo: *Semicosimplicial DGLAs in deformation theory.* Preprint arXiv:0803.0399v1.

[Fu03]   K. Fukaya: *Deformation theory, homological algebra and mirror symmetry.* Geometry and physics of branes (Como, 2001), Ser. High Energy Phys. Cosmol. Gravit., IOP Bristol, (2003), 121-209. Electronic version available at http://www.math.kyoto-u.ac.jp/%7Efukaya/como.dvi

[Get04]   E. Getzler: *Lie theory for nilpotent L_∞-algebras.* Ann. of Math. **170** (1), (2009) 271-301; arXiv:math/0404003v4.
Ann. of Math., **170** (1), (2009), 271-301. arXiv:math/0404003v4.

[GM88]   W.M. Goldman, J.J. Millson: *The deformation theory of representations of fundamental groups of compact kähler manifolds.* Publ. Math. I.H.E.S. **67** (1988) 43-96.

[GM90]   W.M. Goldman, J.J. Millson: *The homotopy invariance of the Kuranishi space.* Illinois J. Math. **34** (1990) 337-367.

[Hin97]   V. Hinich: *Descent of Deligne groupoids.* Internat. Math. Res. Notices 1997, no. 5, 223–239.

[HiS97a]   V. Hinich, V. Schechtman: *Deformation theory and Lie algebra homology. I.* Algebra Colloq. 4 (1997), no. 2, 213–240.

[HiS97b]   V. Hinich, V. Schechtman: *Deformation theory and Lie algebra homology. II.* Algebra Colloq. 4 (1997), no. 3, 291–316.

[Hir78]   F. Hirzebruch: *Topological Methods in Algebraic Geometry.* Classics in Mathematics, Springer-Verlag, New York Berlin, (1978).

[HK91]   J. Huebschmann, T. Kadeishvili: *Small models for chain algebras.* Math. Z. **207** (1991) 245-280.

[Ia06]   D. Iacono: *Differential Graded Lie Algebras and Deformations of Holomorphic Maps.* Phd Thesis (2006) arXiv:math.AG/0701091.

[Ia07]   D. Iacono: *A semiregularity map annihilating obstructions to deforming holomorphic maps.* Canad. Math. Bull. (to appear); arXiv:0707.2454.

[Ia08]   D. Iacono: *L_∞-algebras and deformations of holomorphic maps.* Int. Math. Res. Not. **8** (2008) 36 pp. arXiv:0705.4532.

[Kad82]   T. V. Kadeishvili: *The algebraic structure in the cohomology of $A(\infty)$-algebras.* Soobshch. Akad. Nauk Gruzin. SSR **108** (1982), 249-252.

[Kaw92]   Y. Kawamata: *Unobstructed deformations - a remark on a paper of Z. Ran.* J. Algebraic Geom. **1** (1992), 183-190.

[Kod86]   K. Kodaira: *Complex manifold and deformation of complex structures.* Springer-Verlag (1986).

[Kon03]   M. Kontsevich: *Deformation quantization of Poisson manifolds, I.* Letters in Mathematical Physics **66** (2003) 157-216; arXiv:q-alg/9709040.

[KS00]   M. Kontsevich, Y. Soibelman: *Deformations of algebras over operads and Deligne's conjecture.* In: G. Dito and D. Sternheimer (eds) *Conférence Moshé Flato 1999, Vol. I (Dijon 1999)*, Kluwer Acad. Publ., Dordrecht (2000) 255-307; arXiv:math.QA/0001151.

[KS01]   M. Kontsevich, Y. Soibelman: *Homological mirror symmetry and torus fibrations.* K. Fukaya, (ed.) et al., Symplectic geometry and mirror symmetry. Proceedings of the 4th KIAS annual international conference, Seoul, South Korea, August 14-18, 2000. Singapore: World Scientific. (2001) 203-263; arXiv:math.SG/0011041.

[LM95]   T. Lada, M. Markl: *Strongly homotopy Lie algebras.* Comm. Algebra **23** (1995) 2147-2161; arXiv:hep-th/9406095.

[LS93]   T. Lada, J. Stasheff: *Introduction to sh Lie algebras for physicists.* Int. J. Theor. Phys. **32** (1993) 1087-1104; arXiv:hep-th/9209099.

[Ma99]   M. Manetti: *Deformation theory via differential graded Lie algebras.* In *Seminari di Geometria Algebrica 1998-1999* Scuola Normale Superiore (1999); arXiv:math.AG/0507284.

[Ma02]   M. Manetti: *Extended deformation functors.* Int. Math. Res. Not. **14** (2002) 719-756; arXiv:math.AG/9910071.

[Ma04a]   M. Manetti: *Cohomological constraint to deformations of compact Kähler manifolds.* Adv. Math. **186** (2004) 125-142; arXiv:math.AG/0105175.

[Ma04b]   M. Manetti: *Lectures on deformations of complex manifolds.* Rend. Mat. Appl. (7) **24** (2004) 1-183; arXiv:math.AG/0507286.

[Ma07]   M. Manetti: *Lie description of higher obstructions to deforming submanifolds.* Ann. Sc. Norm. Super. Pisa Cl. Sci. **6** (2007) 631-659; arXiv:math.AG/0507287.

[Ma09]   M. Manetti: *Differential graded Lie algebras and formal deformation theory.* In *Algebraic Geometry: Seattle 2005.* Proc. Sympos. Pure Math. **80** (2009) 785-810.

[Me99]   S.A. Merkulov: *Strong homotopy algebras of a Kähler manifold.* Intern. Math. Res. Notices (1999) 153-164; arXiv:math.AG/9809172.

[NaA87]   V. Navarro Aznar: *Sur la théorie de Hodge-Deligne.* Invent. Math. **90** (1987) 11-76.

[Pr03]   J. P. Pridham: *Deformations via Simplicial Deformation Complexes.* Preprint arXiv:math/0311168v6.

[Qui69]   D. Quillen: *Rational homotopy theory.* Ann. of Math. **90** (1969) 205-295.

[Ra92]   Z. Ran: *Deformations of manifolds with torsion or negative canonical bundle.* J. Algebraic Geom. **1** (1992), 279-291.

[Sch97]   V. Schechtman: *Local structure of moduli spaces.* arXiv:alg-geom/9708008.

[SS79]   M. Schlessinger, J. Stasheff: *Deformation Theory and Rational Homotopy Type.* Preprint (1979).

[Se06]   E. Sernesi: *Deformations of Algebraic Schemes.* Grundlehren der mathematischen Wissenschaften, **334**, Springer-Verlag, New York Berlin, (2006).

[Ti87]   G. Tian: *Smoothness of the universal deformation space of compact Calabi-Yau manifolds and its Petersson-Weil metric.* Mathematical Aspects of String Theory (San Diego, 1986), Adv. Ser. Math. Phys. 1, World Sci. Publishing, Singapore, (1987), 629-646.

[To89]   A.N. Todorov: *The Weil-Petersson geometry of the moduli space of $SU(n \geq 3)$ (Calabi-Yau) Manifolds I.* Commun. Math. Phys., **126**, (1989), 325-346.

[We94]   C.A. Weibel: *An introduction to homological algebra.* Cambridge Studies in Advanced Mathematics **38**, Cambridge Univesity Press, Cambridge, (1994).

[Whi57] H. Whitney: *Geometric integration theory.* Princeton University Press, Princeton, N. J., (1957).

INSTITUT FÜR MATHEMATIK,
JOHANNES GUTENBERG-UNIVERSITÄT,
STAUDINGERWEG 9, D 55128 MAINZ GERMANY.
   E-mail address: iacono@uni-mainz.de

DIPARTIMENTO DI MATEMATICA "GUIDO CASTELNUOVO",
SAPIENZA UNIVERSITÀ DI ROMA,
P.LE ALDO MORO 5, I-00185 ROMA ITALY.
   E-mail address: manetti@mat.uniroma1.it
   URL: www.mat.uniroma1.it/people/manetti/

Quantizing deformation theory

John Terilla

ABSTRACT. We describe a step toward quantizing deformation theory. The L_∞ operad is encoded in a Hochschild cocyle $\circ_1$ in a simple universal algebra $(P, \circ_0)$. This Hochschild cocyle can be extended naturally to a star product $\star = \circ_0 + \hbar \circ_1 + \hbar^2 \circ_2 + \cdots$. The algebraic structure encoded in $\star$ is the properad $\Omega(coFrob)$ which, conjecturally, controls a quantization of deformation theory—a theory for which Frobenius algebras replace ordinary commutative parameter rings.

1. Introduction

There is a well-known philosophy that deformations of anything are controlled by L_∞ algebras. It's more correct to say that deformations of anything *in characteristic zero* over *commutative parameter spaces* are controlled by L_∞ algebras. In this note, I'm not going to say anything about nonzero characteristic, but I will address the parameter spaces. When one refers to deformation theory abstractly, it is an umbrella theory that concerns deformations of a wide variety of mathematical objects. Examples include complex manifolds, associative algebras, vector bundles with connections, group representations, and quantum field theories. What makes deformation theory a theory in itself, what unifies such seemingly different situations, is the consideration of parameter spaces or rings. The idea is this: given a mathematical object X, deformation theory organizes X and its nearby deformations as a family fibered over a local base space $\mathcal{B}$. One can pullback a fibration over a base space $\mathcal{B}$ to a fibration over $\mathcal{B}'$ using a map $\mathcal{B}' \to \mathcal{B}$. The deformation theory of an object X leads to a contravariant functor from a category of parameter spaces to a category of sets:

$$\mathcal{B} \mapsto \{\text{deformations of } X \text{ over } \mathcal{B}\}/\{\text{equivalence}\}.$$

Using the usual correspondence between spaces and rings

$$\text{space } \mathcal{B} \leftrightarrow \text{ring of functions on } \mathcal{B}$$

one passes to a covariant functor from a category of (commutative, associative) rings to a category of sets.

A good example to keep in mind is that of complex manifolds: a deformation of a complex manifold M over a base space $\mathcal{B}$ is a fibration $\pi : \mathcal{N} \to \mathcal{B}$ together with an isomorphism between M and a fixed fiber $\pi^{-1}(b)$. Another example to have in mind is that of associative algebras: a deformation of an associative algebra A over

1991 *Mathematics Subject Classification.* 14B12,14D15,53D55,18D50.

a base ring R is an algebra structure on $A \otimes R$ together with a map of algebras $A \otimes R \to A$. Here, the picture is a family of algebra structures on A fibered over $spec(R)$.

Deformation theory itself is the study of functors $\mathfrak{F} : \textbf{Rings} \to \textbf{Sets}$ that satisfy a set of axioms, first identified by Schlessinger [13], that are observed in the known examples—call them *deformation functors*. One particularly important example of a deformation functor is $\mathfrak{Def}_L$ which arises from considering deformations of the differential in an L_∞ algebra L. It's defined by

$$\mathfrak{Def}_L(R) = \{\text{solutions to the master equation in } L \otimes R\}/\{\text{equivalence}\}.$$

Moreover, we have what Kontsevich calls [7] "the fundamental theorem of deformation theory"

THEOREM 1. *An L_∞ map $L \to L'$ induces a natural transformations of functors $\mathfrak{Def}_L \to \mathfrak{Def}_{L'}$, which is a natural equivalence if and only if $L \to L'$ is a quasi-isomorphism.*

A beautiful theorem of Manetti [9] states that

THEOREM 2. *For every deformation functor $\mathfrak{F}$ there exists an L_∞ algebra L so that $\mathfrak{F}$ is naturally equivalent to $\mathfrak{Def}_L$.*

One says that L *controls* the deformation of an object X if the deformation theory of X leads to a functor $\mathfrak{F}$ that is naturally equivalent to $\mathfrak{Def}_L$. Familiar examples are: deformations of a complex manifold X are controlled by a Dolbeault resolution of the sheaf of polyvectorfields (which obtains an L_∞ structure from the $\bar\partial$ operator and the Schouten-Nijenhuis bracket) and deformations of an associative algebra A are controlled by the Hochschild cochain complex of A (which obtains an L_∞ structure from the Hochschild coboundary operator and the Gerstenhaber bracket). The upshot of Theorems 1 and 2 is that deformation theory itself can be regarded as the homotopy theory of L_∞ algebras. Now, I want to emphasize that the correspondence between deformation theory and L_∞ algebras depends on the use of commutative parameter rings[1]. If R is not commutative, the set $\mathfrak{Def}_L(R)$ is not defined (one reason is that $L \otimes R$ isn't an L_∞ algebra). Many deformation functors can be defined for noncommutative rings. The geometric picture is that of families of structures fibered over noncommutative base spaces. There's strong evidence [8, 5] that deformations over noncommutative rings are controlled by A_∞, instead of L_∞, algebras.

In this note, I discuss generalizing deformation theory in a new, quantum direction. What I will discuss conjecturally corresponds to deformation theory over parameter rings which are Frobenius algebras—that is commutative associative algebras that have a compatible cocommutative coassociative coproduct. What I will do is define a certain simple commutative algebra $(P, \circ_0)$ and identify the L_∞ operad with a Hochschild one-cocycle of $(P, \circ_0)$ which I denote by $\circ_1$. There is a canonical quantization of $(P, \circ_0)$ producing a star product $\star = \circ_0 + \hbar\circ_1 + \hbar^2\circ_2 + \cdots$. This star product is identified with the $\Omega(coFrob)$ properad, which conjecturally controls deformations over Frobenius parameter rings.

---

[1]The setting for Theorem 2 is generalized to differential graded commutative parameter rings.

2. Linear algebra

Let V be a graded vector space, over a field of characteristic zero. Let V^n be shorthand for the n-fold tensor product $T^n V = V \otimes \cdots \otimes V$ and denote the element $v_1 \otimes \cdots \otimes v_1$ by $(v_1, \ldots, v_n)$. There is a simple product, call it $\hat{\circ}_0$,

$$\hat{\circ}_0 : \hom(V^i, V^j) \otimes \hom(V^m, V^n) \to \hom(V^{i+m}, V^{j+n})$$

defined by

$$(f \hat{\circ}_0 g)(v_1, \ldots, v_{i+m}) = f(v_1, \ldots, v_i) \otimes g(v_{i+1}, \ldots, v_{i+m})$$

for $f \in \hom(V^i, V^j)$ and $g \in \hom(V^m, V^n)$. (Don't worry about the confusing notation: I'm really interested in some other products that I'll be denoting without the hat: $\circ_0, \circ_1, \ldots$).

Define $S^n V$, the n-fold symmetric product of V as the quotient of $T^n V$ by a signed action of the symmetric group Σ_n on $T^n V$. To simplify this note, I supress shifts and signs, see [2] for the details of a coherent sign convention. There's a map $\pi : T^n V \to S^n V$ defined by $x \mapsto [x]$ and a map $\mathrm{sym} : S^n V \to T^n V$ defined by $[x] \mapsto \sum_{\sigma \in \Sigma_n} \sigma x$. Using sym and π one obtains a product

$$\circ_0 : \hom(S^i V, S^j V) \otimes \hom(S^m V, S^n V) \to \hom(S^{i+m} V, S^{j+n} V)$$

defined by

$$f \circ_0 g = \pi((\mathrm{sym}\, f\, \pi) \hat{\circ}_0 (\mathrm{sym}\, g\, \pi))\, \mathrm{sym},$$

Let $SV = \oplus_{n \geq 0} S^n V$, $\widehat{SV} = \prod_{n \geq 0} S^n V$, and $P = \hom(SV, \widehat{SV})$. Elements in SV are finite sums, elements of $\widehat{SV}$ are infinite sums, and elements of P are doubly indexed, infinite sums $\sum_{i,j=0}^{\infty} \alpha_j^i$ with each $\alpha_j^i \in \hom(S^i V, S^j V)$. One one has the product $\circ_0 : P \otimes P \to P$ defined term by term:

$$\left(\sum \alpha_j^i \right) \circ_0 \left(\sum \beta_n^m \right) = \sum \gamma_s^r$$

where $\gamma_s^r = \sum \alpha_j^i \circ_0 \beta_n^m$, the sum being the finite sum over indices i, j, m, n with $i + m = r$ and $j + n = s$. The result $(P, \circ_0)$ is a commutative, associative algebra.

Next, we define an operator $\circ_1 : P \otimes P \to P$ by using sym and π to transfer a $\hat{\circ}_1 : \hom(V^i, V^j) \otimes \hom(V^m, V^n) \to \hom(V^{i+m-1}, V^{j+n-1})$ product in the tensor product, as pictured below (in the picture, $i = 5$, $j = 3$, $m = 4$, and $n = 4$)

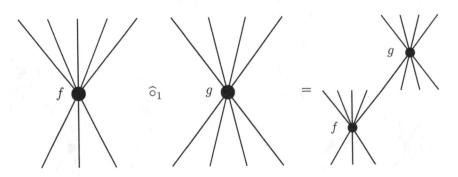

The picture means that $(f \hat{\circ}_1 g)(v_1, \ldots, v_{i+m-1}) = w_1, \ldots, w_{j+n-1}$ if $\{g(v_i, \ldots, v_{i+m-1}) = (y, w_{j+1}, \ldots, w_{j+n-1})$ and $f(v_1, \ldots, v_{i-1}, y) = (w_1, \ldots, w_j)$.

The product $\circ_1 : \hom(S^i V, S^j V) \otimes \hom(S^m V, S^n V) \to \hom(S^{i+m-1}, S^{j+n-1} V)$ is defined by

$$f \circ_1 g = \pi((\operatorname{sym} f \, \pi) \widehat{\circ}_1 (\operatorname{sym} g \, \pi)) \operatorname{sym}.$$

The effect of using sym and π is to sum, with signs, over outputs of g and inputs of f.

The operation $\circ_1$ encodes the theory of L_∞ algebras. Recall that an L_∞ structure on V is given by a degree one element $D \in \hom(SV, V)$ satisfying a certain condition. The condition is sometimes expressed by decomposing D as $D = \sum \mu_i$, each $\mu_i \in \hom(S^i V, V)$, and writing an infinite collection of quadratic relations satisfied by the μ_i. These relations are precisely given by saying that $D \circ_1 D = 0$. Therefore,

PROPOSTION 1. *A degree one map $D \in \hom(SV, V) \subset P$ defines an L_∞ structure on V if and only if $D \circ_1 D = 0$.*

Now we analyze $\circ_1$ as an element of $\hom(P \otimes P, P)$. Let

$$HC(P, P) = \oplus_{n>0} \hom(P^n, P)$$

denote the Hochschild cochain complex of $(P, \circ_0)$ with values in P. A simple but important observation is:

THEOREM 3. *The operation $\circ_1 : P \otimes P \to P$ is a Hochschild cocycle in $HC(P, P)$.*

PROOF. For homogeneous elements $f, g, h \in P = \hom(SV, \widehat{SV})$, we need to show that $(f \circ_1 g) \circ_0 h \pm f \circ_1 (g \circ_0 h) \pm (f \circ_0 g) \circ_1 h \pm f \circ_0 (g \circ_1 h) = 0$. The idea is contained in this picture:

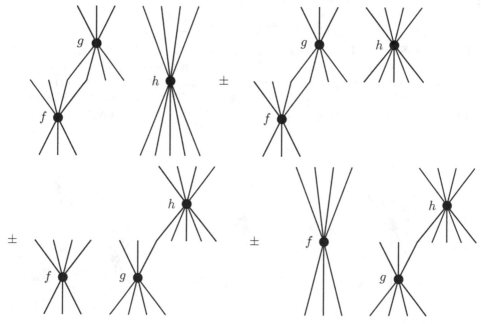

To see that all terms cancel, note that the first term cancels with half of the second term, and the last term cancels with half of the third term. The remaining halves of the second and third term, which consist of expressions that involve an output of f glued to an input of h cancel with each other.

$\square$

So, the triple $(P, \circ_0, \circ_1)$ presents a familiar situation: we have a commutative, associative algebra and a Hochschild one cocycle. One can try to "quantize" in the sense of deformation quantization. That is, deform $\circ_0$ in the direction $\circ_1$ by finding $\circ_2, \circ_3, \circ_4, \dots : P \otimes P \to P$ so that the star product

$$\star : P[[\hbar]] \otimes_{k[[\hbar]]} P[[\hbar]] \to P[[\hbar]]$$

defined for $f, g \in P$ by

$$f \star g = f \circ_0 g + \hbar f \circ_1 g + \hbar^2 f \circ_2 g + \hbar^3 f \circ_3 g \cdots$$

makes $P[[\hbar]]$ into an associative $k[[\hbar]]$ algebra. A canonical quantization exists and it's easy to give an explicit description of the terms $\circ_k$, which is transferred to P from a product $\widehat{\circ}_k : \hom(V^i, V^j) \otimes \hom(V^m, V^n) \to \hom(V^{i+m-k}, V^{n+j-k})$ defined for $f \in \hom(V^i, V^j)$ and $g \in \hom(V^m, V^n)$ by glueing the last k outputs of f to the first k inputs of g as below (in the picture, $i = 5$, $j = 3$, $m = 4$, $n = 4$, and $k = 2$):

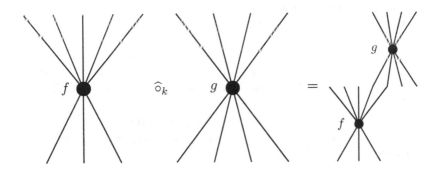

I consider the "quantization" of the concept of an L_∞ algebra as the passage

(1) $\{(V, D) : D \in \hom(SV, V) : D \circ_1 D = 0\}$

$$\rightsquigarrow \{(V, H) : H \in \hom(SV, \widehat{SV})[[\hbar]] : H \star H = 0\}.$$

The L_∞ concept controls a certain type of many-to-one operation algebraic structure. Its quantization controls a certain type of many-to-many, genus-graded, algebraic structure. (One could factor the passage (1) through $\{(V, B) : B \in \hom(SV, SV) : B \circ_1 B = 0\}$, which would control an algebraic structure consisting of many-to-many genus zero operations, bi-Lie infinity algebras in fact, but once you have many-to-many operations, it's more natural to introduce genus as well.) Given an element $H \in \hom(SV, SV)[[\hbar]]$, you can decompose

$$H = \sum_g \sigma_g \hbar^g \quad \text{where } \sigma_g \in \hom(SV, \widehat{SV})$$

and decompose $\sigma_g = \sum_{i,j} (\sigma_g)_i^j$ where $\sigma_i^j : S^i V \to S^j V$. Then, the condition that $H \star H = 0$ is equivalent to a collection of quadratic relations on the $(\sigma_g)_i^j$. In particular, $\{(\sigma_0)_i^1\}$ is an L_∞ algebra. One would like to describe the algebra structure on V defined by a degree one $H \in P[[\hbar]]$ satisfying $H \star H = 0$. Using the properadic language of Vallette [16], one finds that:

THEOREM 4. [2] *A degree one* $H \in \hom(SV, SV)[[\hbar]]$ *satisfying* $H \star H = 0$ *is equivalent to giving* V *the structure of a* $\Omega(coFrob)$ *algebra.*

For a detailed description of properadic algebra, see [16, 10] and for the proof of Theorem 4 see [2]. Here, I want to sketch a high-level picture. The expression *coFrob* denotes a certain, simple coproperad—the coFrobenius coproperad—which I will say more about in a moment. The symbol Ω denotes "cobar". Cobar, and bar, are examples of more general categorical constructions. In this case, the cobar construction assigns a properad to a coproperad, and fits into an adjunction with the bar construction, which assigns a coproperad to a properad. There are functors

$$\Omega : coProperads \to Properads \text{ and } B : Properads \to coProperads$$

satisfying

$$\hom_{properads}(\Omega(C), P) \simeq \hom_{coproperads}(C, B(P)).$$

An antecedent example was identified in the paper of Ginzburg and Kapranov on Koszul duality for operads [4]: the L_∞ operad is obtained as $\Omega(coCom)$, where *coCom* is the co-Commutative cooperad, governing cocommutative algebras. And this brings me back to the parameter rings.

The commutative operad *Com* works as a unit for the tensor product in a large category of operads. For each $n > 1$, $Com(n)$ is one dimensional. In terms of representations, a *Com* algebra is a commutative associative algebra: if $(W, \cdot)$ is a commutative associative algebra, then there is one way to compose $\cdot$ to get a map $W^n \to W$ for $n \geq 1$. Given any operad[2] O, one has $O \otimes Com \simeq O$. An implication of this is that

> If $L = (V, D)$ is an $L_\infty = \Omega(coCom)$ algebra and $C = (W, \cdot)$ is a commutative algebra, then $L \otimes C$ is again an $L_\infty = \Omega(coCom)$ algebra.

In order to define the functor Def_L, it was required that for a parameter ring R, $V \otimes R$ be an L_∞ algebra.

For properads without nullary operations, Frobenius plays the role of the unit. A Frobenius algebra consists of a triple $F = (W, \cdot, \Delta)$ where $\cdot$ is a commutative associative product on W and Δ is a coccommutative coassociative coproduct on W, satisfying a compatibility. The compatibility between the product and the coproduct is such that for each number of input $m > 0$, number of output $n > 0$, and genus $g \geq 0$, all ways to compose the product and the coproduct to get a genus g operation $W^n \to W^m$ are the same. In the Frobenius properad, $Frob(m, n, g)$ is one dimensional for each m, n, g. An implication is

> If $Z = (V, H)$ is a $\Omega(CoFrob)$ algebra and $F = (W, \cdot, \Delta)$ is a Frobenius algebra, then $Z \otimes F$ is again an $\Omega(CoFrob)$ algebra.

The program of quantizing deformation theory is to develop a theory of functors of Frobenius algebras. The important example will be Def_Z for a $\Omega(coFrob)$ algebra $Z = (V, H)$, which will be defined by

$$Def_Z(F) = \{\text{solutions to the master equation in } Z \otimes F\}/\{\text{equivalence}\}.$$

---

[2]There is a detail here: if the operad has n-to-0 operations, then one should replace *Com* with unital *Com*.

The importance of the particular functor Def_Z will be a version of Manetti's theorem: For every set-valued functor $\mathfrak{F}$ of Frobenius algebras satsifying a version of Schlessinger's axioms, there exists a $\Omega(coFrob)$ algebra $Z = (V, H)$ so that $\mathfrak{F}$ is naturally equivalent to Def_Z.

The program would be applied in settings where deformation theories are already established, by finding an algebra over $\Omega(coFrob)$ that extends the L_∞ algebra governing the deformation theory. By extend, I mean that the n-to-1, genus zero part of the $\Omega(coFrob)$ algebra, which is an L_∞ algebra, controls the given deformation theory over commutative parameter rings. Examples: symplectic field theory[**3, 15**], Park's algebraic quantum field theory [**11, 12**], differential BV algebras such as those arising from a CY-category[**1, 6**], and string topology [**14**].

References

[1] K. Costello, T. Tradler, and M. Zeinalian. Closed string TCFT for Hermitian Calabi-Yau elliptic spaces. arXiv:0807.3052, 2008.

[2] G. Drummond-Cole, J. Terilla, and T. Tradler. Algebras over cobar(cofrob). math.arXiv:0807.1241, 2008.

[3] Y. Eliashberg, A. Givental, and H. Hofer. Introduction to symplectic field theory. math.SG/0010059, 2000.

[4] V. Ginzberg and M. Kapranov. Koszul duality for operads. *Duke J. Math*, 76:203–272, 1994.

[5] J. Hirsh. Noncommutative deformation theory. in progress.

[6] L. Katzarkov, M. Kontsevich, and T. Pantev. Hodge theoretic aspects of mirror symmetry, 2008.

[7] M. Kontsevich. Deformation quantization of Poisson manifolds, I. *Lett. Math. Phys.*, 66:157–216, 2003.

[8] M. Kontsevich and Y. Soibelman. Notes on A_∞ algebras, A_∞ categories, and non-commutative geometry, I. arXiv:math/0606241, 2006, 2006.

[9] M. Manetti. Extended deformation functors. *Internat. Math. Res. Not.*, 14:719–756, 2002.

[10] S. Merkulov and B. Vallette. Deformation theory of representations of prop(erad)s. math.arXiv:0707.0889, 2007.

[11] J.-S. Park. Flat family of QFTs and deformations of d-algebras. hep-th/0308130, 2003, 2003.

[12] J. S. Park, J. Terilla, and T. Tradler. Quantum backgrounds and QFT. *Intl. Math. Res. Notices*, 2009.

[13] M. Schlessinger. Functors of Artin rings. *Trans. Am. Math. Soc.*, 130:208–222, 1968.

[14] D. Sullivan and M. Chas. String topology. math.GT/9911159, 1999.

[15] M. Sullivan, J. Terilla, and T. Tradler. Symplectic field theory and quantum backgrounds. *J. Symplectic Geom.*, 6(4):379405, 2008.

[16] B. Vallette. A Koszul duality for props. *Trans. Amer. Math. Soc.*, 359:4865–4943, 2007.

QUEENS COLLEGE AND CUNY GRADUATE CENTER, CITY UNIVERSITY OF NEW YORK
E-mail address: jterilla@qc.cuny.edu

L_∞-interpretation of a classification of deformations of Poisson structures in dimension three

Anne Pichereau

ABSTRACT. We give an L_∞-interpretation of the classification, obtained in [17], of the formal deformations of a family of exact Poisson structures in dimension three. We indeed reobtain the explicit formulas for all the formal deformations of these Poisson structures, together with a classification in the generic case, by constructing a suitable quasi-isomorphism between two L_∞-algebras, which are associated to these Poisson structures.

1. Introduction

In [17], we have exhibited a classification of the formal deformations of the Poisson structures defined on $\mathbf{F}[x, y, z]$ ($\mathbf{F}$ is an arbitrary field of characteristic zero), of the form:

$$(1) \qquad \{\cdot, \cdot\}_\varphi = \frac{\partial \varphi}{\partial x} \frac{\partial}{\partial y} \wedge \frac{\partial}{\partial z} + \frac{\partial \varphi}{\partial y} \frac{\partial}{\partial z} \wedge \frac{\partial}{\partial x} + \frac{\partial \varphi}{\partial z} \frac{\partial}{\partial x} \wedge \frac{\partial}{\partial y},$$

where φ is a weight-homogeneous polynomial of $\mathbf{F}[x, y, z]$, admitting an isolated singularity, in the generic case. In the present paper, following an idea of B. Fresse, we give an L_∞-interpretation of this result, that is to say, we obtain this result again by methods, which are different and which use the theory of L_∞-algebras.

The Poisson structures appear in classical mechanics, where physical systems are described by commutative algebras which are algebras of smooth functions on Poisson manifolds. They generalize the symplectic structures, as for example the natural symplectic structure on $\mathbf{R}^{2r}$, which was introduced by D. Poisson in 1809. On the contrary, in quantum mechanics, physical systems are described by non-commutative algebras, which are algebras of observables on Hilbert spaces, and P. Dirac has observed that, up to a factor depending on the Planck's constant, the commutator of observables appearing in the work of W. Heisenberg is the analogue of the Poisson bracket of classical mechanics. The Poisson structures and their deformations also appear in the theory of deformation quantization (see for instance [2]) with, in particular, the very important result obtained by M. Kontsevich in 1997: given a Poisson manifold (M, π) and the associative algebra $(\mathcal{A} = C^\infty(M), \cdot)$, there is a one-to-one correspondence between the equivalence classes of star products of $\mathcal{A}$,

2000 *Mathematics Subject Classification.* 17B63, 58H15, 16E45.
Key words and phrases. Deformations, L_∞-algebras, Poisson structures.
The author was supported by a grant of the EPDI (European Post-Doctoral Institut).

for which the first term is π, and the equivalence classes of the formal deformations of π.

In a more general context, a *Poisson structure* on an associative commutative algebra $\mathcal{A}$ is a Lie algebra structure on $\mathcal{A}$, $\pi : \mathcal{A} \times \mathcal{A} \to \mathcal{A}$, which is a biderivation of $\mathcal{A}$ (see paragraph 2.2.1). In the case where $\mathcal{A} = C^\infty(M)$ is the algebra of smooth functions over a manifold M, one says that (M, π) is a Poisson manifold. Formally deforming a Poisson structure π defined on an associative commutative algebra $\mathcal{A}$ means considering the Poisson structures π_* defined on the ring $\mathcal{A}[[\nu]]$ of all the formal power series with coefficients in $\mathcal{A}$ and in one parameter ν, which extend the initial Poisson structure (i.e., which are π, modulo ν). In this paper and in [17], we study a classification of formal deformations of Poisson structures modulo equivalence, two formal deformations π_* and π'_* of π being equivalent if there exists a morphism $\Phi : (\mathcal{A}[[\nu]], \pi_*) \to (\mathcal{A}[[\nu]], \pi'_*)$ of Poisson algebras over $\mathbf{F}[[\nu]]$ which is the identity modulo ν. There is a similar definition for the formal deformations of an associative product, the $*$-products being formal deformations of an associative product, for which each coefficient is a bidifferential operator. We refer to [17] for an introduction to the study of formal deformations of Poisson structures and the role played by the Poisson cohomology in this study.

M. Kontsevich proved the one-to-one correspondence mentioned above by using the theory of L_∞-algebras and Maurer-Cartan equations. In fact, he obtained this result by proving his conjecture of *formality* for a certain differential graded Lie algebra. A *differential graded Lie algebra* (*dg Lie algebra*, in short) is a graded Lie algebra $(\mathfrak{g}, [\cdot, \cdot]_\mathfrak{g})$, endowed with a differential $\partial_\mathfrak{g}$, which is a graded derivation with respect to $[\cdot, \cdot]_\mathfrak{g}$. The differential $\partial_\mathfrak{g}$ is a degree 1 map satisfying $\partial_\mathfrak{g} \circ \partial_\mathfrak{g} = 0$, giving rise to a cohomology $H(\mathfrak{g}, \partial_\mathfrak{g})$. A dg Lie algebra is a particular example of an L_∞-*algebra* (also called *strongly homotopy Lie algebra*), which is a graded vector space L, equipped with a collection of skew-symmetric multilinear maps $(\ell_n)_{n \in \mathbf{N}^*}$, satisfying different conditions, which can be viewed as generalized Jacobi identities. (The strongly homotopy algebras were introduced by J. Stasheff in [18] in the associative case, see also [10] and [11].) An $(L_\infty\text{-})$*quasi-isomorphism* between two dg Lie algebras (or between two L_∞-algebras) is an L_∞-morphism between them (that is to say a collection of multilinear maps $(f_n)_{n \in \mathbf{N}^*}$ from one to the other, satisfying a collection of compatibility conditions), which induces an isomorphism between their cohomologies. These notions will be recalled in the paragraph 2.1. A dg Lie algebra is said to be *formal* if there exists a quasi-isomorphism between it and the dg Lie algebra given by its cohomology $H(\mathfrak{g}, \partial_\mathfrak{g})$ (equipped with the trivial differential and the graded Lie bracket induced by $[\cdot, \cdot]_\mathfrak{g}$). To a dg Lie algebra $(\mathfrak{g}, \partial_\mathfrak{g}, [\cdot, \cdot]_\mathfrak{g})$ is associated an equation, called the *Maurer-Cartan equation* and given by:

$$\partial_\mathfrak{g}(\gamma) + \frac{1}{2}[\gamma, \gamma]_\mathfrak{g} = 0,$$

whose solutions $\gamma \in \mathfrak{g}^1$ are degree one homogeneous elements of $\mathfrak{g}$, which can also be considered as depending on a formal parameter ν, $\gamma \in \nu\mathfrak{g}^1[[\nu]]$. The set of all these formal solutions is denoted by $\mathcal{MC}^\nu(\mathfrak{g})$. Notice that there is a also a notion of *generalized Maurer-Cartan equation* associated to an L_∞-algebra, which is more complicated (because it takes into account the whole L_∞-structure). Given a Poisson manifold (M, π) (respectively, a Poisson algebra $(\mathcal{A}, \pi)$), the Poisson cohomology complex $H(M, \pi)$ associated to (M, π) (respectively, $H(\mathcal{A}, \pi)$ associated to $(\mathcal{A}, \pi)$) is defined as follows: the cochains are the polyvector fields (respectively, the skew-symmetric multiderivations of $\mathcal{A}$) and the Poisson coboundary operator is given by $\delta_\pi := -[\cdot, \pi]_S$, where $[\cdot, \cdot]_S$ is the Schouten bracket (which is a graded Lie bracket, obtained by extending the commutator of vector fields to a graded biderivation with respect to the wedge product). One can then associate to π the dg Lie algebra $\mathfrak{g}_\pi$, given by the graded vector space of the Poisson cochains (with a shift of degree), equipped with the Poisson coboundary operator associated to π as differential (up to a sign) and the Schouten bracket as graded Lie bracket. Because a degree one element $\gamma \in \mathfrak{g}_\pi^1$ or $\gamma \in \nu \mathfrak{g}_\pi^1[[\nu]]$ satisfies the Jacobi identity (hence, is a Poisson structure) if and only if $[\gamma, \gamma]_S = 0$, an element $\gamma \in \nu \mathfrak{g}_\pi^1[[\nu]]$ is then a formal solution of the Maurer-Cartan equation associated to $\mathfrak{g}_\pi$ if and only if $\pi + \gamma$ is a formal deformation of π. Similarly, the star products also correspond to the formal solutions of the Maurer-Cartan equation associated to a dg Lie algebra $\mathfrak{g}_H$, constructed from the Hochschild cohomology complex of the associative algebra $(C^\infty(M), \cdot)$. M. Kontsevich showed that the dg Lie algebra $\mathfrak{g}_H$ is formal, by showing that it is quasi-isomorphic to the particular dg Lie algebra $\mathfrak{g}_\pi$, associated to the trivial Poisson bracket $\pi = 0$. This result, together with the fact that a quasi-isomorphism between two dg Lie algebras induces a bijection between the sets of all the formal solutions of the Maurer-Cartan equations modulo a gauge equivalence, leads to the desired one-to-one correspondence.

In this paper, we follow an idea of B. Fresse to reobtain, but with L_∞-methods, the explicit formulas for all the formal deformations (modulo equivalence) of $\{\cdot, \cdot\}_\varphi$ (defined in (1) and sometimes called *exact Poisson structures*), which were obtained in [**17**], when $\varphi \in \mathcal{A} := \mathbf{F}[x, y, z]$ is a weight homogeneous polynomial with an isolated singularity, and in particular, the classification of the formal deformations of these Poisson structures, when φ is generic (i.e., when its weighted degree is different from the sum of the weights of the three variables x, y and z or, equivalently, when $H^1(\mathcal{A}, \{\cdot, \cdot\}_\varphi)$ is zero). To do this, we show that this classification is not a consequence of the formality of a certain dg Lie algebra, but still of the existence of a suitable *quasi-isomorphism between two L_∞-algebras*. In order to explain this, let us consider φ a polynomial as before and $(\mathfrak{g}_\varphi, \partial_\varphi, [\cdot, \cdot]_S)$, the dg Lie algebra associated to the Poisson algebra $(\mathcal{A} := \mathbf{F}[x, y, z], \{\cdot, \cdot\}_\varphi)$ and $H_\varphi := H(\mathcal{A}, \{\cdot, \cdot\}_\varphi)$ its associated Poisson cohomology. As said before, there is a shift of degree implying that H_φ^ℓ, the homogeneous part of H_φ of degree ℓ, is in fact the $(\ell + 1)$-st Poisson cohomology space $H^{\ell+1}(\mathcal{A}, \{\cdot, \cdot\}_\varphi)$, associated to $(\mathcal{A}, \{\cdot, \cdot\}_\varphi)$.

In fact, the classification of the formal deformations of the Poisson bracket $\{\cdot, \cdot\}_\varphi$ obtained in [**17**] was indexed by elements of $H_\varphi^1 \otimes \nu \mathbf{F}[[\nu]] = H^2(\mathcal{A}, \{\cdot, \cdot\}_\varphi) \otimes \nu \mathbf{F}[[\nu]]$ and B. Fresse pointed out to me that it could come from the formality of the dg Lie algebra $\mathfrak{g}_\varphi$, or at least from the existence of a suitable quasi-isomorphism between H_φ and the dg Lie algebra $\mathfrak{g}_\varphi$, where H_φ would be equipped with a suitable L_∞-algebra structure. In our context, having a suitable L_∞-algebra structure on H_φ means that the set of all the formal solutions of the generalized Maurer-Cartan equation associated to H_φ would be exactly $H_\varphi^1 \otimes \nu \mathbf{F}[[\nu]]$. Indeed, as in the case of dg Lie algebras, a quasi-isomorphism between two L_∞-algebras induces an isomorphism between the sets of formal solutions of the corresponding generalized Maurer-Cartan equations, modulo a gauge equivalence, and in our case, the formal solutions of the Maurer-Cartan equation associated to $\mathfrak{g}_\varphi$, modulo the gauge equivalence, correspond exactly to the equivalence classes of the formal deformations of the Poisson structure $\{\cdot, \cdot\}_\varphi$. Using the explicit bases exhibited for the Poisson cohomology associated to $(\mathcal{A}, \{\cdot, \cdot\}_\varphi)$ in [**16**] and the idea of B. Fresse, we indeed obtained the following result (see the theorems 4.3 and 4.5):

THEOREM 1.1. *Let $\varphi \in \mathcal{A} = \mathbf{F}[x, y, z]$ be a weight homogeneous polynomial with an isolated singularity. Consider $(\mathfrak{g}_\varphi, \partial_\varphi, [\cdot, \cdot]_S)$ the dg Lie algebra associated to the Poisson algebra $(\mathcal{A}, \{\cdot, \cdot\}_\varphi)$, as explained above, where $\{\cdot, \cdot\}_\varphi$ is given by (1), and H_φ the cohomology associated to the cochain complex $(\mathfrak{g}_\varphi, \partial_\varphi)$.*
There exist

(1) *an L_∞-algebra structure on H_φ, such that the generalized Maurer-Cartan equation associated to H_φ is trivial (i.e., every element $\gamma \in \nu H_\varphi^1[[\nu]]$ is solution);*

(2) *a quasi-isomorphism $f_\bullet^\varphi$ from the L_∞-algebra H_φ to the dg Lie algebra $(\mathfrak{g}_\varphi, \partial_\varphi, [\cdot, \cdot]_S)$, such that the isomorphism, induced by $f_\bullet^\varphi$ between the formal solutions of the Maurer-Cartan equations, sends $\mathcal{MC}^\nu(H_\varphi)$ to the representatives, exhibited in [**17**], for all the formal deformations of the Poisson bracket $\{\cdot, \cdot\}_\varphi$, modulo equivalence.*

This theorem 1.1 permits us to recover the results obtained in [**17**], concerning the formal deformations of the Poisson structures $\{\cdot, \cdot\}_\varphi$. It also permits us to better understand different phenomena about this result. In particular, we used in [**17**] that, in the generic case, $H^1(\mathcal{A}, \{\cdot, \cdot\}_\varphi)$ is zero and we now know that this fact implies that the gauge equivalence in $\mathcal{MC}^\nu(H_\varphi)$ is trivial. Moreover, in the special case (when the weighted degree of φ is the sum of the weights of the three variables x, y and z or, equivalently, when $H^1(\mathcal{A}, \{\cdot, \cdot\}_\varphi)$ is not zero), we can now better understand the equivalence classes of the formal deformations, as the equivalence relation for the formal deformations of $\{\cdot, \cdot\}_\varphi$ can be obtained by transporting the gauge equivalence in $\mathcal{MC}^\nu(H_\varphi)$ to $\mathcal{MC}^\nu(\mathfrak{g}_\varphi)$.

Finally, notice that, given a dg Lie algebra $(\mathfrak{g}, \partial_\mathfrak{g}, [\cdot, \cdot]_\mathfrak{g})$ and a choice of bases for the cohomology spaces $H^\ell(\mathfrak{g}, \partial_\mathfrak{g})$ associated to the cochain complex $(\mathfrak{g}, \partial_\mathfrak{g})$, and using a theorem of transfer structure (see for instance the "move" (M1) of [**15**]), we know that there always exist an L_∞-algebra structure on $H(\mathfrak{g}, \partial_\mathfrak{g})$, together with a quasi-isomorphism between this L_∞-algebra and the dg Lie algebra $(\mathfrak{g}, \partial_\mathfrak{g}, [\cdot, \cdot]_\mathfrak{g})$. The problem to use this result in our context where $\mathfrak{g} = \mathfrak{g}_\varphi$, is that we do not need only the existence of this L_∞-structure and this quasi-isomorphism, but we also need to be able to control these data, in order:

(1) for the formal solutions of the generalized Maurer-Cartan equation associated to H_φ to be simple (given by $H_\varphi^1 \otimes \nu \mathbf{F}[[\nu]]$),

(2) for the image of the isomorphism, induced by $f_\bullet^\varphi$ between the sets of formal solutions of the Maurer-Cartan equations, to give exactly the representatives of the formal deformations of the Poisson bracket $\{\cdot, \cdot\}_\varphi$ modulo equivalence which were exhibited in [**17**].

To be able to do this, we have proved, in the section 3, a proposition which permits one, given a dg Lie algebra $(\mathfrak{g}, \partial_\mathfrak{g}, [\cdot, \cdot]_\mathfrak{g})$ and a choice of basis for its associated cohomology $H(\mathfrak{g}, \partial_\mathfrak{g})$, to construct step by step, both an L_∞-algebra structure $\ell_\bullet = (\ell_n)_{n \in \mathbf{N}^*}$ on $H(\mathfrak{g}, \partial_\mathfrak{g})$ with $\ell_1 = 0$, and a quasi-isomorphism $f_\bullet = (f_n)_{n \in \mathbf{N}^*}$ from $H(\mathfrak{g}, \partial_\mathfrak{g})$ to $\mathfrak{g}$, such that, at each step, whatever the choices made at the previous steps for the maps $\ell_2, \ldots, \ell_{m-1}$ and $f_1, \ldots, f_{m-1}$, satisfying the conditions required at this step, the collections of maps $(\ell_n)_{1 \le n \le m-1}$ and $(f_n)_{1 \le n \le m-1}$ extend to an L_∞-algebra structure $\ell_\bullet$ on $H(\mathfrak{g}, \partial_\mathfrak{g})$ and a quasi-isomorphism $f_\bullet$ from $H(\mathfrak{g}, \partial_\mathfrak{g})$ to $\mathfrak{g}$. This result is given in the proposition 3.1 and permits us, together with the explicit bases exhibited for the Poisson cohomology associated to $(\mathcal{A}, \{\cdot, \cdot\}_\varphi)$ in [16], to prove the desired theorem 1.1.

Acknowledgments. I am grateful to B. Fresse for having introduced me to the theory of L_∞-algebras anf for pointing out to me a possible L_∞-interpretation of my results. Moreover, I would like to thank J. Stasheff for interesting and useful comments on this work. I also would like to thank H. Abbaspour, M. Marcolli and T. Tradler for giving me the opportunity of participating to this volume. Finally, this work has been done when I was a visitor at the CRM (Centre de Recerca Matemàticà, Barcelona) and at the MPIM (Max-Planck-Institut für Mathematik, Bonn), whose hospitality are also greatly acknowledged.

2. Preliminaries: L_∞-algebras and Poisson algebras

In this first section we recall some definitions in the theory of L_∞-algebras. We indeed need to fix our sign conventions and the notations. The notions of Poisson structures, cohomology and the deformations of Poisson structures are also recalled.

2.1. L_∞-algebras and morphisms, Maurer-Cartan equations.

We first recall the notions of L_∞-algebras, L_∞-morphisms, Maurer-Cartan equations, mainly in order to fix the sign conventions. For these notions and the conventions we choose, we refer to (the appendix A of) [13] (see also [10] and [3]).

In this paper, $\mathbf{F}$ is an arbitrary field of characteristic zero and every algebra, dg Lie algebra, L_∞-algebra, etc, is considered over $\mathbf{F}$. If V is a graded[1] vector space, we denote by $|x| \in \mathbf{Z}$ the degree of a homogeneous element x of V. Let us denote by $\bigwedge^\bullet V$, the graded commutative associative algebra (rather denoted by $\bigodot^\bullet V$ in [13]), obtained by dividing the tensor algebra $T^\bullet V = \bigoplus_{k \in \mathbf{N}} V^{\otimes k}$ of V by the ideal generated by the elements of the form $x \otimes y - (-1)^{|x||y|} y \otimes x$. Denoting by $\wedge$ the product in $\bigwedge^\bullet V$, one then has:

$$x \wedge y = (-1)^{|x||y|} y \wedge x,$$

where x and y are homogeneous elements of V. Then, if $x_1, \ldots, x_k \in V$ are homogeneous elements of V and $\sigma \in \mathfrak{S}_k$ is a permutation of $\{1, \ldots, k\}$, one defines the so-called *Koszul sign* $\varepsilon(\sigma; x_1, \ldots, x_k)$, associated to $x_1, \ldots, x_k$ and σ, by the equality:

$$x_1 \wedge \cdots \wedge x_k = \varepsilon(\sigma; x_1, \ldots, x_k) \, x_{\sigma(1)} \wedge \cdots \wedge x_{\sigma(k)},$$

valid in the algebra $\bigwedge^\bullet V$. Then, one also defines the number $\chi(\sigma; x_1, \ldots, x_k) \in \{-1, 1\}$, by:

$$\chi(\sigma; x_1, \ldots, x_k) := \text{sign}(\sigma) \, \varepsilon(\sigma; x_1, \ldots, x_k),$$

where $\text{sign}(\sigma)$ denotes the sign of the permutation σ. When no confusion can arise, we write $\chi(\sigma)$ for $\chi(\sigma; x_1, \ldots, x_k)$. For $i, j \in \mathbf{N}$, a (i,j)-*shuffle* is a permutation $\sigma \in \mathfrak{S}_{i+j}$ such that $\sigma(1) < \cdots < \sigma(i)$ and $\sigma(i+1) < \cdots < \sigma(i+j)$, and the set of

---

[1] We here consider graded vector spaces as being graded over $\mathbf{Z}$, but for the specific cases which we study in the section 4, the considered graded vector spaces are graded only on $\mathbf{N} \cup \{-1\}$

all (i,j)-shuffles is denoted by $S_{i,j}$.

2.1.1. L_∞-*algebras*. An L_∞-*algebra* L is graded vector space $L = \bigoplus_{n\in\mathbf{Z}} L^n$, equipped with a collection of linear maps

$$\ell_\bullet = \left(\ell_k : \bigotimes^k L \to L \right)_{k\in\mathbf{N}^*},$$

such that:

    a. each map ℓ_k is a graded map of degree $\deg(\ell_k) = 2 - k$,
    b. each map ℓ_k is a skew-symmetric map, which means that

$$\ell_k(\xi_{\sigma(1)}, \dots, \xi_{\sigma(k)}) = \chi(\sigma)\ell_k(\xi_1, \dots, \xi_k),$$

    for all homogeneous elements $\xi_1, \dots, \xi_k \in L$ and all permutation $\sigma \in \mathfrak{S}_k$;
    c. the maps $\ell_k, k \in \mathbf{N}^*$ satisfy the following "generalized Jacobi identity":

$$(\mathfrak{J}_n) \quad \sum_{\substack{i+j=n+1 \\ i,j\geq 1}} \sum_{\sigma\in S_{i,n-i}} \chi(\sigma)(-1)^{i(j-1)}\, \ell_j\left(\ell_i\left(\xi_{\sigma(1)}, \dots, \xi_{\sigma(i)}\right), \xi_{\sigma(i+1)}, \dots, \xi_{\sigma(n)}\right) = 0,$$

    for all $n \in \mathbf{N}^*$ and all homogeneous $\xi_1, \dots, \xi_n \in L$.

The map ℓ_1 (which satisfies $\ell_1^2 = 0$, by $(\mathfrak{J}_1)$) is sometimes called the differential of L and denoted by ∂, while the map ℓ_2 is sometimes denoted by a bracket $[\cdot, \cdot]$. A *differential graded Lie algebra* (*dg Lie algebra* in short) is an L_∞-algebra $(L, \ell_1, \ell_2, \ell_3, \dots)$, with $\ell_k = 0$, for all $k \geq 3$.

Notice that if $\ell_1 = 0$, then the equation $(\mathfrak{J}_n)$ reads as follows:

$$J_n(\mathcal{L}_{n-1}; \xi_1, \dots, \xi_n) = 0,$$

where $J_n(\mathcal{L}_{n-1}; \xi_1, \dots, \xi_n)$ depends only on $\mathcal{L}_{n-1} := (\ell_2, \dots, \ell_{n-1})$ (and not on ℓ_n) and is defined by:

$$J_n(\mathcal{L}_{n-1}; \xi_1, \dots, \xi_n) :=$$

$$(2) \qquad \sum_{\substack{i+j=n+1 \\ i,j\geq 2}} \sum_{\sigma\in S_{i,n-i}} \chi(\sigma)(-1)^{i(j-1)}\, \ell_j\left(\ell_i\left(\xi_{\sigma(1)}, \dots, \xi_{\sigma(i)}\right), \xi_{\sigma(i+1)}, \dots, \xi_{\sigma(n)}\right).$$

When no confusion can arise, we rather write $J_n(\xi_1, \dots, \xi_n)$ for $J_n(\mathcal{L}_{n-1}; \xi_1, \dots, \xi_n)$.

2.1.2. L_∞-*morphisms, quasi-isomorphisms*. There is a notion of (weak) morphism of L_∞-algebras, which we do not need here. We only need the particular case when the considered morphism goes from an L_∞-algebra to a dg Lie algebra. (For the general definition of L_∞-morphisms between L_∞-algebras, see [8].) Let $L = (L, \ell_1, \ell_2, \dots)$ be an L_∞-algebra and let $\mathfrak{g} = (\mathfrak{g}, \partial_\mathfrak{g}, [\cdot, \cdot]_\mathfrak{g})$ be a dg Lie algebra. A *(weak) L_∞-morphism* from L to $\mathfrak{g}$ is a collection of linear maps

$$f_\bullet = \left(f_n : \bigotimes^n L \to \mathfrak{g} \right)_{n\in\mathbf{N}^*},$$

such that:

    a. each map f_n is a graded map of degree $\deg(f_n) = 1 - n$;
    b. each map f_n is skew-symmetric;

c. the following identities hold, for all $n \in \mathbf{N}^*$ and all homogeneous elements $\xi_1, \ldots, \xi_n \in L$:

(3)

$$\partial_{\mathfrak{g}} \left(f_n(\xi_1, \ldots, \xi_n) \right)$$

$$+ \sum_{\substack{j+k=n+1 \\ j,k \geq 1}} \sum_{\sigma \in S_{k,n-k}} \chi(\sigma) \, (-1)^{k(j-1)+1} f_j \left(\ell_k \left(\xi_{\sigma(1)}, \ldots, \xi_{\sigma(k)} \right), \xi_{\sigma(k+1)}, \ldots, \xi_{\sigma(n)} \right)$$

$$+ \sum_{\substack{s+t=n \\ s,t \geq 1}} \sum_{\substack{\tau \in S_{s,n-s} \\ \tau(1) < \tau(s+1)}} \chi(\tau) \, e_{s,t}(\tau) \left[f_s \left(\xi_{\tau(1)}, \ldots, \xi_{\tau(s)} \right), f_t \left(\xi_{\tau(s+1)}, \ldots, \xi_{\tau(n)} \right) \right]_{\mathfrak{g}} = 0,$$

where $e_{s,t}(\tau) := (-1)^{s-1} \cdot (-1)^{(t-1)\left(\sum_{p=1}^{s} |\xi_{\tau(p)}| \right)}$ and where $\chi(\sigma)$ (respectively, $\chi(\tau)$) stands for $\chi(\sigma; \xi_1, \ldots, \xi_n)$ (respectively, $\chi(\tau; \xi_1, \ldots, \xi_n)$).

We point out that, for $1 \leq s \leq n$ and for an $(s, n-s)$-shuffle $\tau \in S_{s,n-s}$, the condition $\tau(1) < \tau(s+1)$ is equivalent to $\tau(1) = 1$. One says that the L_∞-morphism $f_\bullet$ from L to $\mathfrak{g}$ is a *quasi-isomorphism* (or a *(weak) L_∞-equivalence*) if the chain map $f_1 : (L, \ell_1) \to (\mathfrak{g}, \partial_{\mathfrak{g}})$ induces an isomorphism between the cohomologies associated to the cochain complexes (L, ℓ_1) and $(\mathfrak{g}, \partial_{\mathfrak{g}})$.

Notice that if the L_∞-algebra $(L, \ell_1, \ell_2, \ldots)$ satisfies $\ell_1 = 0$, then we write equation (3) rather in the following form:

$$(\mathcal{E}_n) \qquad \partial_{\mathfrak{g}} \left(f_n(\xi_1, \ldots, \xi_n) \right) - f_1 \left(\ell_n \left(\xi_1, \ldots, \xi_n \right) \right) = T_n(\mathcal{F}_n, \mathcal{L}_{n-1}; \xi_1, \ldots, \xi_n)$$

where $T_n(\mathcal{F}_n, \mathcal{L}_{n-1}; \xi_1, \ldots, \xi_n)$ depends on the elements $\mathcal{F}_n := (f_1, \ldots, f_{n-1})$ and $\mathcal{L}_{n-1} := (\ell_2, \ldots, \ell_{n-1})$, and is defined by:

(4)

$$T_n(\mathcal{F}_n, \mathcal{L}_{n-1}; \xi_1, \ldots, \xi_n) :=$$

$$\sum_{\substack{j+k=n+1 \\ j,k \geq 2}} \sum_{\sigma \in S_{k,n-k}} \chi(\sigma) \, (-1)^{k(j-1)} f_j \left(\ell_k \left(\xi_{\sigma(1)}, \ldots, \xi_{\sigma(k)} \right), \xi_{\sigma(k+1)}, \ldots, \xi_{\sigma(n)} \right)$$

$$- \sum_{\substack{s+t=n \\ s,t \geq 1}} \sum_{\substack{\tau \in S_{s,n-s} \\ \tau(1)=1}} \chi(\tau) \, e_{s,t}(\tau) \left[f_s \left(\xi_{\tau(1)}, \ldots, \xi_{\tau(s)} \right), f_t \left(\xi_{\tau(s+1)}, \ldots, \xi_{\tau(n)} \right) \right]_{\mathfrak{g}},$$

for all $n \in \mathbf{N}^*$ and all (homogeneous) elements $\xi_1, \ldots, \xi_n \in L$. When no confusion can arise, we simply write $T_n(\xi_1, \ldots, \xi_n)$ for $T_n(\mathcal{F}_n, \mathcal{L}_{n-1}; \xi_1, \ldots, \xi_n)$.

2.1.3. *Maurer-Cartan equation.* To an L_∞-algebra is associated the so-called generalized Maurer-Cartan equation (or homotopy Maurer-Cartan equation). In our context, we only need a particular case of it, where the solutions depend formally on a parameter ν. Let $L = (L, \ell_1, \ell_2, \ldots)$ be an L_∞-algebra. The *generalized Maurer-Cartan equation* associated to L is written as follows:

$$(5) \quad -\ell_1(\gamma) - \frac{1}{2}\ell_2(\gamma, \gamma) + \frac{1}{3!}\ell_3(\gamma, \gamma, \gamma) + \cdots + \frac{(-1)^{n(n+1)/2}}{n!}\ell_n(\gamma, \ldots, \gamma) + \cdots = 0,$$

for $\gamma \in L^1 \otimes \nu \mathbf{F}[[\nu]] = \nu L^1[[\nu]]$, where ν is a formal parameter. Notice that the maps ℓ_n, $n \in \mathbf{N}^*$ are extended by multilinearity with respect to the parameter ν (and are still denoted by ℓ_n) and that this infinite sum (5) is well-defined because there is no constant term in γ (i.e., γ is zero modulo ν), so that the coefficient of each ν^i, $i \in \mathbf{N}$ is given by a finite sum. The same will hold for the equations (6) and (10).

The set of all the solutions of the generalized Maurer-Cartan equation associated to L and depending formally on a parameter ν is denoted by $\mathcal{MC}^\nu(L)$. One introduces the *gauge equivalence* on this set, which is denoted by $\sim$ and generated by infinitesimal transformations of the form:

$$(6) \qquad \gamma \longmapsto \xi \cdot \gamma := \gamma - \sum_{n \in \mathbf{N}^*} \frac{(-1)^{n(n-1)/2}}{(n-1)!} \ell_n(\xi, \gamma, \gamma, \dots, \gamma),$$

where $\xi \in L^0 \otimes \nu\mathbf{F}[[\nu]]$.

REMARK 2.1. Let us consider the particular case where the L_∞-algebra L is a dg Lie algebra $(\mathfrak{g}, \partial_\mathfrak{g}, [\cdot, \cdot]_\mathfrak{g})$ whose differential is given by $\partial_\mathfrak{g} = [\chi, \cdot]_\mathfrak{g}$, for some degree one element $\chi \in \mathfrak{g}^1$ satisfying $[\chi, \chi]_\mathfrak{g} = 0$. Then, we have:

$$(7) \qquad \begin{aligned} \mathcal{MC}^\nu(\mathfrak{g}) &= \left\{ \gamma \in \mathfrak{g}^1 \otimes \nu\mathbf{F}[[\nu]] \mid [\chi, \gamma]_\mathfrak{g} + \tfrac{1}{2}[\gamma, \gamma]_\mathfrak{g} = 0 \right\} \\ &= \left\{ \gamma \in \mathfrak{g}^1 \otimes \nu\mathbf{F}[[\nu]] \mid [\chi + \gamma, \chi + \gamma]_\mathfrak{g} = 0 \right\}. \end{aligned}$$

Moreover the infinitesimal transformation (6) becomes in this case:

$$(8) \qquad \gamma \longmapsto \xi \cdot \gamma := \gamma + [\xi, \chi + \gamma]_\mathfrak{g},$$

for $\xi \in \mathfrak{g}^0 \otimes \nu\mathbf{F}[[\nu]]$.

We denote by $\mathcal{D}ef^\nu(L)$ the set of all the gauge equivalence classes of the formal solutions of the generalized Maurer-Cartan equation associated to L,

$$\mathcal{D}ef^\nu(L) := \mathcal{MC}^\nu(L)/ \sim .$$

For $\gamma \in \mathcal{MC}^\nu(L)$, we denote by $\mathrm{cl}(\gamma) \in \mathcal{D}ef^\nu(L)$ its equivalence class modulo the gauge equivalence. In the following we will use the theorem (see for instance [9] or [3]):

THEOREM 2.2 ([9], [3], ...). *Let L and L' be two L_∞-algebras and let us suppose that $f_\bullet = (f_n : \bigotimes^n L \to L')_{n \in \mathbf{N}^*}$ is a quasi-isomorphism from L to L'. Then $f_\bullet$ induces an isomorphism $\mathcal{D}ef^\nu(f_\bullet)$ from $\mathcal{D}ef^\nu(L)$ to $\mathcal{D}ef^\nu(L')$. This isomorphism is given, for $\gamma \in \mathcal{MC}^\nu(L)$, by:*

$$(9) \qquad \mathcal{D}ef^\nu(f_\bullet)\,(\mathrm{cl}(\gamma)) := \mathrm{cl}\left(\mathcal{MC}^\nu(f_\bullet)(\gamma)\right),$$

where

$$(10) \qquad \mathcal{MC}^\nu(f_\bullet)(\gamma) := \sum_{n \geq 1} \frac{(-1)^{1+n(n+1)/2}}{n!} f_n(\gamma, \dots, \gamma).$$

Notice that we will only use this theorem in the case L' is a dg Lie algebra.

2.2. Poisson algebras, cohomology and deformations. In this paper, our goal is to apply the theory of L_∞-algebras to the problem of deformations of Poisson structures. We here recall the notions of Poisson algebras, cohomology and deformations, and explain how one can associate a dg Lie algebra to a Poisson algebra.

2.2.1. *Poisson algebra and cohomology.* We recall that a *Poisson structure* $\{\cdot,\cdot\}$ (also denoted by π_0) on an associative commutative algebra $(\mathcal{A},\cdot)$ is a skew-symmetric biderivation of $\mathcal{A}$, i.e., a map $\{\cdot,\cdot\} : \bigwedge^2 \mathcal{A} \to \mathcal{A}$ satisfying the derivation property:

(11) $\qquad \{FG,H\} = F\{G,H\} + G\{F,H\}$, for all $F,G,H \in \mathcal{A}$,

(where FG stands for $F \cdot G$), which is also a Lie structure on $\mathcal{A}$, i.e., which satisfies the Jacobi identity

(12) $\qquad \{\{F,G\},H\} + \{\{G,H\},F\} + \{\{H,F\},G\} = 0$, for all $F,G,H \in \mathcal{A}$.

The couple $(\mathcal{A}, \{\cdot,\cdot\} = \pi_0)$ is then called a *Poisson algebra.*

The Poisson cohomology has been introduced by A. Lichnerowicz in [14]; see also [5] for an algebraic approach. The *Poisson cohomology* complex, associated to a Poisson algebra $(\mathcal{A}, \pi_0)$, is defined as follows. The space of all Poisson cochains is $\mathfrak{X}^\bullet(\mathcal{A}) := \bigoplus_{k\in\mathbf{N}} \mathfrak{X}^k(\mathcal{A})$, where $\mathfrak{X}^0(\mathcal{A})$ is $\mathcal{A}$ and, for all $k \in \mathbf{N}^*$, $\mathfrak{X}^k(\mathcal{A})$ denotes the space of all skew-symmetric k-derivations of $\mathcal{A}$, i.e., the skew-symmetric k-linear maps $\mathcal{A}^k \to \mathcal{A}$ that satisfy the derivation property (12) in each of their arguments. The Poisson coboundary operator $\delta^k_{\pi_0} : \mathfrak{X}^k(\mathcal{A}) \to \mathfrak{X}^{k+1}(\mathcal{A})$ is given by the formula

$$\delta^k_{\pi_0} := -[\cdot, \pi_0]_S,$$

where $[\cdot,\cdot]_S : \mathfrak{X}^p(\mathcal{A}) \times \mathfrak{X}^q(\mathcal{A}) \to \mathfrak{X}^{p+q-1}(\mathcal{A})$ is the so called Schouten bracket. The Schouten bracket is a graded Lie bracket that generalizes the commutator of derivations and that is a graded biderivation with respect to the wedge product of multiderivations (see [12]). It is defined, for $P \in \mathfrak{X}^p(\mathcal{A})$, $Q \in \mathfrak{X}^q(\mathcal{A})$ and for $F_1, \ldots, F_{p+q-1} \in \mathcal{A}$, by:

$$[P,Q]_S [F_1, \ldots, F_{p+q-1}]$$

(13) $$= \sum_{\sigma \in S_{q,p-1}} \text{sign}(\sigma) P\left[Q[F_{\sigma(1)}, \ldots, F_{\sigma(q)}], F_{\sigma(q+1)}, \ldots, F_{\sigma(q+p-1)}\right]$$

$$-(-1)^{(p-1)(q-1)} \sum_{\sigma \in S_{p,q-1}} \text{sign}(\sigma) Q\left[P[F_{\sigma(1)}, \ldots, F_{\sigma(p)}], F_{\sigma(p+1)}, \ldots, F_{\sigma(p+q-1)}\right].$$

It is easy and useful to verify that, given a skew-symmetric biderivation $\pi \in \mathfrak{X}^2(\mathcal{A})$, the Jacobi identity for π is equivalent to $[\pi,\pi]_S = 0$, in other words, if $\pi \in \mathfrak{X}^2(\mathcal{A})$ is a skew-symmetric biderivation of $\mathcal{A}$, then π is a Poisson structure on $\mathcal{A}$ if and only if $[\pi,\pi]_S = 0$.

2.2.2. *The dg Lie algebra associated to the Poisson complex.* The Poisson cohomology complex associated to a Poisson algebra $(\mathcal{A}, \pi_0)$ together with the Schouten bracket give rise to a dg Lie algebra, $(\mathfrak{g}, \partial_{\mathfrak{g}}, [\cdot,\cdot]_{\mathfrak{g}})$, defined as follows.

(1) For all $n \in \mathbf{N}^*$, the degree n homogeneous part of $\mathfrak{g}$ is given by

$$\mathfrak{g}^n := \mathfrak{X}^{n+1}(\mathcal{A}),$$

so that the degree of $P \in \mathfrak{X}^p(\mathcal{A}) = \mathfrak{g}^{p-1}$, viewed as an element of $\mathfrak{g}$, is $|P| := p - 1$,

(2) for all $P \in \mathfrak{X}^p(\mathcal{A}) = \mathfrak{g}^{p-1}$,

$$\partial_{\mathfrak{g}}(P) := (-1)^{|P|} \delta^p_{\pi_0}(P) = (-1)^{p-1} \delta^p_{\pi_0}(P),$$

(3) the graded Lie bracket on $\mathfrak{g}$ is given by the Schouten bracket:

$$[\cdot,\cdot]_{\mathfrak{g}} := [\cdot,\cdot]_S.$$

Notice that, using the skew-symmetry of the Schouten bracket, and the definition of $\delta^p_{\pi_0}$, we can write $\partial_{\mathfrak{g}} = [\pi_0,\cdot]_S$. As $[\pi_0,\pi_0]_S = 0$ (see last paragraph 2.2.1), the dg Lie algebra $(\mathfrak{g},\partial_{\mathfrak{g}},[\cdot,\cdot]_{\mathfrak{g}})$ associated to a Poisson algebra $(\mathcal{A},\pi_0)$ satisfied the conditions of the remark 2.1.

2.2.3. *Formal deformations of Poisson structures.* In this paragraph, we define the notion of formal deformations of Poisson structures. For more details about this, see [**17**]. Let $(\mathcal{A},\cdot)$ be an associative commutative algebra over $\mathbf{F}$ and let π_0 be a Poisson structure on $(\mathcal{A},\cdot)$. We consider the $\mathbf{F}[[\nu]]$-vector space $\mathcal{A}[[\nu]]$ of all formal power series in ν, with coefficients in $\mathcal{A}$. The associative commutative product "$\cdot$", defined on $\mathcal{A}$, is naturally extended to an associative commutative product on $\mathcal{A}[[\nu]]$, still denoted by "$\cdot$". A *formal deformation* of π_0 is a Poisson structure on the associative $\mathbf{F}[[\nu]]$-algebra $\mathcal{A}[[\nu]]$, that extends the initial Poisson structure. In other words, it is given by a map $\pi_* : \mathcal{A}[[\nu]] \times \mathcal{A}[[\nu]] \to \mathcal{A}[[\nu]]$ satisfying the Jacobi identity and of the form:

$$\pi_* = \pi_0 + \pi_1\nu + \cdots + \pi_n\nu^n + \cdots,$$

where the π_i are skew-symmetric biderivations of $\mathcal{A}$ (extended by bilinearity with respect to ν). Notice that given a map $\pi_* = \pi_0 + \pi_1\nu + \cdots + \pi_n\nu^n + \cdots : \mathcal{A}[[\nu]] \times \mathcal{A}[[\nu]] \to \mathcal{A}[[\nu]]$ where for all $i \in \mathbf{N}$, $\pi_i \in \mathfrak{X}^2(\mathcal{A})$ is a skew-symmetric biderivation of $\mathcal{A}$, we have that π_* is a formal deformation of π_0 if and only if $[\pi_*,\pi_*]_S = 0$.

There is a natural notion of equivalence for deformations of a Poisson structure π_0. Two formal deformations π_* and π'_* of π_0 are said to be *equivalent* if there exists an $\mathbf{F}[[\nu]]$-linear map $\Phi : (\mathcal{A}[[\nu]],\pi_*) \to (\mathcal{A}[[\nu]],\pi'_*)$, which is equal to the identity modulo ν and is a Poisson morphism, i.e., it is a morphism of associative algebras $\Phi : (\mathcal{A}[[\nu]],\cdot) \to (\mathcal{A}[[\nu]],\cdot)$, which satisfies:

(14) $$\pi'_*[\Phi(F),\Phi(G)] = \Phi(\pi_*[F,G]),$$

for all $F,G \in \mathcal{A}$ (and therefore, for all $F,G \in \mathcal{A}[[\nu]]$). It is also possible to write such a morphism Φ as the exponential of an element $\xi \in \nu\mathfrak{X}^1(\mathcal{A})[[\nu]] = \mathfrak{X}^1(\mathcal{A}) \otimes \nu\mathbf{F}[[\nu]]$, so that (see for example the lemma 2.1 of [**17**]) the map π'_* given by (14) can also be written as:

(15) $$\pi'_* = e^{\mathrm{ad}_\xi}(\pi_*) = \pi_* + \sum_{k\in\mathbf{N}^*} \frac{1}{k!} \underbrace{[\xi,[\xi,\ldots,[\xi,\pi_*]_S\cdots]_S]_S}_{k \text{ brackets}}.$$

Let us now consider the dg Lie algebra $(\mathfrak{g},\partial_{\mathfrak{g}},[\cdot,\cdot]_{\mathfrak{g}})$ associated to the Poisson algebra $(\mathcal{A},\pi_0)$, as explained in the previous paragraph 2.2.2. According to the remark 2.1, we have:

$$\mathcal{MC}^\nu(\mathfrak{g}) = \left\{ \gamma = \sum_{i\geq 1}\pi_i\nu^i \in \mathfrak{X}^2(\mathcal{A}) \otimes \nu\mathbf{F}[[\nu]] \mid \right.$$

$$\left. \pi_* := \pi_0 + \sum_{i\geq 1}\pi_i\nu^i \text{ is a formal deformation of } \pi_0 \right\},$$

so that, there is a natural one-to-one correspondence between $\mathcal{MC}^\nu(\mathfrak{g})$ and the space of all formal deformations of π_0:

$$
\begin{aligned}
\mathcal{MC}^\nu(\mathfrak{g}) &\rightarrow \{\text{formal deformations of } \pi_0\} \\
\gamma &\mapsto \pi_0 + \gamma
\end{aligned}
$$

Moreover, the infinitesimal transformation (8) on elements of $\mathcal{MC}^\nu(\mathfrak{g})$ can be transposed to an infinitesimal transformation on formal deformations π_* of π_0. It then becomes:

$$
\pi_* \longmapsto \xi \cdot \pi_* := \pi_* + [\xi, \pi_*]_S \,,
$$

for $\xi \in \mathfrak{X}^1(A) \otimes \nu\mathbf{F}[[\nu]]$. We conclude that there is a one-to-one correspondence between the elements of $\mathcal{D}ef^\nu(\mathfrak{g})$ and the equivalence classes of the formal deformations of π_0.

3. Choice in a transfer of L_∞-algebra structure

For $\mathfrak{g} = (\mathfrak{g}, \partial_\mathfrak{g}, [\,\cdot\,,\cdot\,]_\mathfrak{g})$ a dg Lie algebra, we denote by $H(\mathfrak{g}, \partial_\mathfrak{g})$, the graded vector space given by the cohomology of the cochain complex $(\mathfrak{g}, \partial_\mathfrak{g})$. Equipped with the trivial differential, it is a cochain complex $(H(\mathfrak{g}, \partial_\mathfrak{g}), 0)$. Moreover, $Z(\mathfrak{g}, \partial_\mathfrak{g})$ denotes the graded vector space of all the cocycles of the cochain complex $(\mathfrak{g}, \partial_\mathfrak{g})$:

$$
Z(\mathfrak{g}, \partial_\mathfrak{g}) = \ker \partial_\mathfrak{g} \subseteq \mathfrak{g},
$$

and $B(\mathfrak{g}, \partial_\mathfrak{g})$, the graded vector space of all its coboundaries:

$$
B(\mathfrak{g}, \partial_\mathfrak{g}) = \operatorname{Im} \partial_\mathfrak{g} \subseteq \mathfrak{g},
$$

so that $H(\mathfrak{g}, \partial_\mathfrak{g}) = Z(\mathfrak{g}, \partial_\mathfrak{g})/B(\mathfrak{g}, \partial_\mathfrak{g})$. (The grading of $Z(\mathfrak{g}, \partial_\mathfrak{g})$, $B(\mathfrak{g}, \partial_\mathfrak{g})$ and $H(\mathfrak{g}, \partial_\mathfrak{g})$ is naturally induced by the grading of $\mathfrak{g}$.) We denote by p the natural projection from $Z(\mathfrak{g}, \partial_\mathfrak{g})$ to the cohomology of $\mathfrak{g}$, and for every cocycle $x \in Z(\mathfrak{g}, \partial_\mathfrak{g}) \subseteq \mathfrak{g}$, the notations $p(x)$ and $\bar{x}$ both stand for the cohomological class of x,

$$
(16) \qquad
\begin{aligned}
p : Z(\mathfrak{g}, \partial_\mathfrak{g}) &\rightarrow H(\mathfrak{g}, \partial_\mathfrak{g}) \\
x &\mapsto p(x) = \bar{x}.
\end{aligned}
$$

We now define a graded linear map f_1, of degree 0, from $H(\mathfrak{g}, \partial_\mathfrak{g})$ to $\mathfrak{g}$. This definition depends on a choice of a basis $\mathbf{b}^\ell$, for each cohomology space $H^\ell(\mathfrak{g}, \partial_\mathfrak{g})$, and on a choice of representatives $(\vartheta_k^\ell)_k$ of the elements of the basis $\mathbf{b}^\ell$:

$$
\mathbf{b}^\ell = \left(\overline{\vartheta_k^\ell}\right)_k.
$$

(We do not need here to specify the set by which the basis $\mathbf{b}^\ell$ is indexed.) Then the map $f_1 : H(\mathfrak{g}, \partial_\mathfrak{g}) \to \mathfrak{g}$ is defined by

$$
(17) \qquad
\begin{aligned}
f_1 : \quad H^\ell(\mathfrak{g}, \partial_\mathfrak{g}) &\rightarrow Z^\ell(\mathfrak{g}, \partial_\mathfrak{g}) \subseteq \mathfrak{g}^\ell \\
\xi = \textstyle\sum_k \lambda_k^\ell \overline{\vartheta_k^\ell} &\mapsto \textstyle\sum_k \lambda_k^\ell \vartheta_k^\ell,
\end{aligned}
$$

for all $\ell \in \mathbf{Z}$, and where $\xi = \sum_k \lambda_k^\ell \overline{\vartheta_k^\ell}$ is the unique decomposition of $\xi \in H^\ell(\mathfrak{g}, \partial_\mathfrak{g})$ in the fixed basis $\mathbf{b}^\ell$ (the λ_k^ℓ are constants). We deduce from the definition of f_1 that we have:

$$
(18) \qquad Z(\mathfrak{g}, \partial_\mathfrak{g}) \simeq \operatorname{Im} f_1 \oplus B(\mathfrak{g}, \partial_\mathfrak{g}),
$$

and

$$
(19) \qquad x - f_1 \circ p(x) \in B(\mathfrak{g}, \partial_\mathfrak{g}), \quad \text{for all } x \in Z(\mathfrak{g}, \partial_\mathfrak{g}).
$$

Also the map f_1 is a chain map between the two cochain complexes $(H(\mathfrak{g}, \partial_\mathfrak{g}), 0)$ and $(\mathfrak{g}, \partial_\mathfrak{g})$, which induces an isomorphism between their cohomologies. This implies in particular that if one extends f_1 to a (weak) L_∞-morphism

$$f_\bullet = \left(f_n : \bigotimes^n H(\mathfrak{g}, \partial_\mathfrak{g}) \to \mathfrak{g} \right)_{n \in \mathbf{N}^*}$$

(where $H(\mathfrak{g}, \partial_\mathfrak{g})$ is equipped with an L_∞-algebra structure), then $f_\bullet$ is automatically a quasi-isomorphism.

We indeed want to construct an L_∞-algebra structure on $H(\mathfrak{g}, \partial_\mathfrak{g})$ together with a quasi-isomorphism from it to the dg Lie algebra $\mathfrak{g}$. We know that, by using a theorem of L_∞-algebra structure transfer, (see for instance the "move" (M1) of [15]), there exists such a L_∞-algebra structure on $H(\mathfrak{g}, \partial_\mathfrak{g})$ and such a quasi-isomorphism from it to the dg Lie algebra $\mathfrak{g}$, which extends f_1, but, as explained in the introduction, the point here is that we need to construct a *specific* L_∞-algebra structure on $H(\mathfrak{g}, \partial_\mathfrak{g})$ and a *specific* quasi-isomorphism. This prevents one to express the transfer structure in terms of a homotopy map (as usually done with the pertubation lemma) because it seems to the author that such a map cannot be explicitly written in general and especially in the context we will use in the section 4. In order to have as much control in this contruction as possible, we show the following:

PROPOSITION 3.1. *Let* $\mathfrak{g} = (\mathfrak{g}, \partial_\mathfrak{g}, [\cdot, \cdot]_\mathfrak{g})$ *be a dg Lie algebra, let* $H(\mathfrak{g}, \partial_\mathfrak{g})$ *denote the graded space given by the cohomology associated to the cochain complex* $(\mathfrak{g}, \partial_\mathfrak{g})$. *We fix* $f_1 : H(\mathfrak{g}, \partial_\mathfrak{g}) \to \mathfrak{g}$ *as being the map defined in (17), associated to a choice of bases* $(\mathbf{b}^\ell)_\ell$ *for the cohomology spaces* $(H^\ell(\mathfrak{g}, \partial_\mathfrak{g}))_\ell$. *We also fix* $\ell_1 : H(\mathfrak{g}, \partial_\mathfrak{g}) \to H(\mathfrak{g}, \partial_\mathfrak{g})$ *as being trivial* ($\ell_1 = 0$) *so that the equations* $(\mathcal{E}_1)$ *and* $(\mathfrak{J}_1)$ *are automatically satisfied.*

(a) *There exist skew-symmetric graded linear maps*

$$\ell_2 : H(\mathfrak{g}, \partial_\mathfrak{g}) \otimes H(\mathfrak{g}, \partial_\mathfrak{g}) \to H(\mathfrak{g}, \partial_\mathfrak{g}), \quad and \quad f_2 : H(\mathfrak{g}, \partial_\mathfrak{g}) \otimes H(\mathfrak{g}, \partial_\mathfrak{g}) \to \mathfrak{g},$$

of degrees $\deg(\ell_2) = 0$ *and* $\deg(f_2) = -1$, *such that the equations* $(\mathcal{E}_2)$ *and* $(\mathfrak{J}_2)$ *are satisfied. Moreover, such a map* ℓ_2 *satisfies also the equation* $(\mathfrak{J}_3)$.

(b) *Let* $m \geq 3$ *be an integer. For any skew-symmetric graded linear maps*

$$\ell_k : \bigotimes^k H(\mathfrak{g}, \partial_\mathfrak{g}) \to H(\mathfrak{g}, \partial_\mathfrak{g}), \quad for \ 2 \leq k \leq m-1,$$

$$f_k : \bigotimes^k H(\mathfrak{g}, \partial_\mathfrak{g}) \to \mathfrak{g}, \quad for \ 2 \leq k \leq m-1,$$

of degrees $\deg(\ell_k) = 2-k$ *and* $\deg(f_k) = 1-k$, *for all* $2 \leq k \leq m-1$, *and such that the equations* $(\mathfrak{J}_2) - (\mathfrak{J}_m)$ *and* $(\mathcal{E}_2) - (\mathcal{E}_{m-1})$ *are satisfied, there exist skew-symmetric graded linear maps*

$$\ell_m : \bigotimes^m H(\mathfrak{g}, \partial_\mathfrak{g}) \to H(\mathfrak{g}, \partial_\mathfrak{g}) \quad and \quad f_m : \bigotimes^m H(\mathfrak{g}, \partial_\mathfrak{g}) \to \mathfrak{g},$$

with $\deg(f_m) = 1-m$, $\deg(\ell_m) = 2-m$ *and satisfying the equation* $(\mathcal{E}_m)$. *Moreover, such a map* ℓ_m *necessarily satisfies also the equation* $(\mathfrak{J}_{m+1})$.

REMARK 3.2. This proposition implies in particular that there exist an L_∞-algebra structure $\ell_\bullet$ on $H(\mathfrak{g}, \partial_\mathfrak{g})$ with the trivial differential $\ell_1 = 0$ and a quasi-isomorphism $f_\bullet$ from $H(\mathfrak{g}, \partial_\mathfrak{g})$ to $\mathfrak{g}$ that extends f_1 (defined in (17)). But, this proposition implies morever that, whatever the choices made for the first $m-1$ maps $\ell_1, \ldots, \ell_{m-1}$ and $f_1, \ldots, f_{m-1}$ (m is an arbitrary integer), with $\ell_1 = 0$ and f_1 given by (17), if these maps satisfy the first m equations defining an L_∞-algebra structure (equations $(\mathfrak{J}_1) - (\mathfrak{J}_m)$) and the first $m - 1$ equations defining an L_∞-morphism (equations $(\mathcal{E}_1) - (\mathcal{E}_{m-1})$), then they still extend to an L_∞-algebra structure $\ell_\bullet$ on $H(\mathfrak{g}, \partial_\mathfrak{g})$ and a quasi-isomorphism $f_\bullet$ from $H(\mathfrak{g}, \partial_\mathfrak{g})$ to $\mathfrak{g}$.

PROOF. The idea of this proof is similar to the one used by T. Kadeishvili in [7] (where he considers A_∞-algebras) to prove his theorem 1. Let us first prove the part (a) of this proposition. To do this, we first show (*Step 1*) that the identity $\ell_1 = 0$ and the definition (17) of f_1 imply that $\partial_\mathfrak{g}(T_2(\xi_1, \xi_2)) = 0$, for all $\xi_1, \xi_2 \in H(\mathfrak{g}, \partial_\mathfrak{g})$. By (18), the cocycle $T_2(\xi_1, \xi_2)$ then decomposes as a coboundary (element in the image of $\partial_\mathfrak{g}$) plus an element in the image of f_1, which permit us to conclude the existence of both maps f_2 and ℓ_2, satisfying the equation $(\mathcal{E}_2)$. Secondly (*Step 2*), we show that the obtained map ℓ_2, satisfying $(\mathcal{E}_2)$, also necessarily satisfies $(\mathfrak{J}_3)$.

(a) - *Step 1.* The skew-symmetric graded linear maps f_1 (given by (17)) and $\ell_1 := 0$ are of degree 0 and -1 respectively, and satisfy both equations:

$$(\mathfrak{J}_1) \qquad\qquad \ell_1 \circ \ell_1 = 0,$$

and

$$(\mathcal{E}_1) \qquad\qquad \partial_\mathfrak{g} \circ f_1 = 0.$$

Let $\xi_1, \xi_2 \in H(\mathfrak{g}, \partial_\mathfrak{g})$. We have $T_2(\xi_1, \xi_2) = -[f_1(\xi_1), f_1(\xi_2)]_\mathfrak{g}$. As $(\mathfrak{g}, \partial_\mathfrak{g}, [\cdot, \cdot]_\mathfrak{g})$ is a dg Lie algebra, $\partial_\mathfrak{g}$ is a (graded) derivation for $[\cdot, \cdot]_\mathfrak{g}$, hence:

$$(20) \quad \partial_\mathfrak{g}(T_2(\xi_1, \xi_2)) = -[\partial_\mathfrak{g}(f_1(\xi_1)), f_1(\xi_2)]_\mathfrak{g} - (-1)^{|\xi_1|}[f_1(\xi_1), \partial_\mathfrak{g}(f_1(\xi_2))]_\mathfrak{g} = 0,$$

by $(\mathcal{E}_1)$. We now define a skew-symmetric graded linear map $\ell_2 : \bigwedge^2 H(\mathfrak{g}, \partial_\mathfrak{g}) \to H(\mathfrak{g}, \partial_\mathfrak{g})$ of degree 0, by:

$$(21) \qquad\qquad \ell_2(\xi_1, \xi_2) := -p \circ T_2(\xi_1, \xi_2),$$

for all $\xi_1, \xi_2 \in H(\mathfrak{g}, \partial_\mathfrak{g})$. This map is well-defined because, according to (20), $T_2(\xi_1, \xi_2)$ is a cocycle for the cochain complex $(\mathfrak{g}, \partial_\mathfrak{g})$, and it trivially satisfies the equation $(\mathfrak{J}_2)$, because $\ell_1 = 0$. It is also possible, according to (19), to define a skew-symmetric graded linear map $f_2 : \bigwedge^2 H(\mathfrak{g}, \partial_\mathfrak{g}) \to \mathfrak{g}$, of degree -1, with the following formula:

$$(22) \qquad\qquad \partial_\mathfrak{g}(f_2(\xi_1, \xi_2)) = T_2(\xi_1, \xi_2) - f_1 \circ p(T_2(\xi_1, \xi_2)),$$

for all $\xi_1, \xi_2 \in H(\mathfrak{g}, \partial_\mathfrak{g})$. The maps ℓ_2 and f_2 then satisfy the equation $(\mathcal{E}_2)$, because $-f_1 \circ p(T_2(\xi_1, \xi_2)) = f_1 \circ \ell_2(\xi_1, \xi_2)$. Notice that, for every $\xi_1, \xi_2 \in H(\mathfrak{g}, \partial_\mathfrak{g})$, the choice of the element $f_2(\xi_1, \xi_2) \in \mathfrak{g}$ is unique, up to a cocycle.

(a) - *Step 2.* Now, let us prove the second part of (a), by showing that the map ℓ_2, defined in (21), satisfies the equation

$$(\mathfrak{J}_3) \qquad\qquad \sum_{\sigma \in S_{2,1}} \chi(\sigma)\, \ell_2\left(\ell_2\left(\xi_{\sigma(1)}, \xi_{\sigma(2)}\right), \xi_{\sigma(3)}\right) = 0,$$

for all homogeneous $\xi_1, \xi_2, \xi_3 \in H(\mathfrak{g}, \partial_\mathfrak{g})$, where $\chi(\sigma)$ stands for $\chi(\sigma; \xi_1, \xi_2, \xi_3)$. We prove this, by using the equations $(\mathcal{E}_1)$ and $(\mathcal{E}_2)$ and the graded Jacobi identity satisfied by $[\cdot, \cdot]_\mathfrak{g}$. Let $\xi_1, \xi_2, \xi_3 \in H(\mathfrak{g}, \partial_\mathfrak{g})$ and let $\sigma \in S_{2,1}$. By the definition (21) of

ℓ_2, we have $\ell_2\left(\ell_2\left(\xi_{\sigma(1)}, \xi_{\sigma(2)}\right), \xi_{\sigma(3)}\right) = -p\left(T_2\left(\ell_2\left(\xi_{\sigma(1)}, \xi_{\sigma(2)}\right), \xi_{\sigma(3)}\right)\right)$. Moreover, by definition of T_2,

$$
\begin{aligned}
T_2\left(\ell_2\left(\xi_{\sigma(1)}, \xi_{\sigma(2)}\right), \xi_{\sigma(3)}\right) &= -\left[f_1\left(\ell_2\left(\xi_{\sigma(1)}, \xi_{\sigma(2)}\right)\right), f_1\left(\xi_{\sigma(3)}\right)\right]_{\mathfrak{g}} \\
&= \left[T_2\left(\xi_{\sigma(1)}, \xi_{\sigma(2)}\right), f_1\left(\xi_{\sigma(3)}\right)\right]_{\mathfrak{g}} - \left[\partial_{\mathfrak{g}}\left(f_2\left(\xi_{\sigma(1)}, \xi_{\sigma(2)}\right)\right), f_1\left(\xi_{\sigma(3)}\right)\right]_{\mathfrak{g}} \\
&= -\left[\left[f_1\left(\xi_{\sigma(1)}\right), f_1\left(\xi_{\sigma(2)}\right)\right]_{\mathfrak{g}}, f_1\left(\xi_{\sigma(3)}\right)\right]_{\mathfrak{g}} - \left[\partial_{\mathfrak{g}}\left(f_2\left(\xi_{\sigma(1)}, \xi_{\sigma(2)}\right)\right), f_1\left(\xi_{\sigma(3)}\right)\right]_{\mathfrak{g}},
\end{aligned}
$$

where we have used $(\mathcal{E}_2)$ (i.e., $\partial_{\mathfrak{g}} \circ f_2 - f_1 \circ \ell_2 = T_2$) in the second step. As $\partial_{\mathfrak{g}}$ is a derivation for $[\cdot, \cdot]_{\mathfrak{g}}$ and using the fact that $\partial_{\mathfrak{g}} \circ f_1 = 0$, one obtains:

$$
\left[\partial_{\mathfrak{g}}\left(f_2\left(\xi_{\sigma(1)}, \xi_{\sigma(2)}\right)\right), f_1\left(\xi_{\sigma(3)}\right)\right]_{\mathfrak{g}} = \partial_{\mathfrak{g}}\left(\left[f_2\left(\xi_{\sigma(1)}, \xi_{\sigma(2)}\right), f_1\left(\xi_{\sigma(3)}\right)\right]_{\mathfrak{g}}\right).
$$

Finally, because $p \circ \partial_{\mathfrak{g}} = 0$,

$$
-\sum_{\sigma \in S_{2,1}} \chi(\sigma)\, p \circ T_2\left(\ell_2\left(\xi_{\sigma(1)}, \xi_{\sigma(2)}\right), \xi_{\sigma(3)}\right) =
$$

$$
p\left(\sum_{\sigma \in S_{2,1}} \chi(\sigma)\left[\left[f_1\left(\xi_{\sigma(1)}\right), f_1\left(\xi_{\sigma(2)}\right)\right]_{\mathfrak{g}}, f_1\left(\xi_{\sigma(3)}\right)\right]_{\mathfrak{g}}\right) = 0,
$$

where we have used the graded Jacobi identity satisfied by $[\cdot, \cdot]_{\mathfrak{g}}$, to obtain the last line. This shows that the map ℓ_2 satisfies $(\mathfrak{J}_3)$.

REMARK 3.3. The skew-symmetric graded linear map ℓ_2 of degree 0 which satisfies $(\mathcal{E}_2)$ is unique and given by (21). Using (21) and the definition of T_2, we obtain that, for all $\xi_1, \xi_2 \in H(\mathfrak{g}, \partial_{\mathfrak{g}})$,

$$
\ell_2(\xi_1, \xi_2) = -p \circ T_2(\xi_1, \xi_2) = p\left([f_1(\xi_1), f_1(\xi_2)]_{\mathfrak{g}}\right).
$$

In other words, the map $\ell_2 : \bigwedge^2 H(\mathfrak{g}, \partial_{\mathfrak{g}}) \to H(\mathfrak{g}, \partial_{\mathfrak{g}})$ is the map induced by the graded Lie bracket $[\cdot, \cdot]_{\mathfrak{g}}$ on $H(\mathfrak{g}, \partial_{\mathfrak{g}})$. For this reason, we sometimes denote ℓ_2 also by $[\cdot, \cdot]_{\mathfrak{g}}$.

Let us now prove the part (b) of the proposition. To do this, we suppose that $m \geq 3$ and that $f_2, \ldots, f_{m-1}$ and $\ell_2, \ldots, \ell_{m-1}$ are skew-symmetric graded linear maps, of degrees $\deg(\ell_k) = 2 - k$ and $\deg(f_k) = 1 - k$, which satisfy the equations $(\mathfrak{J}_2) - (\mathfrak{J}_m)$ and $(\mathcal{E}_2) - (\mathcal{E}_{m-1})$. Then, we show (*Step 1*), that

$$
\partial_{\mathfrak{g}}\left(T_m(\xi_1, \ldots, \xi_m)\right) = 0, \quad \text{for all } \xi_1, \ldots, \xi_m \in H(\mathfrak{g}, \partial_{\mathfrak{g}}).
$$

This indeed implies, by (18), that the cocycle $T_m(\xi_1, \ldots, \xi_m)$ decomposes as a coboundary (element in the image of $\partial_{\mathfrak{g}}$) plus an element in the image of f_1, which leads to the existence of both maps f_m and ℓ_m, satisfying the equation $(\mathcal{E}_m)$.

Then (*Step 2*), we show that the obtained map ℓ_m, satisfying $(\mathcal{E}_m)$, necessarily also satisfies the equation $(\mathfrak{J}_m)$.

(b) - *Step 1.* Let $\xi_1, \ldots, \xi_m \in H(\mathfrak{g}, \partial_{\mathfrak{g}})$ be homogeneous elements. Recall that we have:

$$
(23) \qquad\qquad T_m(\xi_1, \ldots, \xi_m) = S_m(\xi_1, \ldots, \xi_m) - U_m(\xi_1, \ldots, \xi_m),
$$

where we define, for all $n \in \mathbf{N}^*$, and all $\zeta_1, \ldots, \zeta_n \in H(\mathfrak{g}, \partial_\mathfrak{g})$:

$$S_n(\zeta_1, \ldots, \zeta_n) :=$$

(24)
$$\sum_{\substack{j+k=n+1 \\ j,k \geq 2}} \sum_{\sigma \in S_{k,n-k}} \chi(\sigma) (-1)^{k(j-1)} f_j \left(\ell_k \left(\zeta_{\sigma(1)}, \ldots, \zeta_{\sigma(k)} \right), \zeta_{\sigma(k+1)}, \ldots, \zeta_{\sigma(n)} \right)$$

and

$$U_n(\zeta_1, \ldots, \zeta_n) :=$$

(25)
$$\sum_{\substack{s+t=n \\ s,t \geq 1}} \sum_{\substack{\tau \in S_{s,n-s} \\ \tau(1)=1}} \chi(\tau) \, e_{s,t}(\tau) \left[f_s \left(\zeta_{\tau(1)}, \ldots, \zeta_{\tau(s)} \right), f_t \left(\zeta_{\tau(s+1)}, \ldots, \zeta_{\tau(n)} \right) \right]_\mathfrak{g},$$

with $e_{s,t}(\tau) = (-1)^{s-1} \cdot (-1)^{(t-1)\left(\sum\limits_{p=1}^{s} |\zeta_{\tau(p)}| \right)}$ and where $\chi(\sigma)$ (respectively, $\chi(\tau)$) stands for $\chi(\sigma; \zeta_1, \ldots, \zeta_n)$ (respectively, $\chi(\tau; \zeta_1, \ldots, \zeta_n)$). For $j = 2, \ldots, m-1$, the equation $(\mathcal{E}_j)$ can be written as $\partial_\mathfrak{g} \circ f_j = T_j + f_1 \circ \ell_j$, so that

$$\partial_\mathfrak{g} \left(S_m(\xi_1, \ldots, \xi_m) \right) =$$

$$\sum_{\substack{j+k=m+1 \\ j,k \geq 2}} \sum_{\sigma \in S_{k,m-k}} \chi(\sigma) (-1)^{k(j-1)} T_j \left(\ell_k \left(\xi_{\sigma(1)}, \ldots, \xi_{\sigma(k)} \right), \xi_{\sigma(k+1)}, \ldots, \xi_{\sigma(m)} \right)$$

$$+ f_1 \left(J_m(\xi_1, \ldots, \xi_m) \right) =$$

$$\sum_{\substack{j+k=m+1 \\ j,k \geq 2}} \sum_{\sigma \in S_{k,m-k}} \chi(\sigma) (-1)^{k(j-1)} T_j \left(\ell_k \left(\xi_{\sigma(1)}, \ldots, \xi_{\sigma(k)} \right), \xi_{\sigma(k+1)}, \ldots, \xi_{\sigma(m)} \right),$$

where we have used the equation $(\mathfrak{J}_m)$ (in the case $\ell_1 = 0$, see (2)), in the second step. Now, using the writing of T_j, for $2 \leq j \leq m-1$, we get:

$$\partial_\mathfrak{g} \left(S_m(\xi_1, \ldots, \xi_m) \right) = \mathfrak{a}_m(\xi_1, \ldots, \xi_m) + \mathfrak{b}_m(\xi_1, \ldots, \xi_m) + \mathfrak{c}_m(\xi_1, \ldots, \xi_m),$$

where, for all $n \in \mathbf{N}^*$ and all homogeneous $\zeta_1, \ldots, \zeta_n \in H(\mathfrak{g}, \partial_\mathfrak{g})$, we have defined:

$$\mathfrak{a}_n(\zeta_1, \ldots, \zeta_n) :=$$

$$\sum_{\substack{p+q+k=n+2 \\ p,q,k \geq 2}} \sum_{\substack{\alpha \in S_{q-1,p-1}^{k+1} \\ \sigma \in S_{k,n-k}}} \chi(\sigma; \zeta_1, \ldots, \zeta_n) \, \chi(\alpha; \zeta_{\sigma(k+1)}, \ldots, \zeta_{\sigma(n)}) \cdot (-1)^{k(p+q)+q(p-1)}.$$

$$f_p \left(\ell_q \left(\ell_k \left(\zeta_{\sigma(1)}, \ldots, \zeta_{\sigma(k)} \right), \zeta_{\sigma\alpha(k+1)}, \ldots, \zeta_{\sigma\alpha(k+q-1)} \right), \zeta_{\sigma\alpha(k+q)}, \ldots, \zeta_{\sigma\alpha(n)} \right),$$

and

$$\mathfrak{b}_n(\zeta_1,\ldots,\zeta_n) :=$$

$$\sum_{\substack{p+q+k=n+2 \\ p,q,k\geq 2}} \sum_{\substack{\alpha\in S_{q,p-2}^{k+1} \\ \sigma\in S_{k,n-k}}} \chi(\sigma;\zeta_1,\ldots,\zeta_n)\,\chi(\alpha;\zeta_{\sigma(k+1)},\ldots,\zeta_{\sigma(n)})\cdot$$

$$(-1)^{k(p+q)}(-1)^{q(p-1)}\cdot(-1)^{q+\left(\sum_{r=1}^{k}|\zeta_{\sigma(r)}|+k\right)\cdot\left(\sum_{s=k+1}^{k+q}|\zeta_{\sigma\alpha(s)}|\right)}.$$

$$f_p\left(\ell_q\left(\zeta_{\sigma\alpha(k+1)},\ldots,\zeta_{\sigma\alpha(k+q)}\right),\ell_k\left(\zeta_{\sigma(1)},\ldots,\zeta_{\sigma(k)}\right),\zeta_{\sigma\alpha(k+q+1)},\ldots,\zeta_{\sigma\alpha(n)}\right),$$

and finally

$$\mathfrak{c}_n(\zeta_1,\ldots,\zeta_n) :=$$

$$-\sum_{\substack{j+k=n+1 \\ j,k\geq 2}}\sum_{\sigma\in S_{k,n-k}}\sum_{\substack{a+b=j \\ a,b\geq 1}}\sum_{\beta\in S_{a-1,b}^{k+1}} \chi(\sigma;\zeta_1,\ldots,\zeta_n)\,\chi(\beta;\zeta_{\sigma(k+1)},\ldots,\zeta_{\sigma(n)})\cdot$$

$$(-1)^{k(j-1)}(-1)^{a-1}\cdot(-1)^{(b-1)\left(\sum_{r=1}^{k}|\zeta_{\sigma(r)}|+k+\sum_{s=k+1}^{k+a-1}|\zeta_{\sigma\beta(s)}|\right)}.$$

$$\left[f_a\left(\ell_k\left(\zeta_{\sigma(1)},\ldots,\zeta_{\sigma(k)}\right),\zeta_{\sigma\beta(k+1)},\ldots,\zeta_{\sigma\beta(k+a-1)}\right),f_b\left(\zeta_{\sigma\beta(k+a)},\ldots,\zeta_{\sigma\beta(n)}\right)\right]_{\mathfrak{g}}.$$

Here, for $r,s,t\in\mathbf{N}$, we have denoted by $S_{s,t}^{r+1}$ the set of all the permutations σ of $\{r+1,\ldots,r+s+t\}$, such that $\sigma(r+1)<\cdots<\sigma(r+s)$ and $\sigma(r+s+1)<\cdots<\sigma(r+s+t)$. A permutation $\sigma\in S_{s,t}^{r+1}$ can also be seen as a permutation of $\{1,\ldots,r+s+t\}$, simply by fixing $\sigma_{|\{1,\ldots,r\}}=\mathrm{id}_{|\{1,\ldots,r\}}$.

REMARK 3.4. Let us justify how one obtains that the sum

$$\mathfrak{d}(\xi_1,\ldots,\xi_m) :=$$

$$\sum_{\substack{j+k=m+1 \\ j,k\geq 2}}\sum_{\sigma\in S_{k,m-k}}\chi(\sigma)\,(-1)^{k(j-1)}T_j\left(\ell_k\left(\xi_{\sigma(1)},\ldots,\xi_{\sigma(k)}\right),\xi_{\sigma(k+1)},\ldots,\xi_{\sigma(m)}\right)$$

is given by $\mathfrak{a}_m(\xi_1,\ldots,\xi_m)+\mathfrak{b}_m(\xi_1,\ldots,\xi_m)+\mathfrak{c}_m(\xi_1,\ldots,\xi_m)$, using only the definition of the T_j. Let $\xi_1,\ldots,\xi_m\in H(\mathfrak{g},\partial_{\mathfrak{g}})$ be homogeneous elements and let $j,k\geq 2$ with $j+k=m+1$, and $\sigma\in S_{k,m-k}$. In order to simplify the notation, we denote by $\eta_1:=\ell_k\left(\xi_{\sigma(1)},\ldots,\xi_{\sigma(k)}\right)$ and $\eta_2:=\xi_{\sigma(k+1)},\ldots,\eta_j:=\xi_{\sigma(m)}$ and write:

$$T_j\left(\eta_1,\eta_2,\ldots,\eta_j\right)=$$

$$\sum_{\substack{p+q=j+1 \\ p,q\geq 2}}\sum_{\gamma\in S_{q,j-q}}\chi(\gamma;\eta_1,\ldots,\eta_j)(-1)^{q(p-1)}f_p\left(\ell_q\left(\eta_{\gamma(1)},\ldots,\eta_{\gamma(q)}\right),\eta_{\gamma(q+1)},\ldots,\eta_{\gamma(j)}\right)$$

$$-\sum_{\substack{a+b=j \\ a,b\geq 1}}\sum_{\substack{\gamma'\in S_{a,j-a} \\ \gamma'(1)=1}}\chi(\gamma';\xi_{\sigma(k+1)},\ldots,\xi_{\sigma(m)})(-1)^{a-1+(b-1)\left(\sum_{r=1}^{a}|\eta_{\gamma'(r)}|\right)}.$$

$$\left[f_a\left(\eta_{\gamma'(1)},\ldots,\eta_{\gamma'(a)}\right),f_b\left(\eta_{\gamma'(a+1)},\ldots,\eta_{\gamma'(j)}\right)\right]_{\mathfrak{g}}.$$

Then, the second sum leads easily to $\mathfrak{c}_m(\xi_1,\ldots,\xi_m)$ and for the first sum, one has to separate the two cases where the permutation $\gamma\in S_{q,j-q}$, which appears in the sum, satisfies $\gamma(1) = 1$ or $\gamma(q+1) = 1$, to obtain respectively the terms $\mathfrak{a}_m(\xi_1, \ldots, \xi_m)$ and $\mathfrak{b}_m(\xi_1, \ldots, \xi_m)$. Indeed, if $\gamma(1) = 1$, then there exists $\alpha \in S^{k+1}_{q-1,p-1}$ such that:

$$
\begin{aligned}
\eta_{\gamma(1)} &= \ell_k(\xi_{\sigma(1)}, \ldots, \xi_{\sigma(k)}), \\
\eta_{\gamma(2)} &= \xi_{\sigma\alpha(k+1)}, \qquad\qquad \eta_{\gamma(q+1)} = \xi_{\sigma\alpha(k+q)} \\
&\vdots \qquad\qquad\qquad\qquad\qquad\qquad\quad \vdots \\
\eta_{\gamma(q)} &= \xi_{\sigma\alpha(k+q-1)}, \qquad\quad \eta_{\gamma(j)} = \xi_{\sigma\alpha(m)}.
\end{aligned}
$$

By checking that $\chi(\gamma; \eta_1, \ldots, \eta_j) = \chi(\alpha; \xi_{\sigma(k+1)}, \ldots, \xi_{\sigma(m)})$, one obtains the sum $\mathfrak{a}_m(\xi_1, \ldots, \xi_m)$. In the case $\gamma(q+1) = 1$, one can rather write:

$$
\begin{aligned}
\eta_{\gamma(1)} &= \xi_{\sigma\alpha(k+1)}, \qquad\qquad \eta_{\gamma(q+2)} = \xi_{\sigma\alpha(k+q+1)}, \\
&\vdots \qquad\qquad\qquad\qquad\qquad\qquad\quad \vdots \\
\eta_{\gamma(q)} &= \xi_{\sigma\alpha(k+q)}, \qquad\qquad \eta_{\gamma(j)} = \xi_{\sigma\alpha(m)}, \\
\eta_{\gamma(q+1)} &= \ell_k(\xi_{\sigma(1)}, \ldots, \xi_{\sigma(k)}),
\end{aligned}
$$

with $\alpha \in S^{k+1}_{q,p-2}$. It is then possible to compute that $\operatorname{sign}(\gamma) = \operatorname{sign}(\alpha) \cdot (-1)^q$ and

$$
\varepsilon(\gamma; \eta_1, \ldots, \eta_j) = \varepsilon(\alpha; \xi_{\sigma(k+1)}, \ldots, \xi_{\sigma(m)}) \cdot (-1)^{\left(\sum\limits_{s=1}^{k} |\xi_{\sigma(s)}|+k\right) \cdot \left(\sum\limits_{r=k+1}^{k+q} |\xi_{\sigma\alpha(r)}|\right)}. \text{ This}
$$

permits one to obtain the sum $\mathfrak{b}_m(\xi_1, \ldots, \xi_m)$.

Now, we will successively show that both sums $\mathfrak{a}_m(\xi_1, \ldots, \xi_m)$ and $\mathfrak{b}_m(\xi_1, \ldots, \xi_m)$ are equal to zero. To do this, we prove the following lemmas.

LEMMA 3.5. *Let $n \in \mathbf{N}^*$. Suppose that the equations $(\mathfrak{J}_j)$ for $1 \leq j \leq n-1$ are satisfied by the maps $\ell_1 = 0, \ell_2, \ldots, \ell_{n-1}$, then*

$$
\mathfrak{a}_n(\zeta_1, \ldots, \zeta_n) = 0, \quad \text{for all } \zeta_1, \ldots, \zeta_n \in H(\mathfrak{g}, \partial_\mathfrak{g}).
$$

PROOF OF LEMMA 3.5. Let $\zeta_1, \ldots, \zeta_n \in H(\mathfrak{g}, \partial_\mathfrak{g})$. For $p, q, k \geq 2$ such that $p+q+k = n+2$, and for $\sigma \in S_{k,n-k}$ and $\alpha \in S^{k+1}_{q-1,p-1}$, the permutation $\sigma \circ \alpha \in \mathfrak{S}_n$ can be uniquely written as $\sigma \circ \alpha = \rho \circ \beta$, with $\rho \in S_{n-p+1,p-1}$ and $\beta \in S_{k,q-1}$. Using this, one obtains:

$$
\mathfrak{a}_n(\zeta_1, \ldots, \zeta_n) =
$$
$$
\sum_{p=2}^{n-2} \sum_{\rho \in S_{n-p+1,p-1}} \chi(\rho)(-1)^{(n-p)(p-1)} f_p\left(J_p(\zeta_{\rho(1)}, \ldots, \zeta_{\rho(n-p+1)}), \zeta_{\rho(n-p+2)}, \ldots, \zeta_{\rho(n)}\right),
$$

where $\chi(\rho)$ stands for $\chi(\rho; \zeta_1, \ldots, \zeta_n)$ and J_p is defined in (2). For every $2 \leq p \leq n-2$ and every $\rho \in S_{n-p+1,p-1}$, one has $J_p(\zeta_{\rho(1)}, \ldots, \zeta_{\rho(n-p+1)}) = 0$, by $(\mathfrak{J}_{n-p+1})$, where $n-p+1 = k+q-1$ runs through all integers between 3 and $n-1$. Hence $\mathfrak{a}_n(\zeta_1, \ldots, \zeta_n) = 0$. $\qquad\square$

According to this lemma, and because the maps $\ell_1 = 0, \ell_2, \ldots, \ell_{m-1}$ are supposed to satisfy the equations $(\mathfrak{J}_1) - (\mathfrak{J}_{m-1})$, we have $\mathfrak{a}_m(\xi_1, \ldots, \xi_m) = 0$. Let us now consider the sum $\mathfrak{b}_m(\xi_1, \ldots, \xi_m)$. It is also zero, according to the following:

LEMMA 3.6. *Let $n \in \mathbf{N}^*$. For all $\zeta_1, \ldots, \zeta_n \in H(\mathfrak{g}, \partial_\mathfrak{g})$, we have*

$$
\mathfrak{b}_n(\zeta_1, \ldots, \zeta_n) = 0.
$$

PROOF OF LEMMA 3.6. This result follows from the skew-symmetry of the maps $f_1, \ldots, f_n$, making the sum $\mathfrak{b}_n(\zeta_1, \ldots, \zeta_n)$ equal to minus itself. $\qquad\square$

Now, we consider the term $\partial_{\mathfrak{g}} \left(U_m(\xi_1, \ldots, \xi_m) \right)$. As $\partial_{\mathfrak{g}}$ is a graded derivation for $[\cdot, \cdot]_{\mathfrak{g}}$ and because $[\cdot, \cdot]_{\mathfrak{g}}$ is skew-symmetric, one has, for all $\zeta_1, \ldots, \zeta_m \in H(\mathfrak{g}, \partial_{\mathfrak{g}})$ and all $s, t \in \{1, \ldots m - 1\}$ such that $s + t = m$:

$$\partial_{\mathfrak{g}} \left([f_s(\zeta_1, \ldots, \zeta_s), f_t(\zeta_{s+1}, \ldots, \zeta_m)]_{\mathfrak{g}} \right) =$$

$$[\partial_{\mathfrak{g}}(f_s(\zeta_1, \ldots, \zeta_s)), f_t(\zeta_{s+1}, \ldots, \zeta_m)]_{\mathfrak{g}}$$

$$-(-1)^{|f_s(\zeta_1, \ldots, \zeta_s)|(1 + |\partial_{\mathfrak{g}}(f_t(\zeta_{s+1}, \ldots, \zeta_m))|)} [\partial_{\mathfrak{g}}(f_t(\zeta_{s+1}, \ldots, \zeta_m)), f_s(\zeta_1, \ldots, \zeta_s)]_{\mathfrak{g}}.$$

Using this, the one-to-one correspondence between the set $\{\tau \in S_{s,m-s} \mid \tau(1) = 1\}$ and the set $\{\tau' \in S_{t,m-t} \mid \tau'(t+1) = 1\}$ and finally the fact that $S_{s,m-s} = \{\tau \in S_{s,m-s} \mid \tau(1) = 1\} \sqcup \{\tau \in S_{s,m-s} \mid \tau(s+1) = 1\}$, we obtain that:

$$\partial_{\mathfrak{g}} \left(U_m(\xi_1, \ldots, \xi_m) \right) =$$

$$\sum_{\substack{s+t=m \\ s,t \geq 1}} \sum_{\tau \in S_{s,m-s}} \chi(\tau) e_{s,t}(\tau) \left[\partial_{\mathfrak{g}} \left(f_s(\xi_{\tau(1)}, \ldots, \xi_{\tau(s)}) \right), f_t(\xi_{\tau(s+1)}, \ldots, \xi_{\tau(m)}) \right]_{\mathfrak{g}}.$$

Finally, it remains for $\partial_{\mathfrak{g}} \left(T_m(\xi_1, \ldots, \xi_m) \right)$:

$$\partial_{\mathfrak{g}} \left(T_m(\xi_1, \ldots, \xi_m) \right) = \partial_{\mathfrak{g}} \left(S_m(\xi_1, \ldots, \xi_m) \right) - \partial_{\mathfrak{g}} \left(U_m(\xi_1, \ldots, \xi_m) \right)$$

$$= c_m(\xi_1, \ldots, \xi_m)$$

$$- \sum_{\substack{s+t=m \\ s,t \geq 1}} \sum_{\tau \in S_{s,m-s}} \chi(\tau) e_{s,t}(\tau) \left[\partial_{\mathfrak{g}} \left(f_s(\xi_{\tau(1)}, \ldots, \xi_{\tau(s)}) \right), f_t(\xi_{\tau(s+1)}, \ldots, \xi_{\tau(m)}) \right]_{\mathfrak{g}}.$$

We now point out that, for all $n \in \mathbf{N}^*$ and for all $\zeta_1, \ldots, \zeta_n \in H(\mathfrak{g}, \partial_{\mathfrak{g}})$,
(26)

$$c_n(\zeta_1, \ldots, \zeta_n) =$$

$$\sum_{\substack{s+t=n \\ s,t \geq 1}} \sum_{\tau \in S_{s,n-s}} \chi(\tau) e_{s,t}(\tau) \left[(f_1 \circ \ell_s + S_s)(\zeta_{\tau(1)}, \ldots, \zeta_{\tau(s)}), f_t(\zeta_{\tau(s+1)}, \ldots, \zeta_{\tau(n)}) \right]_{\mathfrak{g}}.$$

We use once more the equation $(\mathcal{E}_s)$ and (23) to write $\partial_{\mathfrak{g}} \circ f_s = f_1 \circ \ell_s + T_s = f_1 \circ \ell_s + S_s - U_s$, for $s = 1, \ldots, m - 1$, and to obtain:

$$\partial_{\mathfrak{g}} \left(T_m(\xi_1, \ldots, \xi_m) \right) =$$

$$\sum_{\substack{s+t=m \\ s,t \geq 1}} \sum_{\tau \in S_{s,m-s}} \chi(\tau) e_{s,t}(\tau) \left[U_s(\xi_{\tau(1)}, \ldots, \xi_{\tau(s)}), f_t(\xi_{\tau(s+1)}, \ldots, \xi_{\tau(m)}) \right]_{\mathfrak{g}}.$$

Written differently, this reads as follows:
(27)
$$\partial_{\mathfrak{g}} \left(T_m(\xi_1, \ldots, \xi_m) \right) = R_m(\xi_1, \ldots, \xi_m),$$

where we have introduced the following notation (because we will need this notation later): for all $n \in \mathbf{N}^*$ and all $\zeta_1, \ldots, \zeta_n \in H(\mathfrak{g}, \partial_{\mathfrak{g}})$,

$$R_n(\zeta_1, \ldots, \zeta_n) :=$$

$$\sum_{\substack{a+b+t=n \\ a,b,t \geq 1}} \sum_{\substack{\tau \in S_{a+b,t} \\ \sigma \in S_{a,b} \\ \sigma(1)=1}} \chi(\tau; \zeta_1, \ldots, \zeta_n) \chi(\sigma; \zeta_{\tau(1)}, \ldots, \zeta_{\tau(a+b)}) e_{a+b,t}(\tau) e_{a,b}(\tau \circ \sigma) \cdot$$

$$\left[\left[f_a(\zeta_{\tau\sigma(1)}, \ldots, \zeta_{\tau\sigma(a)}), f_b(\zeta_{\tau\sigma(a+1)}, \ldots, \zeta_{\tau\sigma(a+b)}) \right]_{\mathfrak{g}}, f_t(\zeta_{\tau(a+b+1)}, \ldots, \zeta_{\tau(n)}) \right]_{\mathfrak{g}}.$$

It is then possible to show that this is zero, using the graded Jacobi identity satisfied by $[\cdot,\cdot]_{\mathfrak{g}}$. Because we will need this result in another context, we show the following:

LEMMA 3.7. *For $n \in \mathbf{N}^*$ and all $\zeta_1, \ldots, \zeta_n \in H(\mathfrak{g}, \partial_{\mathfrak{g}})$, one has:*

$$R_n(\zeta_1, \ldots, \zeta_n) = 0.$$

PROOF OF LEMMA 3.7. Let $\zeta_1, \ldots, \zeta_n \in H(\mathfrak{g}, \partial_{\mathfrak{g}})$. One first can show that

$$2\, R_n(\zeta_1, \ldots, \zeta_n) =$$

$$\sum_{\substack{a+b+t=n \\ a,b,t\geq 1}} \sum_{\rho \in S_{a,b,t}} \chi(\rho; \zeta_1, \ldots, \zeta_n)\, e_{a+b,t}(\rho)\, e_{a,b}(\rho) \cdot$$

$$\left[\left[f_a\left(\zeta_{\rho(1)}, \ldots, \zeta_{\rho(a)}\right), f_b\left(\zeta_{\rho(a+1)}, \ldots, \zeta_{\rho(a+b)}\right) \right]_{\mathfrak{g}}, f_t\left(\zeta_{\rho(a+b+1)}, \ldots, \zeta_{\rho(n)}\right) \right]_{\mathfrak{g}},$$

where for $a, b, t \in \mathbf{N}$, $S_{a,b,t}$ is the set of all the permutations $\sigma \in \mathfrak{S}_{a+b+t}$ of $\{1, \ldots, a+b+t\}$, satisfying: $\sigma(1) < \cdots < \sigma(a)$, $\sigma(a+1) < \cdots < \sigma(a+b)$ and $\sigma(a+b+1) < \cdots < \sigma(a+b+t)$. It is now possible to check that one has:

$$6\, R_n(\zeta_1, \ldots, \zeta_n) = \sum_{\substack{a+b+t=n \\ a,b,t\geq 1}} \sum_{\rho \in S_{a,b,t}} \chi(\rho)\,(-1)^e \cdot$$

$$\mathrm{Jac}_{\mathfrak{g}}\left(f_a\left(\zeta_{\rho(1)}, \ldots, \zeta_{\rho(a)}\right), f_b\left(\zeta_{\rho(a+1)}, \ldots, \zeta_{\rho(a+b)}\right), f_t\left(\zeta_{\rho(a+b+1)}, \ldots, \zeta_{\rho(n)}\right) \right),$$

where $e \in \mathbf{Z}$ is an integer depending on $\zeta_1, \ldots, \zeta_n$ and on the permutation ρ, and where, for all $x, y, z \in \mathfrak{g}$,

$$\mathrm{Jac}_{\mathfrak{g}}(x, y, z) := (-1)^{|x||z|} \left[[x, y]_g, z \right]_{\mathfrak{g}} + (-1)^{|y||x|} \left[[y, z]_g, x \right]_{\mathfrak{g}} + (-1)^{|z||y|} \left[[z, x]_g, y \right]_{\mathfrak{g}},$$

which is zero because of the graded Jacobi identity satisfied by $[\cdot,\cdot]_{\mathfrak{g}}$. We now conclude that $R_n(\zeta_1, \ldots, \zeta_n) = 0$. $\qquad \square$

This lemma, together with (27), imply that $\partial_{\mathfrak{g}}(T_m(\xi_1, \ldots, \xi_m)) = 0$. This fact means that, for all $\xi_1, \ldots, \xi_m \in H(\mathfrak{g}, \partial_{\mathfrak{g}})$, the element $T_m(\xi_1, \ldots, \xi_m)$ is a cocycle for the cochain complex $(\mathfrak{g}, \partial_{\mathfrak{g}})$. This allows us to define a skew-symmetric graded linear map $\ell_m : \bigwedge^m H(\mathfrak{g}, \partial_{\mathfrak{g}}) \to H(\mathfrak{g}, \partial_{\mathfrak{g}})$, of degree $2 - m$, with the following formula:

$$(28) \qquad \ell_m(\xi_1, \ldots, \xi_m) := -p \circ T_m(\xi_1, \ldots, \xi_m),$$

for all $\xi_1, \ldots, \xi_m \in H(\mathfrak{g}, \partial_{\mathfrak{g}})$. As in the case $m = 2$ and according to (19), we also have the existence of a skew-symmetric graded linear map $f_m : \bigwedge^m H(\mathfrak{g}, \partial_{\mathfrak{g}}) \to \mathfrak{g}$, of degree $1 - m$, which satisfies the equation $(\mathcal{E}_m)$:

$$T_m(\xi_1, \ldots, \xi_m) = \partial_{\mathfrak{g}}(f_m(\xi_1, \ldots, \xi_m)) - f_1(\ell_m(\xi_1, \ldots, \xi_m)),$$

for all $\xi_1, \ldots, \xi_m \in H(\mathfrak{g}, \partial_{\mathfrak{g}})$.

(b) - *Step 2.* It remains to show, using the equations $(\mathfrak{J}_1) - (\mathfrak{J}_m)$ and $(\mathcal{E}_1) - (\mathcal{E}_{m-1})$, satisfied by the maps $\ell_1, \ldots, \ell_{m-1}$ and $f_1, \ldots, f_{m-1}$ and the equation $(\mathcal{E}_m)$ also satisfied by the maps ℓ_m and f_m, that the map ℓ_m, defined in (28), satisfies necessarily, for all $\xi_1, \ldots, \xi_{m+1} \in H(\mathfrak{g}, \partial_{\mathfrak{g}})$, the equation:

$(\mathfrak{J}_{m+1})$

$$\sum_{\substack{j+k=m+2 \\ j,k\geq 2}} \sum_{\sigma \in S_{k,m+1-k}} \chi(\sigma)(-1)^{k(j-1)}\, \ell_j\left(\ell_k\left(\xi_{\sigma(1)}, \ldots, \xi_{\sigma(k)}\right), \xi_{\sigma(k+1)}, \ldots, \xi_{\sigma(m)}\right) = 0.$$

Let us fix $\xi_1, \ldots, \xi_{m+1} \in H(\mathfrak{g}, \partial_\mathfrak{g})$. By equations $(\mathcal{E}_1) - (\mathcal{E}_m)$, we know that the maps ℓ_j, for $1 \leq j \leq m$, can be written as $\ell_j = -p \circ T_j$. Using the notation of the remark 3.4, this implies that $(\mathfrak{J}_{m+1})$ is equivalent to:

$$p\left(\partial_{m+1}(\xi_1, \ldots, \xi_{m+1})\right) = 0.$$

We also use the same reasoning as the one explained in the remark 3.4 to obtain:

$$\partial_{m+1}(\xi_1, \ldots, \xi_{m+1}) = (\mathfrak{a}_{m+1} + \mathfrak{b}_{m+1} + \mathfrak{c}_{m+1})(\xi_1, \ldots, \xi_{m+1}).$$

Then, the lemma 3.5, together with the fact that the maps $\ell_2, \ldots, \ell_m$ satisfy the equations $(\mathfrak{J}_j)$ for $1 \leq j \leq m$, imply that $\mathfrak{a}_{m+1}(\xi_1, \ldots, \xi_{m+1}) = 0$. Secondly, the lemma 3.6 also says that $\mathfrak{b}_{m+1}(\xi_1, \ldots, \xi_{m+1}) = 0$. Finally it remains that:

$$(\mathfrak{J}_{m+1}) \text{ is equivalent to: } p\left(\mathfrak{c}_{m+1}(\xi_1, \ldots, \xi_{m+1})\right) = 0,$$

which is also equivalent to say that $\mathfrak{c}_{m+1}(\xi_1, \ldots, \xi_{m+1})$ is a coboundary for the cochain complex $(\mathfrak{g}, \partial_\mathfrak{g})$. As (26) can be obtained without using anything but the definitions of $\mathfrak{c}_m$ and S_s, we also have:

$$\mathfrak{c}_{m+1}(\xi_1, \ldots, \xi_{m+1}) =$$
$$\sum_{\substack{p+q=m+1 \\ q \geq 1, p \geq 2}} \sum_{\alpha \in S_{p,q}} \chi(\alpha) e_{p,q}(\alpha) \left[(S_p + f_1 \circ \ell_p)\left(\xi_{\alpha(1)}, \ldots, \xi_{\alpha(p)}\right), f_q\left(\xi_{\alpha(p+1)}, \ldots, \xi_{\alpha(m+1)}\right)\right]_\mathfrak{g}.$$

Now, we use $S_p = T_p + U_p$ and the equations $(\mathcal{E}_p)$, satisfied by the maps ℓ_p and f_p, for $1 \leq p \leq m$, to write $S_p + f_1 \circ \ell_p = \partial_\mathfrak{g} \circ f_p + U_p$ and:

$$\mathfrak{c}_{m+1}(\xi_1, \ldots, \xi_{m+1}) =$$
$$\sum_{\substack{p+q=m+1 \\ q \geq 1, p \geq 2}} \sum_{\alpha \in S_{p,q}} \chi(\alpha) e_{p,q}(\alpha) \left[\partial_\mathfrak{g}\left(f_p\left(\xi_{\alpha(1)}, \ldots, \xi_{\alpha(p)}\right)\right), f_q\left(\xi_{\alpha(p+1)}, \ldots, \xi_{\alpha(m+1)}\right)\right]_\mathfrak{g}$$
$$+ \ R_{m+1}(\xi_1, \ldots, \xi_{m+1}).$$

By lemma 3.7, $R_{m+1}(\xi_1, \ldots, \xi_{m+1}) = 0$, and using the bijection between $S_{p,q}$ and $S_{q,p}$, given by:

$$S_{p,q} \ \rightarrow \ S_{q,p}$$
$$\alpha \ \mapsto \ \alpha' := \begin{pmatrix} 1 & \cdots & q & q+1 & \cdots & p+q \\ \alpha(p+1) & \cdots & \alpha(p+q) & \alpha(1) & \cdots & \alpha(p) \end{pmatrix},$$

for which

$$\mathrm{sign}(\alpha') \ = \ \mathrm{sign}(\alpha) \cdot (-1)^{pq},$$

$$\varepsilon(\alpha'; \xi_1, \ldots, \xi_{p+q}) \ = \ \varepsilon(\alpha; \xi_1, \ldots, \xi_{p+q}) \cdot (-1)^{\left(\sum_{r=1}^{q} |\xi_{\alpha(r)}|\right) \cdot \left(\sum_{r=q+1}^{p+q} |\xi_{\alpha(r)}|\right)},$$

and also using the skew-symmetry of $[\cdot, \cdot]_\mathfrak{g}$ and the fact that $\partial_\mathfrak{g}$ is a graded derivation for $[\cdot, \cdot]_\mathfrak{g}$, we finally obtain:

$$2\,\mathfrak{c}_{m+1}(\xi_1, \ldots, \xi_{m+1}) =$$
$$\sum_{\substack{p+q=m+1 \\ q, p \geq 2}} \sum_{\alpha \in S_{p,q}} \chi(\alpha) e_{p,q}(\alpha) \, \partial_\mathfrak{g}\left(\left[f_p\left(\xi_{\alpha(1)}, \ldots, \xi_{\alpha(p)}\right), f_q\left(\xi_{\alpha(p+1)}, \ldots, \xi_{\alpha(m+1)}\right)\right]_\mathfrak{g}\right).$$

We have then obtained that $\mathfrak{c}_{m+1}(\xi_1, \ldots, \xi_{m+1})$ is a coboundary for the cochain complex $(\mathfrak{g}, \partial_\mathfrak{g})$, so that $\partial_\mathfrak{g}\left(\mathfrak{c}_{m+1}(\xi_1, \ldots, \xi_{m+1})\right) = 0$ and the equation $(\mathfrak{J}_{m+1})$ is satisfied. This finishes the proof of the proposition 3.1. $\qquad \square$

4. Deformations of Poisson structures via L_∞-algebras

In this section, we consider a family of dg Lie algebras, constructed from a family of Poisson structures in dimension three. We will then use the proposition 3.1, to obtain a classification of all formal deformations of these Poisson structures in the generic case, together with an explicit formula for the representative of each equivalence classes of these deformations.

4.1. Poisson structures in dimension three and their cohomology.

In the following, $\mathcal{A}$ denotes the polynomial algebra in three generators $\mathcal{A} := \mathbf{F}[x, y, z]$, where $\mathbf{F}$ is an arbitrary field of characteristic zero. To each polynomial $\varphi \in \mathcal{A}$, one associates a Poisson structure $\{\cdot, \cdot\}_\varphi$ defined by:

$$(29) \qquad \{\cdot, \cdot\}_\varphi := \frac{\partial \varphi}{\partial x} \frac{\partial}{\partial y} \wedge \frac{\partial}{\partial z} + \frac{\partial \varphi}{\partial y} \frac{\partial}{\partial z} \wedge \frac{\partial}{\partial x} + \frac{\partial \varphi}{\partial z} \frac{\partial}{\partial x} \wedge \frac{\partial}{\partial y}.$$

In this context, the Poisson cohomology of $(\mathcal{A}, \{\cdot, \cdot\}_\varphi)$ is denoted by $H(\mathcal{A}, \{\cdot, \cdot\}_\varphi)$. We also denote by $(\mathfrak{g}_\varphi, \partial_\varphi, [\cdot, \cdot]_S)$, the dg Lie algebra associated to the Poisson algebra $(\mathcal{A}, \{\cdot, \cdot\}_\varphi)$, as explained in the paragraph 2.2.2. Notice that $\mathfrak{g}_\varphi^k \simeq \{0\}$, for all $k \geq 3$. With these notations, and those of the previous section, we have: $H^n(\mathfrak{g}_\varphi, \partial_\varphi) = H^{n+1}(\mathcal{A}, \{\cdot, \cdot\}_\varphi)$, for all $n \in \mathbf{Z}$ (in fact, $n \in \mathbf{N} \cup \{-1\}$). As previously, for every cocycle P of the cochain complex $(\mathfrak{g}_\varphi, \partial_\varphi)$, $\bar{P}$ denotes its cohomology class in $H(\mathfrak{g}_\varphi, \partial_\varphi)$. As we want to use the result of the previous section (proposition 3.1), we need to choose representatives $(\vartheta_k^n)_k$ of an $\mathbf{F}$-basis of $H^n(\mathfrak{g}_\varphi, \partial_\varphi)$, for $n \in \mathbf{Z}$. To do this, we use the results of [16], in which the polynomial φ is supposed to be weight-homogeneous and with an isolated singularity (at the origin). Let us recall that a polynomial $\varphi \in \mathbf{F}[x, y, z]$ is said to be *weight homogeneous* of (weighted) degree $\varpi(\varphi) \in \mathbf{N}$, if there exists (unique) positive integers $\varpi_1, \varpi_2, \varpi_3 \in \mathbf{N}^*$ (the *weights* of the variables x, y and z), without any common divisor, such that:

$$(30) \qquad \varpi_1 x \frac{\partial \varphi}{\partial x} + \varpi_2 y \frac{\partial \varphi}{\partial y} + \varpi_3 z \frac{\partial \varphi}{\partial z} = \varpi(\varphi)\varphi.$$

This equation is called the *Euler Formula* and can also be written as: $\vec{e}_\varpi[\varphi] = \varpi(\varphi)\varphi$, where $\vec{e}_\varpi$ is the so-called *Euler derivation* (associated to the weights of the variables), defined by:

$$\vec{e}_\varpi := \varpi_1 x \frac{\partial}{\partial x} + \varpi_2 y \frac{\partial}{\partial y} + \varpi_3 z \frac{\partial}{\partial z}.$$

Recall that a weight homogeneous polynomial $\varphi \in \mathbf{F}[x, y, z]$ is said to admit an *isolated singularity* (at the origin) if the vector space

$$(31) \qquad \mathcal{A}_{sing}(\varphi) := \mathbf{F}[x, y, z] / \langle \frac{\partial \varphi}{\partial x}, \frac{\partial \varphi}{\partial y}, \frac{\partial \varphi}{\partial z} \rangle$$

is finite-dimensional. Its dimension is then denoted by μ and called the *Milnor number* associated to φ. When $\mathbf{F} = \mathbf{C}$, this amounts, geometrically, to saying that the surface $\mathcal{F}_\varphi : \{\varphi = 0\}$ has a singular point only at the origin.

From now on, the polynomial φ will always be a weight homogeneous polynomial with an isolated singularity. The corresponding weights of the three variables $(\varpi_1, \varpi_2 \text{ and } \varpi_3)$ are then fixed and the weight homogeneity of any polynomial in $\mathcal{A} = \mathbf{F}[x, y, z]$ has now to be understood as associated to these weights. In the following, $|\varpi|$ denotes the sum of the weights of the three variables x, y and z: $|\varpi| := \varpi_1 + \varpi_2 + \varpi_3$ and we fix $u_0 := 1, u_1, \ldots, u_{\mu-1} \in \mathcal{A}$, a family composed of weight homogeneous polynomials in $\mathcal{A}$ whose images in $\mathcal{A}_{sing}(\varphi)$ give a basis of this $\mathbf{F}$-vector space (and $u_0 = 1$). (For example, one can choose the polynomials $u_0, \ldots, u_{\mu-1}$ as being monomials of $\mathbf{F}[x, y, z]$).

PROPOSITION 4.1 ([**16**]). *Let $\varphi \in \mathcal{A}$ be a weight-homogeneous polynomial with an isolated singularity. Let $(\mathfrak{g}_\varphi, \partial_\varphi, [\cdot, \cdot]_S)$ denote the dg Lie algebra associated to the Poisson algebra $(\mathcal{A}, \{\cdot, \cdot\}_\varphi)$, as explained in the paragraph 2.2.2, and where $\{\cdot, \cdot\}_\varphi$ is defined in (29). Here we give explicit representatives for $\mathbf{F}$-bases of the Poisson cohomology spaces associated to $(\mathcal{A}, \{\cdot, \cdot\}_\varphi)$ or equivalently to $(\mathfrak{g}_\varphi, \partial_\varphi)$.*

(1) *An $\mathbf{F}$-basis of the first cohomology space $H^{-1}(\mathfrak{g}_\varphi, \partial_\varphi) = H^0(\mathcal{A}, \{\cdot, \cdot\}_\varphi)$ is given by:*

$$\mathbf{b}_\varphi^{-1} := \left(\overline{\varphi^i}, \, i \in \mathbf{N} \right);$$

(2) *An $\mathbf{F}$-basis of the space $H^0(\mathfrak{g}_\varphi, \partial_\varphi) = H^1(\mathcal{A}, \{\cdot, \cdot\}_\varphi)$ is given by:*

$$\mathbf{b}_\varphi^0 := \begin{cases} (0) & \text{if } \varpi(\varphi) \neq |\varpi|, \\ \left(\overline{\varphi^i \, \vec{e}_\varpi}, \, i \in \mathbf{N} \right) & \text{if } \varpi(\varphi) = |\varpi|; \end{cases}$$

(3) *An $\mathbf{F}$-basis of the space $H^1(\mathfrak{g}_\varphi, \partial_\varphi) = H^2(\mathcal{A}, \{\cdot, \cdot\}_\varphi)$ is given by:*

$$\mathbf{b}_\varphi^1 := \left(\overline{\varphi^i \, u_q \{\cdot, \cdot\}_\varphi}, \, i \in \mathbf{N}, q \in \mathcal{E}_\varphi \right) \cup \left(\overline{\{\cdot, \cdot\}_{u_r}}, \, 1 \leq r \leq \mu - 1 \right),$$

where

$$\mathcal{E}_\varphi := \begin{cases} \{1, \ldots, \mu - 1\} & \text{if } \varpi(\varphi) \neq |\varpi|, \\ \{0, \ldots, \mu - 1\} & \text{if } \varpi(\varphi) = |\varpi|, \end{cases}$$

and where the skew-symmetric biderivation $\{\cdot, \cdot\}_{u_q}$ is naturally obtained by replacing φ by u_q in (29);

(4) *An $\mathbf{F}$-basis of the space $H^2(\mathfrak{g}_\varphi, \partial_\varphi) = H^3(\mathcal{A}, \{\cdot, \cdot\}_\varphi)$ is given by:*

$$\mathbf{b}_\varphi^2 := \left(\overline{\varphi^i \, u_s \mathcal{D}}, \, i \in \mathbf{N}, 0 \leq s \leq \mu - 1 \right),$$

where $\mathcal{D}$ is the skew-symmetric triderivation of $\mathcal{A}$, defined by:

$$\mathcal{D} := \frac{\partial}{\partial x} \wedge \frac{\partial}{\partial y} \wedge \frac{\partial}{\partial z};$$

(5) *For $k \geq 3$,*

$$H^k(\mathfrak{g}_\varphi, \partial_\varphi) = H^{k+1}(\mathcal{A}, \{\cdot, \cdot\}_\varphi) \simeq \{0\}.$$

REMARK 4.2. More precisely, the basis of $H^2(\mathcal{A}, \{\cdot, \cdot\}_\varphi)$ given here is obtained by using the proposition 4.8 and the equality (27) of [**16**].

4.2. A suitable quasi-isomorphism between $H(\mathfrak{g}_\varphi, \partial_\varphi)$ and $\mathfrak{g}_\varphi$. Similarly to the definition (17), we now have a linear graded map f_1^φ of degree 0, associated to the bases $\mathbf{b}_\varphi^{-1}, \mathbf{b}_\varphi^0, \mathbf{b}_\varphi^1, \mathbf{b}_\varphi^2$:

(32)
$$f_1^\varphi: \quad H^\ell(\mathfrak{g}_\varphi, \partial_\varphi) \quad \rightarrow \quad Z^\ell(\mathfrak{g}_\varphi, \partial_\varphi)$$
$$\xi = \sum_k \lambda_k^\ell \overline{\vartheta_k^\ell} \quad \mapsto \quad \sum_k \lambda_k^\ell \vartheta_k^\ell,$$

where $\xi = \sum_k \lambda_k^\ell \overline{\vartheta_k^\ell}$ is the unique decomposition of ξ in the basis $\mathbf{b}_\varphi^\ell$, $\ell = -1, 0, 1, 2$, for which the elements $(\vartheta_k^\ell)_k$ denote here the representatives, chosen in the previous proposition 4.1, of the basis $\mathbf{b}_\varphi^\ell$.

Using the proposition 3.1 and the bases $\mathbf{b}_\varphi{}^1, \mathbf{b}_\varphi^0, \mathbf{b}_\varphi^1, \mathbf{b}_\varphi^2$ of the Poisson cohomology spaces associated to $(\mathfrak{g}_\varphi, \partial_\varphi)$, we construct an L_∞-algebra structure on $H(\mathfrak{g}_\varphi, \partial_\varphi)$: $(H(\mathfrak{g}_\varphi, \partial_\varphi), \ell_1 = 0, \ell_2 = [\cdot, \cdot]_S, \ell_3, \dots)$, and a (weak) L_∞-morphism

$$f_\bullet^\varphi = \left(f_n^\varphi : \bigoplus{}^n H(\mathfrak{g}_\varphi, \partial_\varphi) \to \mathfrak{g}_\varphi\right)_{n \in \mathbf{N}^*},$$

which extends f_1^φ, thus is a quasi-isomorphism. We indeed prove the following:

THEOREM 4.3. *Let* $\varphi \in \mathcal{A} = \mathbf{F}[x, y, z]$ *be a weight-homogeneous polynomial, with an isolated singularity and let* $\{\cdot, \cdot\}_\varphi$ *be the associated Poisson bracket defined in (29). Let* $(\mathfrak{g}_\varphi, \partial_\varphi, [\cdot, \cdot]_S)$ *be the dg Lie algebra associated to the Poisson cohomology complex of* $(\mathcal{A}, \{\cdot, \cdot\}_\varphi)$, *as explained in the paragraph 2.2.2. For simplicity, we denote by* H_φ *the space* $H(\mathfrak{g}_\varphi, \partial_\varphi)$, *and for all* $i \in \mathbf{N}^*$, H_φ^i *the space* $H^i(\mathfrak{g}_\varphi, \partial_\varphi)$ *(the i-th cohomology space associated to* $(\mathfrak{g}_\varphi, \partial_\varphi)$). *We fix* f_1^φ *as being the map defined in (32) and* $\ell_1^\varphi : H(\mathfrak{g}, \partial_\mathfrak{g}) \to H(\mathfrak{g}, \partial_\mathfrak{g})$ *as being the trivial map. We also fix the map* ℓ_2^φ *as being the bracket induced by the Schouten bracket* $[\cdot, \cdot]_S$, *i.e.,*

$$\ell_2^\varphi(\overline{x}, \overline{y}) := \overline{[x, y]_S}, \quad \text{for all } x, y \in \mathfrak{g}_\varphi.$$

There exist an L_∞-*algebra structure on* $H_\varphi := H(\mathfrak{g}_\varphi, \partial_\varphi)$, *denoted by* $\ell_\bullet^\varphi := (\ell_i^\varphi)_{i \in \mathbf{N}^*}$ *(with* ℓ_1^φ *and* ℓ_2^φ *given previously) and a quasi-isomorphism* $f_\bullet^\varphi := (f_i^\varphi)_{i \in \mathbf{N}^*}$ *(extending* f_1^φ) *from* H_φ *to the dg Lie algebra* $(\mathfrak{g}_\varphi, \partial_\varphi, [\cdot, \cdot]_S)$, *satisfying the following properties:*

(P_1) *The map* f_2^φ *is defined by the values given in the table 1, for the case* $\varpi(\varphi) \neq |\varpi|$, *and in the table 2, for the case* $\varpi(\varphi) = |\varpi|$;

(P_2) *For all* $i \geq 2$, *the map* ℓ_i^φ *is zero on* H_φ^1:

$$\ell_i^\varphi\big|_{(H_\varphi^1)^{\otimes i}} = 0, \quad \text{for all } i \geq 2;$$

(P_3) *For all* $i \geq 3$, *the map* f_i^φ *is zero on* H_φ^1:

$$f_i^\varphi\big|_{(H_\varphi^1)^{\otimes i}} = 0, \quad \text{for all } i \geq 3.$$

PROOF OF THEOREM 4.3. One can check (by a direct computation) that the following hold:

(33)
$$\begin{aligned}
\left[F(\varphi) u_k \{\cdot, \cdot\}_\varphi, G(\varphi) u_l \{\cdot, \cdot\}_\varphi\right]_S &= 0, \\
\left[F(\varphi) u_k \{\cdot, \cdot\}_\varphi, \{\cdot, \cdot\}_{u_t}\right]_S &= -\partial_\varphi\left(F(\varphi) u_k \{\cdot, \cdot\}_{u_t}\right), \\
\left[\{\cdot, \cdot\}_{u_s}, \{\cdot, \cdot\}_{u_t}\right]_S &= 0,
\end{aligned}$$

for all $0 \leq k, l \leq \mu - 1$ and all $1 \leq s, t \leq \mu - 1$ and for arbitrary elements $F(\varphi)$ and $G(\varphi)$ of $\mathbf{F}[\varphi]$. Because of (3) of proposition 4.1, this implies that the map ℓ_2^φ, which is the map induced by the Schouten bracket on the cohomology H_φ (and also denoted by $[\cdot, \cdot]_S$), is zero when restricted to $H_\varphi^1 \otimes H_\varphi^1$.

Now, by $\ell_1^\varphi = 0$ and the definition (32) of f_1^φ, it is straightforward to show that the skew-symmetric graded linear map $f_2^\varphi : \bigotimes^2 H(\mathfrak{g}_\varphi, \partial_\varphi) \to \mathfrak{g}_\varphi$, defined by the tables 1 and 2, together with $\ell_2^\varphi = [\cdot, \cdot]_S$, satisfy the equation $(\mathcal{E}_2)$. In particular, let us check this on $H_\varphi^1 \otimes H_\varphi^1$. Indeed, for all $0 \leq k, l \leq \mu - 1$ and for arbitrary

TABLE 1. Case $\varpi(\varphi) \neq |\varpi|$. The values of the linear map f_2^φ on the elements of the bases $\mathbf{b}_\varphi^i$ and $\mathbf{b}_\varphi^j$ of the spaces H_φ^i and H_φ^j, for $i, j = -1, 1, 2$. Notice that in this case, $H_\varphi^0 = \{0\}$. In this table, $F(\varphi), G(\varphi)$ are arbitrary elements of $\mathbf{F}[\varphi]$ and $1 \leq k, l, s, t \leq \mu - 1$.

$H_\varphi^i \times H_\varphi^j$	$(\overline{\vartheta^i}, \overline{\vartheta^j}) \in \mathbf{b}_\varphi^i \times \mathbf{b}_\varphi^j$	$f_2^\varphi(\overline{\vartheta^i}, \overline{\vartheta^j}) \in \mathfrak{g}_\varphi^{i+j-1}$
$H_\varphi^{-1} \times H_\varphi^1$	$\left(\overline{F(\varphi)}, \, \overline{G(\varphi) u_l \{\cdot,\cdot\}_\varphi}\right)$	0
	$\left(\overline{F(\varphi)}, \, \overline{\{\cdot,\cdot\}_{u_s}}\right)$	$F'(\varphi) u_s$
$H_\varphi^{-1} \times H_\varphi^2$	$\left(\overline{F(\varphi)}, \, \overline{G(\varphi) u_l \mathcal{D}}\right)$	0
	$\left(\overline{F(\varphi)}, \, \overline{G(\varphi)\mathcal{D}}\right)$	$\frac{1}{\varpi(\varphi)-|\varpi|} G(\varphi)F'(\varphi)\vec{e}_\varpi$
$H_\varphi^1 \times H_\varphi^1$	$\left(\overline{F(\varphi) u_k \{\cdot,\cdot\}_\varphi}, \, \overline{G(\varphi) u_l \{\cdot,\cdot\}_\varphi}\right)$	0
	$\left(\overline{F(\varphi) u_k \{\cdot,\cdot\}_\varphi}, \, \overline{\{\cdot,\cdot\}_{u_s}}\right)$	$F(\varphi) u_k \{\cdot,\cdot\}_{u_s}$
	$\left(\overline{\{\cdot,\cdot\}_{u_s}}, \, \overline{\{\cdot,\cdot\}_{u_t}}\right)$	0

TABLE 2. Case $\varpi(\varphi) = |\varpi|$. The values of the linear map f_2^φ on the elements of the bases $\mathbf{b}_\varphi^i$ and $\mathbf{b}_\varphi^j$ of the spaces H_φ^i and H_φ^j, for $i, j = -1, 0, 1, 2$. In this table, $F(\varphi), G(\varphi)$ are arbitrary elements of $\mathbf{F}[\varphi]$ and $0 \leq k, l \leq \mu - 1$ and $1 \leq s, t \leq \mu - 1$.

$H_\varphi^i \times H_\varphi^j$	$(\overline{\vartheta^i}, \overline{\vartheta^j}) \in \mathbf{b}_\varphi^i \times \mathbf{b}_\varphi^j$	$f_2^\varphi(\overline{\vartheta^i}, \overline{\vartheta^j}) \in \mathfrak{g}_\varphi^{i+j-1}$
$H_\varphi^{-1} \times H_\varphi^0$	$\left(\overline{F(\varphi)}, \, \overline{G(\varphi)\vec{e}_\varpi}\right)$	0
$H_\varphi^{-1} \times H_\varphi^1$	$\left(\overline{F(\varphi)}, \, \overline{G(\varphi) u_l \{\cdot,\cdot\}_\varphi}\right)$	0
	$\left(\overline{F(\varphi)}, \, \overline{\{\cdot,\cdot\}_{u_s}}\right)$	$F'(\varphi) u_s$
$H_\varphi^{-1} \times H_\varphi^2$	$\left(\overline{F(\varphi)}, \, \overline{G(\varphi) u_l \mathcal{D}}\right)$	0
$H_\varphi^0 \times H_\varphi^0$	$\left(\overline{F(\varphi)\vec{e}_\varpi}, \, \overline{G(\varphi)\vec{e}_\varpi}\right)$	0
$H_\varphi^0 \times H_\varphi^1$	$\left(\overline{F(\varphi)\vec{e}_\varpi}, \, \overline{G(\varphi) u_l \{\cdot,\cdot\}_\varphi}\right)$	0
	$\left(\overline{F(\varphi)\vec{e}_\varpi}, \, \overline{\{\cdot,\cdot\}_{u_s}}\right)$	$\left(\frac{\varpi(u_s)-|\varpi|}{|\varpi|} \frac{F(\varphi)-F(0)}{\varphi} - F'(\varphi)\right) u_s \vec{e}_\varpi$
$H_\varphi^0 \times H_\varphi^2$	$\left(\overline{F(\varphi)\vec{e}_\varpi}, \, \overline{G(\varphi) u_l \mathcal{D}}\right)$	0
$H_\varphi^1 \times H_\varphi^1$	$\left(\overline{F(\varphi) u_k \{\cdot,\cdot\}_\varphi}, \, \overline{G(\varphi) u_l \{\cdot,\cdot\}_\varphi}\right)$	0
	$\left(\overline{F(\varphi) u_k \{\cdot,\cdot\}_\varphi}, \, \overline{\{\cdot,\cdot\}_{u_s}}\right)$	$F(\varphi) u_k \{\cdot,\cdot\}_{u_s}$
	$\left(\overline{\{\cdot,\cdot\}_{u_s}}, \, \overline{\{\cdot,\cdot\}_{u_t}}\right)$	0

elements $F(\varphi)$ and $G(\varphi)$ of $\mathbf{F}[\varphi]$, the equation ($\mathcal{E}_2$) for $\xi_1 = \overline{F(\varphi) u_k \{\cdot,\cdot\}_\varphi}$ and

$\xi_2 = \overline{G(\varphi) \, u_l \, \{\cdot, \cdot\}_\varphi}$ becomes, using (33),

$$\partial_\varphi \left(f_2^\varphi \left(\overline{F(\varphi) \, u_k \, \{\cdot, \cdot\}_\varphi}, \overline{G(\varphi) \, u_l \, \{\cdot, \cdot\}_\varphi} \right) \right)$$

$$= \; f_1^\varphi \left(\left[\overline{F(\varphi) \, u_k \, \{\cdot, \cdot\}_\varphi}, \overline{G(\varphi) \, u_l \, \{\cdot, \cdot\}_\varphi} \right]_S \right) - \left[\overline{F(\varphi) \, u_k \, \{\cdot, \cdot\}_\varphi}, \overline{G(\varphi) \, u_l \, \{\cdot, \cdot\}_\varphi} \right]_S$$

$$= \; 0.$$

Similarly, one also obtains $\partial_\varphi \left(f_2^\varphi \left(\overline{\{\cdot, \cdot\}_{u_s}}, \overline{\{\cdot, \cdot\}_{u_t}} \right) \right) = 0$, for all $1 \le s, t \le \mu - 1$. Finally, for any arbitrary element $F(\varphi)$ of $\mathbf{F}[\varphi]$, and for all $0 \le k \le \mu - 1$ and $1 \le t \le \mu - 1$, the identities (33) imply that the equation ($\mathcal{E}_2$) for $\xi_1 = \overline{F(\varphi) \, u_k \, \{\cdot, \cdot\}_\varphi}$ and $\xi_2 = \overline{\{\cdot, \cdot\}_{u_t}}$ reads as follows

$$\partial_\varphi \left(f_2^\varphi \left(\overline{F(\varphi) \, u_k \, \{\cdot, \cdot\}_\varphi}, \overline{\{\cdot, \cdot\}_{u_t}} \right) \right)$$

$$= \; f_1^\varphi \left(\left[\overline{F(\varphi) \, u_k \, \{\cdot, \cdot\}_\varphi}, \overline{\{\cdot, \cdot\}_{u_t}} \right]_S \right) - \left[\overline{F(\varphi) \, u_k \, \{\cdot, \cdot\}_\varphi}, \overline{\{\cdot, \cdot\}_{u_t}} \right]_S$$

$$= \; \partial_\varphi \left(\overline{F(\varphi) \, u_k \, \{\cdot, \cdot\}_{u_t}} \right),$$

where we have used that $f_1^\varphi \circ \partial_\varphi = 0$. This implies that if the map f_2^φ takes, on $H_\varphi^1 \otimes H_\varphi^1$, the values given in the tables 1 and 2, then the previous equations are satisfied, i.e., the equation ($\mathcal{E}_2$) is satisfied on $H_\varphi^1 \otimes H_\varphi^1$. From now on, we fix f_2^φ to take, on $H_\varphi^1 \otimes H_\varphi^1$, the values given in the tables 1 and 2.

We have obtained the existence of the maps ℓ_1^φ, ℓ_2^φ and f_1^φ, f_2^φ, satisfying the equations ($\mathcal{E}_1$), ($\mathcal{E}_2$) and ($\mathfrak{J}_1$), ($\mathfrak{J}_2$), ($\mathfrak{J}_3$). By the proposition 3.1, this implies that there exist skew-symmetric graded linear maps

$$f_3^\varphi : \bigotimes^3 H(\mathfrak{g}_\varphi, \partial_\varphi) \to \mathfrak{g}_\varphi \quad \text{and} \quad \ell_3^\varphi : \bigotimes^3 H(\mathfrak{g}_\varphi, \partial_\varphi) \to H(\mathfrak{g}_\varphi, \partial_\varphi)$$

with $\deg(f_3^\varphi) = -2$, $\deg(\ell_3^\varphi) = -1$ and satisfying the equation ($\mathcal{E}_3$). Moreover, the proposition 3.1 also says that such a map ℓ_3^φ necessarily satisfies the equation ($\mathfrak{J}_4$).

In the equation ($\mathcal{E}_n$), we denote $T_n(\mathcal{F}_n^\varphi, \mathcal{L}_{n-1}^\varphi; \xi_1, \dots, \xi_n)$ by $T_n^\varphi(\xi_1, \dots, \xi_n)$, for $n \in \mathbf{N}^*$ and $\xi_1, \dots, \xi_n \in H(\mathfrak{g}, \partial_\mathfrak{g})$, when $\mathcal{F}_n^\varphi$ and $\mathcal{L}_{n-1}^\varphi$ denote the elements $\mathcal{F}_n^\varphi := (f_1^\varphi, \dots, f_n^\varphi)$ and $\mathcal{L}_{n-1}^\varphi := (\ell_1^\varphi, \dots, \ell_{n-1}^\varphi)$. By ($\mathcal{E}_3$), we have $\ell_3^\varphi := -p \circ T_3^\varphi$. Moreover, given the maps $\ell_1^\varphi, \ell_2^\varphi, f_1^\varphi, f_2^\varphi$ as previously, one can also verify that:

$$T_3^\varphi \big|_{\left(H_\varphi^1 \right)^{\otimes 3}} = 0,$$

so that, $\ell_3^\varphi \big|_{\left(H_\varphi^1 \right)^{\otimes 3}} = -p \circ T_3^\varphi \big|_{\left(H_\varphi^1 \right)^{\otimes 3}} = 0$, and the equation ($\mathcal{E}_3$) is still satisfied if we choose $f_3^\varphi \big|_{\left(H_\varphi^1 \right)^{\otimes 3}} := 0$, what we do from now on. Let us for example show that

$$T_3^\varphi \left(\overline{F(\varphi) u_l \, \{\cdot, \cdot\}_\varphi}, \overline{\{\cdot, \cdot\}_{u_s}}, \overline{\{\cdot, \cdot\}_{u_t}} \right) = 0,$$

for any arbitrary element $F(\varphi)$ of $\mathbf{F}[\varphi]$, for all $0 \le l \le \mu - 1$ and all $1 \le s, t \le \mu - 1$. First, let us point out that, by the definition (4) of T_3^φ, and because $\ell_2^\varphi \big|_{H_\varphi^1 \otimes H_\varphi^1} = 0$, we simply get, for any $\xi_1, \xi_2, \xi_3 \in H_\varphi^1$:

$$T_3^\varphi \left(\xi_1, \xi_2, \xi_3 \right) = \left[f_1^\varphi(\xi_1), f_2^\varphi(\xi_2, \xi_3) \right]_S + \left[f_2^\varphi(\xi_1, \xi_2), f_1^\varphi(\xi_3) \right]_S + \left[f_2^\varphi(\xi_1, \xi_3), f_1^\varphi(\xi_2) \right]_S.$$

Now, by the tables 1 and 2, we obtain:

$$T_3^\varphi \left(\overline{F(\varphi)u_l \{\cdot,\cdot\}_\varphi}, \overline{\{\cdot,\cdot\}_{u_s}}, \overline{\{\cdot,\cdot\}_{u_t}} \right) =$$

$$\left[F(\varphi)u_l \{\cdot,\cdot\}_{u_s}, \{\cdot,\cdot\}_{u_t} \right]_S + \left[F(\varphi)u_l \{\cdot,\cdot\}_{u_t}, \{\cdot,\cdot\}_{u_s} \right]_S.$$

To conclude that this is equal to zero, it suffices to show (by a direct computation) that, for any $f, g, h, l \in \mathcal{A}$, we have:

$$\begin{aligned} [f\{\cdot,\cdot\}_l, g\{\cdot,\cdot\}_h]_S &= f \left(\frac{\partial l}{\partial x} \left(\frac{\partial g}{\partial y} \frac{\partial h}{\partial z} - \frac{\partial g}{\partial z} \frac{\partial h}{\partial y} \right) + \circlearrowleft (x,y,z) \right) D \\ &+ g \left(\frac{\partial h}{\partial x} \left(\frac{\partial f}{\partial y} \frac{\partial l}{\partial z} - \frac{\partial f}{\partial z} \frac{\partial l}{\partial y} \right) + \circlearrowleft (x,y,z) \right) D \\ &= - [f\{\cdot,\cdot\}_h, g\{\cdot,\cdot\}_l]_S, \end{aligned}$$

where we recall that D denotes the skew-symmetric triderivation of $\mathcal{A}$ defined by $D := \frac{\partial}{\partial x} \wedge \frac{\partial}{\partial y} \wedge \frac{\partial}{\partial z}$, and where "$+ \circlearrowleft (x,y,z)$" means that we consider the other terms with cyclically permuted variables x, y, z.

Now we have chosen the maps $\ell_1^\varphi, \ell_2^\varphi, \ell_3^\varphi$ and $f_1^\varphi, f_2^\varphi, f_3^\varphi$ such that the equations $(\mathcal{E}_1), (\mathcal{E}_2), (\mathcal{E}_3)$ and $(\mathfrak{J}_1), (\mathfrak{J}_2), (\mathfrak{J}_3), (\mathfrak{J}_4)$ are satisfied, $\ell_i^\varphi\big|_{(H_\varphi^1)^{\otimes i}} = 0$ for $i = 2, 3$, f_2^φ is given by the tables 1 and 2, and $f_3^\varphi\big|_{(H_\varphi^1)^{\otimes 3}} = 0$. The proposition 3.1 once more gives us the existence of skew-symmetric graded linear maps

$$f_4^\varphi : \bigotimes^4 H(\mathfrak{g}_\varphi, \partial_\varphi) \to \mathfrak{g}_\varphi \quad \text{and} \quad \ell_4^\varphi : \bigotimes^4 H(\mathfrak{g}_\varphi, \partial_\varphi) \to H(\mathfrak{g}_\varphi, \partial_\varphi)$$

with $\deg(f_4^\varphi) = -3$ and $\deg(\ell_4^\varphi) = -2$ and satisfying the equation $(\mathcal{E}_4)$. Moreover, according to the proposition 3.1, such a map ℓ_4^φ satisfies also the equation $(\mathfrak{J}_5)$. It is also straightforward, with the choices made previously, to show that

$$T_4^\varphi\big|_{(H_\varphi^1)^{\otimes 4}} = 0.$$

This implies that $\ell_4^\varphi\big|_{(H_\varphi^1)^{\otimes 4}} = -p \circ T_4^\varphi\big|_{(H_\varphi^1)^{\otimes 4}} = 0$ and that it is possible to choose $f_4^\varphi\big|_{(H_\varphi^1)^{\otimes 4}} = 0$ (what we do from now on), so that $(\mathcal{E}_4)$ is still satisfied. Finally, because $\ell_i^\varphi\big|_{(H_\varphi^1)^{\otimes i}} = 0$, for $i = 2, 3, 4$, and $f_i^\varphi\big|_{(H_\varphi^1)^{\otimes i}} = 0$, for $i = 3, 4$, one has necessarily that:

$$T_j^\varphi\big|_{(H_\varphi^1)^{\otimes j}} = 0, \text{ for all } j \geq 5.$$

This fact, together with the proposition 3.1, imply that there finally exist skew-symmetric graded linear maps

$$\ell_k^\varphi : \bigotimes^k H(\mathfrak{g}_\varphi, \partial_\varphi) \to H(\mathfrak{g}_\varphi, \partial_\varphi), \quad \text{with } k \geq 5,$$

$$f_k^\varphi : \bigotimes^k H(\mathfrak{g}_\varphi, \partial_\varphi) \to \mathfrak{g}_\varphi, \quad \text{with } k \geq 5,$$

of degrees $2-k$ and $1-k$ respectively, and satisfying $\ell_k^\varphi\big|_{(H_\varphi^1)^{\otimes k}} = 0$, and $f_k^\varphi\big|_{(H_\varphi^1)^{\otimes k}} = 0$, for all $k \geq 5$, such that the maps $(\ell_1^\varphi, \ell_2^\varphi, \ell_3^\varphi, \ldots)$ and $(f_1^\varphi, f_2^\varphi, f_3^\varphi, \ldots)$ satisfy the conditions $(P_1) - (P_3)$, and

- $(\ell_k^\varphi)_{k \in \mathbf{N}^*}$ is an L_∞-algebra structure on H_φ,

- $(f_k^\varphi)_{k \in \mathbf{N}^*}$ is a quasi-isomorphism from H_φ to $\mathfrak{g}_\varphi$,

hence the theorem 4.3.   □

REMARK 4.4. There is a natural question concerning this theorem 4.3, which is: is it possible that $\ell_k^\varphi = 0$, for all $k \geq 3$? In other words, is it possible that the theorem extends to a result of *formality* for $\mathfrak{g}_\varphi$? Indeed, a dg Lie algebra $(\mathfrak{g}, \partial_\mathfrak{g}, [\cdot, \cdot]_\mathfrak{g})$ is said to be *formal* if it is linked to the dg Lie algebra $(H(\mathfrak{g}, \partial_\mathfrak{g}), 0, [\cdot, \cdot]_\mathfrak{g})$ (endowed with the trivial differential and the graded Lie bracket induced by $[\cdot, \cdot]_\mathfrak{g}$) by a quasi-isomorphism.

In fact, we can show that, except maybe if we change the definition of f_1^φ (i.e., if we consider another choice of bases $\mathbf{b}_\varphi^{-1}, \mathbf{b}_\varphi^0, \mathbf{b}_\varphi^1, \mathbf{b}_\varphi^2$), the map ℓ_3^φ cannot be zero. In the case $\varpi(\varphi) \neq |\varpi|$, one indeed has for example:

$$T_3^\varphi\left(\bar{\varphi}, \bar{\varphi}, \bar{D}\right) = 2\left[\varphi, f_2^\varphi\left(\bar{\varphi}, \bar{D}\right)\right]_S.$$

We know that the choice we made for the value $f_2^\varphi\left(\bar{\varphi}, \bar{D}\right)$ is unique, up to a 1-cocycle for the Poisson cohomology associated to $(\mathcal{A}, \{\cdot, \cdot\}_\varphi)$. According to the fact that $H^1(\mathcal{A}, \{\cdot, \cdot\}_\varphi) \simeq \{0\}$, when $\varpi(\varphi) \neq |\varpi|$, a 1-cocycle is a 1-coboundary, that is to say an element of the form $\mathcal{V} = \{\cdot, F\}_\varphi$, with $F \in \mathcal{A}$ (called an hamiltonian derivation). For such an element, $[\varphi, \mathcal{V}]_S = -\mathcal{V}[\varphi] = 0$. This implies that the value of $T_3^\varphi\left(\bar{\varphi}, \bar{\varphi}, \bar{D}\right)$ does not depend on the choice for $f_2^\varphi\left(\bar{\varphi}, \bar{D}\right)$ and, using the table 1,

$$T_3^\varphi\left(\bar{\varphi}, \bar{\varphi}, \bar{D}\right) = 2\left[\varphi, \frac{1}{\varpi(\varphi) - |\varpi|}\bar{e}_\varpi\right]_S = 2\frac{\varpi(\varphi)}{|\varpi|}\frac{}{\varpi(\varphi)}\varphi.$$

Because $\ell_3^\varphi = -p \circ T_3^\varphi$, we have $\ell_3^\varphi\left(\bar{\varphi}, \bar{\varphi}, \bar{D}\right) = 2\frac{\varpi(\varphi)}{|\varpi| - \varpi(\varphi)}\bar{\varphi}$, which is not zero.

4.3. Classification of the formal deformations of $\{\cdot, \cdot\}_\varphi$.

To obtain the theorem 1.1, we fix an L_∞-algebra structure $\ell_\bullet^\varphi$ on H_φ and a quasi-isomorphism $f_\bullet^\varphi$ from H_φ to $\mathfrak{g}_\varphi$, as in theorem 4.3. By the paragraph 2.2.3, we know that $\mathcal{D}ef^\nu(\mathfrak{g}_\varphi)$ corresponds to the set of all the equivalence classes of the formal deformations of $\{\cdot, \cdot\}_\varphi$. Let us now consider the set $\mathcal{D}ef^\nu(H_\varphi)$. By definition of the generalized Maurer-Cartan equation (5) and because the L_∞-algebra structure $\ell_\bullet^\varphi = (\ell_k^\varphi)_{k \in \mathbf{N}^*}$ satisfies $\ell_1^\varphi = 0$ and the property (P_2) of the theorem 4.3, we have:

$$MC^\nu(H_\varphi) = H_\varphi^1 \otimes \nu\mathbf{F}[[\nu]] = H^2(\mathcal{A}, \{\cdot, \cdot\}_\varphi) \otimes \nu\mathbf{F}[[\nu]].$$

In the generic case, that is to say when $\varpi(\varphi) \neq |\varpi|$, according to proposition 4.1, one has $H_\varphi^0 \simeq \{0\}$, so that the gauge equivalence in $MC^\nu(H_\varphi)$ is trivial and

$$\mathcal{D}ef^\nu(H_\varphi) \simeq MC^\nu(H_\varphi) = H_\varphi^1 \otimes \nu\mathbf{F}[[\nu]] = H^2(\mathcal{A}, \{\cdot, \cdot\}_\varphi) \otimes \nu\mathbf{F}[[\nu]].$$

Moreover, in the special case where $\varpi(\varphi) = |\varpi|$, then according to proposition 4.1, one has $H_\varphi^0 = \mathbf{F}[\varphi]\bar{e}_\varpi$ and in this case:

$$\mathcal{D}ef^\nu(H_\varphi) = H_\varphi^1 \otimes \nu\mathbf{F}[[\nu]]/\sim = H^2(\mathcal{A}, \{\cdot, \cdot\}_\varphi) \otimes \nu\mathbf{F}[[\nu]]/\sim,$$

where $\sim$ is the gauge equivalence in $MC^\nu(H_\varphi)$, generated by the infinitesimal transformations of the form:

(34)
$$\gamma \longmapsto \gamma - \sum_{k \geq 1}\frac{(-1)^{k(k-1)/2}}{(k-1)!}\ell_k^\varphi(\xi, \gamma, \dots, \gamma),$$

for $\xi = \sum\limits_{i \geq 1} \overline{F_i(\varphi)\, \vec{e}_\varpi}\, \nu^i \in H^0_\varphi \otimes \nu \mathbf{F}[[\nu]]$, where the $F_i(\varphi)$ are elements of $\mathbf{F}[\varphi]$. We now are able to show the following:

THEOREM 4.5. *Let $\varphi \in \mathcal{A} = \mathbf{F}[x, y, z]$ be a weight-homogeneous polynomial, with an isolated singularity. To φ is associated the Poisson structure defined by:*

$$\{\cdot, \cdot\}_\varphi := \frac{\partial \varphi}{\partial x} \frac{\partial}{\partial y} \wedge \frac{\partial}{\partial z} + \frac{\partial \varphi}{\partial y} \frac{\partial}{\partial z} \wedge \frac{\partial}{\partial x} + \frac{\partial \varphi}{\partial z} \frac{\partial}{\partial x} \wedge \frac{\partial}{\partial y}.$$

We consider the dg Lie algebra $(\mathfrak{g}_\varphi, \partial_\varphi, [\cdot, \cdot]_S)$, associated to φ and defined in the paragraph 4.1, of all skew-symmetric multiderivations of $\mathcal{A}$, equipped with the Schouten bracket $[\cdot, \cdot]_S$ and the differential $\partial_\varphi := \left[\{\cdot, \cdot\}_\varphi, \cdot \right]_S$.

We denote by $\mathfrak{C}$, the set of all $(\mathbf{c}, \bar{\mathbf{c}})$, where $\mathbf{c} := \left(c^k_{l,i} \in \mathbf{F} \right)_{\substack{(l,i) \in \mathbf{N} \times \mathcal{E}_\varphi \\ k \in \mathbf{N}^*}}$ is a family of constants indexed by $\mathbf{N} \times \mathcal{E}_\varphi \times \mathbf{N}^*$ and $\bar{\mathbf{c}} := \left(\bar{c}^k_r \in \mathbf{F} \right)_{\substack{1 \leq r \leq \mu-1 \\ k \in \mathbf{N}^*}}$ is a family of constants indexed by $\{1, \dots, \mu - 1\} \times \mathbf{N}^*$, such that, for every $k_0 \in \mathbf{N}^*$, the sequences $(c^{k_0}_{l,i})_{(l,i) \in \mathbf{N} \times \mathcal{E}_\varphi}$ and $(\bar{c}^{k_0}_r)_{1 \leq r \leq \mu-1}$ have finite supports. Now, for every element $(\mathbf{c}, \bar{\mathbf{c}}) = \left((c^k_{l,i}), (\bar{c}^k_r) \right) \in \mathfrak{C}$, we associate an element $\gamma^{\mathbf{c}, \bar{\mathbf{c}}}$ of $\mathfrak{g}^1_\varphi \otimes \nu \mathbf{F}[[\nu]]$, by the following formula:

$$(35) \qquad \gamma^{\mathbf{c}, \bar{\mathbf{c}}} := \sum_{n \in \mathbf{N}^*} \gamma^{\mathbf{c}, \bar{\mathbf{c}}}_n \nu^n,$$

with, for all $n \in \mathbf{N}^*$, $\gamma^{\mathbf{c}, \bar{\mathbf{c}}}_n$ given by:

$$
(36) \qquad
\begin{aligned}
\gamma^{\mathbf{c}, \bar{\mathbf{c}}}_n &:= \sum_{\substack{(l,i) \in \mathbf{N} \times \mathcal{E}_\varphi \\ 1 \leq r \leq \mu-1}} \ \sum_{\substack{a+b=n \\ a,b \in \mathbf{N}^*}} c^a_{l,i}\, \bar{c}^b_r\, \varphi^l\, u_i\, \{\cdot, \cdot\}_{u_r} \\[2mm]
&+ \sum_{(m,j) \in \mathbf{N} \times \mathcal{E}_\varphi} c^n_{m,j}\, \varphi^m\, u_j\, \{\cdot, \cdot\}_\varphi + \sum_{1 \leq s \leq \mu-1} \bar{c}^n_s\, \{\cdot, \cdot\}_{u_s},
\end{aligned}
$$

where the u_j, for $0 \leq j \leq \mu - 1$, are weight homogeneous polynomials of $\mathcal{A} = \mathbf{F}[x, y, z]$, whose images in $\mathcal{A}_{sing}(\varphi) = \mathbf{F}[x, y, z]/\langle \frac{\partial \varphi}{\partial x}, \frac{\partial \varphi}{\partial y}, \frac{\partial \varphi}{\partial z} \rangle$ give a basis of the $\mathbf{F}$-vector space $\mathcal{A}_{sing}(\varphi)$, and $u_0 = 1$. Then, one has:

(1) *The set of all the gauge equivalence classes of the solutions of the Maurer-Cartan equation associated to the dg Lie algebra $(\mathfrak{g}_\varphi, \partial_\varphi, [\cdot, \cdot]_S)$ is then given by:*

$$\mathcal{D}ef^\nu(\mathfrak{g}_\varphi) = \{\gamma^{\mathbf{c}, \bar{\mathbf{c}}} \mid (\mathbf{c}, \bar{\mathbf{c}}) \in \mathfrak{C}\}/ \sim,$$

where $\sim$ still denotes the gauge equivalence;

(2) *In the generic case where $\varpi(\varphi) \neq |\varpi|$, this set is exactly given by:*

$$\mathcal{D}ef^\nu(\mathfrak{g}_\varphi) = \{\gamma^{\mathbf{c}, \bar{\mathbf{c}}} \mid (\mathbf{c}, \bar{\mathbf{c}}) \in \mathfrak{C}\}.$$

PROOF. To show this theorem, we fix an L_∞-algebra structure $\ell^\varphi_\bullet$ on H_φ and a quasi-isomorphism $f^\varphi_\bullet$, as in theorem 4.3. According to the theorem 2.2, we know that

$$\mathcal{D}ef^\nu(\mathfrak{g}_\varphi) = \mathcal{D}ef^\nu(f^\varphi_\bullet)\,(\mathcal{D}ef^\nu(H_\varphi)).$$

We also have seen at the beginning of this paragraph that, because $\ell^\varphi_1 = 0$ and because of the property (P_2) of theorem 4.3, $\mathcal{D}ef^\nu(H_\varphi) = H^1_\varphi \otimes \nu \mathbf{F}[[\nu]]/ \sim$. Now, by definition of $f_\bullet^\varphi$, and because it satisfies the property (P_3) of the theorem 4.3, and by definition (9) (and (10)) of $\mathcal{D}ef^\nu(f_\bullet^\varphi)$, we have:

$$\mathcal{D}ef^\nu(\mathfrak{g}_\varphi) = \mathcal{D}ef^\nu(f_\bullet^\varphi)\left(\mathcal{D}ef''(H_\varphi)\right)$$

$$= \left(f_1^\varphi + \frac{1}{2}f_2^\varphi\right)(H_\varphi^1 \otimes \nu\mathbf{F}[[\nu]])/\sim .$$

Let $\gamma = \sum_{n\in\mathbf{N}^*}\gamma_n\,\nu^n$ be an element of $H_\varphi^1 \otimes \nu\mathbf{F}[[\nu]]$, where each γ_n is an element of H_φ^1. For $n \in \mathbf{N}^*$, every element γ_n can be decomposed in the basis $\mathbf{b}_\varphi^1$ (see the proposition 4.1), i.e., there exist families of constants $(\mathbf{c}, \bar{\mathbf{c}}) = \left((c_{m,j}^n), (\bar{c}_s^n)\right) \in \mathfrak{C}$ satisfying:

$$\gamma_n = \sum_{(m,j)\in\mathbf{N}\times\mathcal{E}_\varphi} c_{m,j}^n\,\varphi^m\,u_j\,\{\cdot,\cdot\}_\varphi + \sum_{1\le s\le\mu-1}\bar{c}_s^n\,\{\cdot,\cdot\}_{u_s},$$

for all $n \in \mathbf{N}$. Now, using the tables 1 and 2, we obtain exactly that $\left(f_1^\varphi + \frac{1}{2}f_2^\varphi\right)(\gamma) = \gamma^{\mathbf{c},\bar{\mathbf{c}}}$, hence the result. For the case where $\varpi(\varphi) \ne |\varpi|$, it only remains to recall that in this case, the gauge equivalence $\sim$ is trivial, as explained at the beginning of this paragraph. □

According to what we have seen in the paragraph 2.2.3, the previous theorem can be translated into a result concerning the formal deformations of the family of Poisson brackets $\{\cdot,\cdot\}_{\varphi}$, for $\varphi \in \mathbf{F}[x,y,z]$, a weight-homogenous polynomial with an isolated singularity. It then becomes exactly the parts (a), (b) and (c) of the proposition 3.3 of [17] and replacing $\nu\mathbf{F}[[\nu]]$ by $\nu\mathbf{F}[[\nu]]/\langle\nu^{m+1}\rangle$ (with $m \in \mathbf{N}^*$) in everything we have done leads to the part (d) of this proposition 3.3 of [17], which we write once more here:

PROPOSITION 4.6 ([17]). *Let* $\varphi \in \mathcal{A} = \mathbf{F}[x,y,z]$ *be a weight homogeneous polynomial with an isolated singularity. Consider the Poisson algebra* $(\mathcal{A}, \{\cdot,\cdot\}_\varphi)$ *associated to* φ*, where* $\{\cdot,\cdot\}_\varphi$ *is the Poisson bracket given by (29). Then we have the following:*

(a) *For all families of constants* $\left(c_{l,i}^k \in \mathbf{F}\right)_{\substack{(l,i)\in\mathbf{N}\times\mathcal{E}_\varphi \\ k\in\mathbf{N}^*}}$ *and* $\left(\bar{c}_r^k \in \mathbf{F}\right)_{\substack{1\le r\le\mu-1 \\ k\in\mathbf{N}^*}}$, *such that, for every* $k_0 \in \mathbf{N}^*$*, the sequences* $(c_{l,i}^{k_0})_{(l,i)\in\mathbf{N}\times\mathcal{E}_\varphi}$ *and* $(\bar{c}_r^{k_0})_{1\le r\le\mu-1}$ *have finite supports, the formula*

$$(37) \qquad \pi_* = \{\cdot,\cdot\}_\varphi + \sum_{n\in\mathbf{N}^*}\pi_n\nu^n,$$

where, for all $n \in \mathbf{N}^*$*,* π_n *is given by:*

$$(38) \qquad \pi_n = \sum_{\substack{(l,i)\in\mathbf{N}\times\mathcal{E}_\varphi \\ 1\le r\le\mu-1}} \sum_{\substack{a+b=n \\ a,b\in\mathbf{N}^*}} c_{l,i}^a\,\bar{c}_r^b\,\varphi^l\,u_i\,\{\cdot,\cdot\}_{u_r}$$

$$+ \sum_{(m,j)\in\mathbf{N}\times\mathcal{E}_\varphi} c_{m,j}^n\,\varphi^m\,u_j\,\{\cdot,\cdot\}_\varphi + \sum_{1\le s\le\mu-1}\bar{c}_s^n\,\{\cdot,\cdot\}_{u_s},$$

defines a formal deformation of $\{\cdot,\cdot\}_\varphi$*, where the* u_j $(0 \le j \le \mu - 1)$ *are weight homogeneous polynomials of* $\mathcal{A} = \mathbf{F}[x,y,z]$*, whose images in* $\mathcal{A}_{sing}(\varphi) = \mathbf{F}[x,y,z]/\langle\frac{\partial\varphi}{\partial x}, \frac{\partial\varphi}{\partial y}, \frac{\partial\varphi}{\partial z}\rangle$ *give a basis of the* $\mathbf{F}$*-vector space* $\mathcal{A}_{sing}(\varphi)$*, and* $u_0 = 1$.

(b) *For any formal deformation π'_* of $\{\cdot,\cdot\}_\varphi$, there exist families of constants*
$\left(c_{l,i}^k\right)_{\substack{(l,i)\in\mathbf{N}\times\mathcal{E}_\varphi \\ k\in\mathbf{N}^*}}$ *and* $\left(\bar{c}_r^k\right)_{\substack{1\leq r\leq\mu-1 \\ k\in\mathbf{N}^*}}$ *(such that, for every $k_0\in\mathbf{N}^*$, only a*
finite number of $c_{l,i}^{k_0}$ and $\bar{c}_r^{k_0}$ are non-zero), for which π'_* is equivalent to
the formal deformation π_* given by the above formulas (37) and (38).

(c) *Moreover, if the (weighted) degree of the polynomial φ is not equal to*
the sum of the weights: $\varpi(\varphi)\neq|\varpi|$, then for any formal deformation
π'_* of $\{\cdot,\cdot\}_\varphi$, there exist unique families of constants $\left(c_{l,i}^k\right)_{\substack{(l,i)\in\mathbf{N}\times\mathcal{E}_\varphi \\ k\in\mathbf{N}^*}}$ and
$\left(\bar{c}_r^k\right)_{\substack{1\leq r\leq\mu-1 \\ k\in\mathbf{N}^*}}$ (with, for every $k_0\in\mathbf{N}^*$, only a finite number of non-zero
$c_{l,i}^{k_0}$ and $\bar{c}_r^{k_0}$), such that π'_* is equivalent to the formal deformation π_* given
by the formulas (37) and (38).

This means that formulas (37) and (38) give a system of representa-
tives for all formal deformations of $\{\cdot,\cdot\}_\varphi$, modulo equivalence.

(d) *Analogous results hold if we replace formal deformations by m-th order*
deformations $(m\in\mathbf{N}^*)$ and impose in (c) that $c_{l,i}^k=0$ and $\bar{c}_r^k=0$, as
soon as $k\geq m+1$.

References

[1] François Bayen, Moshé Flato, Christian Fronsdal, André Lichnerowicz, and Daniel Stern-heimer. Deformation theory and quantization. I and II. *Annals of Physics*, 111(1):61–110, 1978.

[2] Alberto Cattaneo, Bernhard Keller, Charles Torossian, and Alain Bruguières. Déformation, quantification, théorie de Lie. *Panoramas et Synthèses*, 20:viii+186, 2005.

[3] Martin Doubek, Martin Markl, and Petr Zima. Deformation theory (lecture notes). *Universitatis Masarykianae Brunensis. Facultas Scientiarum Naturalium. Archivum Mathematicum*, 43(5):333–371, 2007.

[4] Johannes Huebschmann. Origins and breadth of the theory of higher homotopies. *Festschrift in honor of M. Gerstenhaber's 80-th and Jim Stashe's 70-th birthday, Progress in Math. (to appear), arXiv:math/0710.2645.*

[5] Johannes Huebschmann. Poisson cohomology and quantization. *Journal für die Reine und Angewandte Mathematik*, 408:57–113, 1990.

[6] Johannes Huebschmann and Jim Stasheff. Formal solution of the master equation via HPT and deformation theory. *Forum Mathematicum*, 14(6):847–868, 2002.

[7] Tornike V. Kadeishvili. On the theory of homology of fiber spaces. *International Topology Conference (Moscow State Univ., Moscow, 1979), kademiya Nauk SSSR i Moskovskoe Matematicheskoe Obshchestvo. Uspekhi Matematicheskikh Nauk.*, 35(3(213)):183–188, 1980. See also arXiv:math/0504437.

[8] Hiroshige Kajiura and Jim Stasheff. Homotopy algebras inspired by classical open-closed string field theory. *Communications in Mathematical Physics*, 263(3):553–581, 2006.

[9] Maxim Kontsevich. Deformation quantization of Poisson manifolds. *Letters in Mathematical Physics*, 66(3):157–216, 2003.

[10] Tom Lada and Martin Markl. Strongly homotopy Lie algebras. *Communications in Algebra*, 23(6):2147–2161, 1995.

[11] Tom Lada and Jim Stasheff. Introduction to SH Lie algebras for physicists. *International Journal of Theoretical Physics*, 32(7):1087–1103, 1993.

[12] Camille Laurent-Gengoux, Anne Pichereau, and Pol Vanhaecke. An invitation to poisson structures. *monograph to appear in Springer.*

[13] Calin I. Lazaroiu. String field theory and brane superpotentials. *The Journal of High Energy Physics*, Paper 18(10):40 p, 2001.

[14] André Lichnerowicz. Les variétés de Poisson et leurs algèbres de Lie associées. *Journal of Differential Geometry*, 12(2):253–300, 1977.

[15] Martin Markl. Homotopy algebras are homotopy algebras. *Forum Mathematicum*, 16(1):129–160, 2004.

[16] Anne Pichereau. Poisson (co)homology and isolated singularities. *Journal of Algebra*, 299(2):747–777, 2006.

[17] Anne Pichereau. Formal deformations of poisson structures in low dimensions. *Pacific Journal of Mathematics*, 239(01):105 – 133, 2009.

[18] James Dillon Stasheff. Homotopy associativity of H-spaces. I, II. *Transactions of the American Mathematical Society*, 108:293–312, 1963.

MAX-PLANCK-INSTITUT FÜR MATHEMATIK VIVATSGASSE 7, 53 111 BONN, GERMANY
E-mail address: pichereau@mpim-bonn.mpg.de

Aspects of Mathematics

Edited by Klas Diederich

*A Publication of the Max-Planck-Institute for Mathematics, Bonn

www.viewegteubner.de

The Bridge between Discrete Mathematics and Statistical Physics

Martin Loebl

Discrete Mathematics in Statistical Physics

Introductory Lectures

2010. X, 187 S. (Advanced Lectures in Mathematics) Br. EUR 34,90
ISBN 978-3-528-03219-7

Basic concepts - Introduction to Graph Theory - Trees and electrical networks – Matroids - Geometric representations of graphs - Game of dualities - The zeta function and graph polynomials – Knots - 2D Ising and dimer models

The book first describes connections between some basic problems and technics of combinatorics and statistical physics. The discrete mathematics and physics terminology are related to each other. Using the established connections, some exciting activities in one field are shown from a perspective of the other field. The purpose of the book is to emphasize these interactions as a strong and successful tool. In fact, this attitude has been a strong trend in both research communities recently.

It also naturally leads to many open problems, some of which seem to be basic. Hopefully, this book will help making these exciting problems attractive to advanced students and researchers.

**VIEWEG+
TEUBNER**

Abraham-Lincoln-Straße 46
65189 Wiesbaden
Fax 0611.7878-400
www.viewegteubner.de

Stand Januar 2010.
Änderungen vorbehalten.
Erhältlich im Buchhandel oder im Verlag.